To my friends and colleagues;
Kamal Bawa, who introduced me to conservation biology,
and Les Kaufman, who generously shared
his knowledge of marine biology.

Contents

Chapter 1
Defining Coservation Biology 3

Chapter 2
What is Biodiversity? 19

A Primer of

Conservation Biology

FIFTH EDITION

A Primer of Conservation Biology

FIFTH EDITION

Richard B. Primack
Boston University

Sinauer Associates, Inc. Publishers
Sunderland, MA U.S.A.

The Cover
A black rhino (*Diceros bicornis*) silhouetted against the fiery sunset at Masai Mara National Reserve in Kenya. Rhinos are protected inside the reserve, but their numbers have declined severely elsewhere in Africa due to habitat destruction, farming, and illegal poaching for their horns. (Photograph © Alex Bernasconi.)

A Primer of Conservation Biology, Fifth Edition

Sinauer Associates
23 Plumtree Road
Sunderland, MA 01375 U.S.A.
E-mail: orders.sinauer.com; publish@sinauer.com
Internet: www.sinauer.com

ISBN 978-0-87893-623-6

Printed in China
5 4 3 2 1

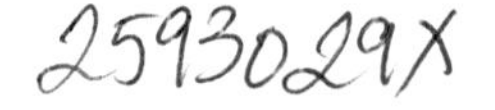

Chapter 3
The Value of Biodiversity 47

Chapter 4
Threats to Biodiversity 79

Chapter 6
Conserving Populations and Species 169

Chapter 7
Protected Areas 213

Chapter 8
Conservation Outside Protected Areas 255

Chapter 9
The Challenge of Sustainable Development 283

Preface

After decades of public interest in nature and the environment, the United Nations focused worldwide attention on conservation by declaring 2010 to be the International Year of Biodiversity and 2011 to be the Year of Forests. The general public has absorbed this message and is asking its political leaders to provide the policy changes needed to address issues of conservation. Conservation biology is the field that seeks to study and protect the living world and its biological diversity (or biodiversity in its shortened form). The field emerged during the last 35 years as a major new discipline to address the alarming loss of biological diversity. The threats to biodiversity are all too real, as demonstrated by the recent recognition that fully one-third of amphibian species are in danger of extinction. At the same time, our need to remain hopeful is highlighted, for example, by increasing sea turtle populations at many locations throughout the world following comprehensive conservation efforts. Many examples described in this book show that governments, individuals, and conservation organizations can work together to make the world a better place for all living things.

University students continue to enroll enthusiastically and in large numbers in conservation biology courses. The first (1995), second (2000), third (2004), and fourth (2008) editions of *A Primer of Conservation Biology* sought to fill the need for a "quick" guide for those who wanted a basic familiarity with conservation biology. Like its predecessors, this fifth edition of the *Primer* is designed for use in non-majors courses and short courses in conservation biology and can also be used as a supplemental text for general biology, ecology, wildlife biology, and environmental policy courses. It is also intended to serve as a concise guide for professionals who require a well-documented overview of the subject but do not require in-depth case studies or lengthy scientific discussion. *Essentials of Conservation Biology*, now in its fifth edition, is recommended to readers who want this more comprehensive treatment of the subject.

This fifth edition of the *Primer* reflects the excitement and new developments in the field. It provides coverage of the latest information available on a number of topics, including the expanding system of marine protected areas, funding for conservation projects, and linkages between conservation and global climate change. It also highlights new approaches culled from the literature on topics such as species reintroductions, population viability analysis, protected areas management, and payments for ecosystem services.

In keeping with the international approach of conservation biology, I feel it is important to make the field accessible to as wide an audience as possible. With the assistance of Marie Scavotto and the staff of Sinauer Associates, I have arranged an active translation program, beginning in 1995 with a translation of *Essentials* into German, followed by a Chinese translation in 1997. It became clear to me that the best way to make the material accessible was to create regional or country-specific translations, identifying local scientists to become coauthors, and to add case studies, examples, and illustrations from their own countries and regions that would be more relevant to the intended audience. To that end, in the past 12 years editions of the *Primer* have appeared in Brazilian Portuguese, Chinese (two editions), Czech (two editions), Estonian, French with a Madagascar focus, Greek, Indonesian (two editions), Italian, Japanese (two editions), Korean (two editions), Mongolian, Romanian, Russian, Spanish, and Vietnamese. An English version of the *Primer* with a South Asian focus has also been published. Editions of *Essentials* have appeared in Arabic, Hungarian, Romanian, and Spanish with a Latin American focus. New editions of the *Primer* for France, Pakistan, Turkey, Bangladesh, and Germany, and the *Essentials* in Chinese are currently in production. It is my hope that these translations will help conservation biology develop as a discipline with a global scope. At the same time, examples from these translations find their way back into the English language editions, thereby enriching the presentation.

I hope that readers of this book will want to find out more about the extinction crisis facing species and ecosystems and how they can take action to halt it. I encourage readers to take the field's activist spirit to heart—use the Appendix to find organizations and sources of information on how to help. If readers gain a greater appreciation for the goals, methods, and importance of conservation biology, and if they are moved to make a difference in their everyday lives, this textbook will have served its purpose.

Available to qualified adopting instructors is an Instructor's Resource CD that includes electronic versions of all the figures, photos, and tables from the textbook.

Acknowledgments

I sincerely appreciate the contribution of everyone who helped make this book accurate and clear. Individual chapters in this edition were reviewed by Dana Bauer, Linus Chen, Richard Corlett, Elizabeth Freeman, Lian Pin Koh, David Lindenmayer, Meg Lowman, Michael Reed, and Anna Sher. Les Kaufman of Boston University provided expertise on marine systems.

Numerous people offered specialized input that helped make the examples and figures current and accurate. I would particularly like to recognize the contributions of Miguel Ferrar, Lucas Joppa, Jay Odell, Ben Phelan, Craig Packer, and Kent Redford.

Jessica Susser was the principal research assistant and organizer for the project, with additional help from Nicole Chabaneix, Libby Ellwood, Caitlin McDonough, Michelle McInnis, Caroline Polgar, and Jasper Primack. Danna Niedzwiecki provided invaluable help in the production of the book, with numerous suggestions on how to make the book friendlier to student readers. Thanks to Lou Doucette for her skillful copyediting and to Andy Sinauer, Chris Small, David McIntyre, Ann Chiara, Janice Holabird, and the rest of the Sinauer staff who helped to transform the manuscript into a finished book.

Special thanks are due to my wife Margaret and my children Dan, Will, and Jasper for encouraging me to fulfill an important personal goal by completing this book. I would like to recognize Boston University for providing me with the facilities and environment that made this project possible and the many Boston University students who have taken my conservation biology courses over the years. Their enthusiasm and suggestions have helped me to find new ways to present this material. And lastly, I would like to express my great appreciation to my coauthors in other countries who have worked with me to produce conservation biology textbooks in their own languages, which are critical for spreading the message of conservation biology to a wider audience.

Richard Primack
Boston, Massachusetts
March, 2012

Media and Supplements

to accompany

A Primer of Conservation Biology

FIFTH EDITION

eBook

(ISBN 978-0-87893-896-4)

www.coursesmart.com

A Primer of Conservation Biology, Fifth Edition is available as an eBook via CourseSmart, at a substantial discount off the price of the printed textbook. The CourseSmart eBook reproduces the look of the printed book exactly, and includes convenient tools for searching the text, highlighting, and note-taking. The eBook is viewable in any Web browser, and via free apps for iPhone/iPad, Android, and Kindle Fire.

Instructor's Resource Library

(ISBN 978-0-87893-897-1)

Available to qualified adopters, the Instructor's Resource Library to accompany *A Primer of Conservation Biology* includes all of the textbook's figures and tables in a variety of formats, making it easy for instructors to incorporate figures into lectures and other course materials. All of the figures have been optimized for use in the classroom and are provided as both low-resolution and high-resolution JPEGs, as well as ready-to-use PowerPoint slides.

A Primer of
Conservation Biology
FIFTH EDITION

A primatologist observes endangered gelada baboons in Ethiopia.

Chapter 1
Defining Conservation Biology

Popular interest in protecting the world's biological diversity—including its amazing range of species, its complex biological communities and associated ecosystem processes, and the genetic variation within species—has intensified during the last few decades. It has become increasingly evident to both scientists and the general public that we are living in a period of unprecedented **biodiversity*** loss. Around the globe, biological communities that took millions of years to develop—including tropical rain forests, coral reefs, old-growth forests, prairies, and coastal wetlands—have been devastated as a result of human actions. Biologists predict that tens of thousands of species and millions of unique populations will go extinct in the coming decades (Millennium Ecosystem Assessment 2005; Barnosky et al. 2011). The overwhelming cause of all this loss is the rapidly expanding human population.

During the last 150 years, the human population has exploded. It took more than 10,000 years for the number of *Homo sapiens* to reach 1 billion, an event that occurred sometime around the year 1850. Estimates for 2011 put the number of humans at 7 billion; at this size, even a modest rate of population increase adds tens of millions of

*__Biological diversity__ is often shortened to *biodiversity*; it includes all species, genetic variation, and biological communities and their ecosystem-level interactions.

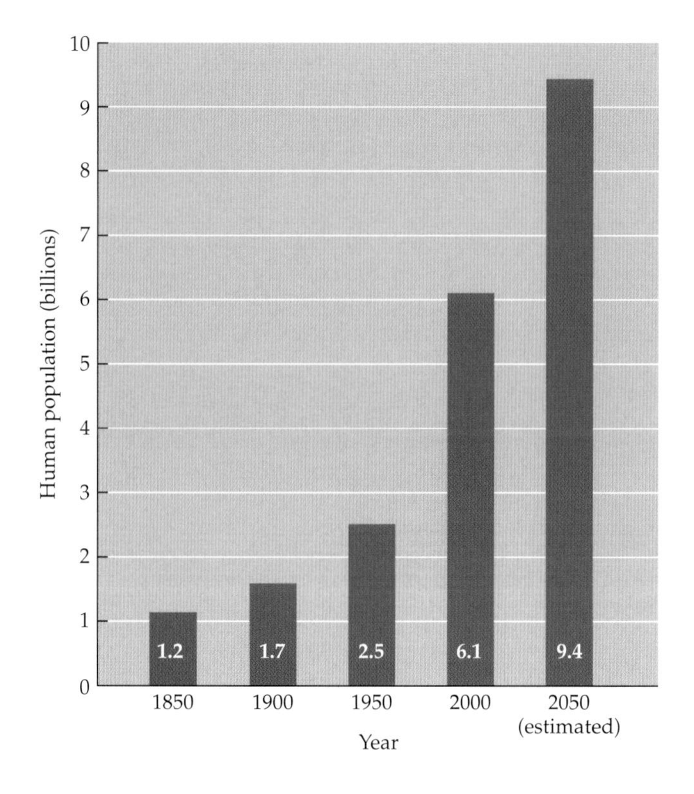

FIGURE 1.1 The human population in 2011 stood at around 7 billion. The World Resources Institute estimates current annual population growth at 1.1%, but even this modest growth rate will add more than 70 million people to the planet in the next year. This number will escalate each year as the increase is compounded. (Data from U.S. Census Bureau, www.census.gov)

individuals each year (**Figure 1.1**). The threats to biodiversity are accelerating because of the demands of the rapidly increasing human population and its rising material consumption. People use natural resources such as firewood, coal, oil, timber, fish, and game, and they convert natural habitats to land dominated by agriculture, cities, housing developments, logging, mining, industrial plants, and other human activities (Caro et al. 2012).

Worsening the situation is the fact that as countries develop and industrialize, the consumption of resources by their citizens increases. For example, the average citizen of the United States uses 5 times more energy than the average global citizen, 10 times more than the average Chinese citizen, and 28 times more than the average Indian citizen (Worldwatch Institute 2008; Encyclopedia of the Nations 2009). The ever-increasing number of human beings and their intensifying use of natural resources have direct and harmful consequences for the diversity of the living world.

The New Science of Conservation Biology

Many of us feel discouraged by the avalanche of species extinctions and the wholesale habitat destruction occurring in the world today. But it is possible—and indeed necessary—to feel challenged in order to find ways to stop the destruction (Stearns and Stearns 2010). Actions taken—or bypassed—during the next few decades will determine how many of the world's species and natural areas will survive. It is quite likely that people will someday look back on the first decades of the twenty-first century as an extraordinarily exciting time, when a collaboration of determined people acting locally and internationally saved large numbers of species from extinction and even entire ecosystems from destruction (Sodhi et al. 2011). Examples of such conservation efforts and positive outcomes are described later in this chapter and throughout this book.

Conservation biology is an integrated, multidisciplinary scientific field that has developed in response to the challenge of preserving species and ecosystems. It has three goals:

- To document the full range of biological diversity on Earth
- To investigate human impact on species, genetic variation, and ecosystems
- To develop practical approaches to prevent the extinction of species, maintain genetic diversity within species, and protect and restore biological communities and their associated ecosystem functions

The first two of these goals involve the dispassionate search for factual knowledge typical of scientific research. The third goal, however, defines conservation biology as a **normative discipline**—that is, it embraces certain values and attempts to apply scientific methods to achieving those values (Lindenmayer and Hunter 2010). Like medical science, which applies knowledge gleaned from physiology, anatomy, biochemistry, and genetics to the goal of achieving human health and eliminating illness, conservation biologists intervene to prevent the human-caused loss of biodiversity, because they believe the preservation of species and ecosystems to be an ultimate good.

Conservation biology arose in the 1980s because the traditional applied disciplines of resource management alone were not comprehensive enough to address the critical threats to biological diversity. In the past, agriculture, forestry, wildlife management, and fisheries biology were concerned primarily with developing methods to manage a small range of species for the marketplace and for recreation. These applied disciplines have gradually been expanding to encompass a broader range of species and ecosystem processes. Conservation biology complements the applied disciplines and provides a more general theoretical approach to the protection of biological diversity. It differs from these disciplines in its primary goal of long-term preservation of biodiversity, with economic factors secondary.

The academic disciplines of population biology, taxonomy, ecology, and genetics constitute the core of conservation biology, and many conservation biologists have been drawn from these ranks. Others come from backgrounds in the applied disciplines, such as forestry and wildlife management. In addition, many leaders in conservation biology have come from

Conservation biology merges applied and theoretical biology and incorporates ideas and expertise from a broad range of fields outside the natural sciences, toward the goal of preserving biodiversity.

zoos and botanical gardens, bringing with them experience in locating rare and endangered species in the wild and then maintaining and propagating them in captivity.

Conservation biology is also closely associated with **environmentalism**, a widespread movement characterized by political and educational activism with the goal of protecting the natural environment from destruction and pollution. In contrast, conservation biology is a scientific discipline based in biological research whose findings often contribute to the environmental movement (Hall and Fleishman 2010).

Because much of the biodiversity crisis arises from human pressures, conservation biology also incorporates ideas and expertise from a broad range of other fields (**Figure 1.2**) (Reyers et al. 2010). For example, environmental law and policy provide the basis for government protection of rare and endangered species and critical habitats. Environmental ethics provides a rationale for preserving species. Ecological economists provide analyses of the economic value of biological diversity to support arguments for preservation. Ecosystem ecologists and climatologists monitor the biological and physical characteristics of the environment and develop models to predict environmental responses to disturbance. Social sciences, such as anthropology, sociology, and geography, provide methods to involve local

FIELD EXPERIENCE AND RESEARCH NEEDS

Basic Sciences

Anthropology
Biogeography
Climatology
Ecology:
  Community ecology
  Ecosystem ecology
  Landscape ecology
Environmental studies:
  Ecological economics
  Environmental ethics
  Environmental law
Ethnobotany
Evolutionary biology
Genetics
Population biology
Sociology
Taxonomy
Other biological, physical, and social sciences

Resource Management

Agriculture
Community education and development
Fisheries management
Forestry
Land-use planning and regulation
Management of captive populations:
  Zoos
  Aquariums
  Botanical gardens
  Seed banks
Management of protected areas
Sustainable development
Wildlife management
Other resource conservation and management activities

FIGURE 1.2 Conservation biology represents a synthesis of many basic sciences (left) that provide principles and new approaches for the applied fields of resource management (right). The experiences gained in the field, in turn, influence the direction of the basic sciences. (After Temple 1991.)

people in actions to protect their immediate environment. Conservation education links academic study and fieldwork to solve environmental problems, teaching people about science and helping them realize the value of the natural environment. Because it draws on the ideas and skills of so many separate fields, conservation biology can be considered a truly multidisciplinary approach.

The philosophical roots of conservation biology

Conservation biology can be grounded in religious and philosophical beliefs that describe the relationship between human societies and the natural world (Dudley et al. 2009). In many of the world's religions, people are seen as both physically and spiritually connected to the plants and animals in the surrounding environment. Many Christian monasteries and religious centers protect the surrounding nature as an important part of their mission. Similarly in Eastern philosophies such as Taoism, Hinduism, and Buddhism, wilderness areas and natural settings are protected and valued for their capacity to provide intense spiritual experiences. These philosophies see a direct connection between the natural world and the spiritual world, a connection that breaks when the natural world is altered or destroyed by human activity. In Islamic teachings, people are given the sacred responsibility to be guardians of nature. Christian and Jewish leaders also encourage preservation of nature as God's creation as part of humanity's moral duties and spiritual practice. Healing damaged environments is seen as an obligation by many religions.

Biological diversity often has immediate significance to traditional societies whose people live close to the land and water. In Native American tribes of the Pacific Northwest, hunters undergo purification rituals in order to be considered worthy of hunting animals. The Iroquois, a Native American group, considered how their actions would affect the lives of their descendants after seven generations. Hunting and gathering societies, such as the Penan of Borneo, give thousands of names to individual trees, animals, and places in their surroundings to create a cultural landscape that is vital to the well-being of the tribe.

In the United States, nineteenth-century philosophers such as Ralph Waldo Emerson and Henry David Thoreau saw wild nature as an important element in human moral and spiritual development (Merchant 2002). Sharing this viewpoint during that century were the prolific artists of the Hudson River School (**Figure 1.3**). These painters were noted for their romantic depictions of "scenes of solitude from which the hand of nature has never been lifted" (Cole 1965). Later conservationist John Muir (1838–1914) also used spiritual arguments for preserving natural landscapes, and wildlife biologist Aldo Leopold (1887–1948) was a vocal advocate for setting aside wilderness areas and maintaining the health of natural ecosystems.

In the latter half of the twentieth century, Rachel Carson's 1962 book *Silent Spring*, which described the linkages between pesticide use and the

FIGURE 1.3 *Looking Down the Yosemite Valley, California* (1865) was the first of many Yosemite paintings by Albert Bierstadt, one of the most noted of the Hudson River School of "wilderness" artists. In 1864, U.S. President Abraham Lincoln signed the Yosemite Land Grant, the first law to set aside territory for public use and preservation. The area is now part of Yosemite National Park, internationally known for its towering granite cliffs and domes, numerous waterfalls, clear streams, and groves of giant Sequoias.

decline in wildlife populations, was instrumental in triggering the modern environmental movement. Later, metaphysical and scientific perspectives merged in the **Gaia hypothesis**, which views the Earth as having the properties of a superorganism whose biological, physical, and chemical components interact to regulate characteristics of the atmosphere and climate (Lovelock 1988). Modern wilderness proponents, such as members of the deep ecology movement that will be discussed in Chapter 3, often advocate the reduction or complete cessation of practices and industries that disrupt the natural interaction of Earth's components. This American focus on untouched wilderness stands in contrast to conservation efforts in other countries that seek to protect nature within a landscape that encompasses traditional uses by humans.

Paralleling these preservationist and ecology-focused orientations are the influential ideas of Gifford Pinchot (1865–1946), the first head of the U.S. Forest Service. Pinchot defined **natural resources** as the commodities and qualities found in nature—including timber, fodder, clean water, wildlife, and even beautiful landscapes. Pinchot believed that the goal of land management is to use these natural resources for the greatest good of the greatest number of people for the longest time. Pinchot's and Leopold's ideas have been combined and extended in the concept of **ecosystem management**, which places the highest management priority on the health of ecosystems and wild species. The current paradigm of **sustainable development** also advocates an integrated approach, calling for development that meets present and future human needs without damaging the environment or diminishing biodiversity.

Discussions of natural resources, ecosystem management, and sustainable development are major themes in conservation biology.

Conservation biology's international scope

In Europe, concern for the protection of wildlife began to spread widely in the late nineteenth century, when many species began to go locally extinct (Galbraith et al. 1998). The combination of both an increasing area of land under cultivation and more widespread use of firearms for hunting led to a marked reduction in the numbers of wild animals.

In the United Kingdom, one of the early leaders in nature conservation, these dramatic changes stimulated the formation of the British conservation movement, leading to the founding of the Commons, Open Spaces and Footpaths Preservation Society in 1865, the National Trust for Places of Historic Interest or Natural Beauty in 1895, and the Royal Society for the Protection of Birds in 1899. Altogether, these groups have preserved over 500,000 hectares (ha) of open land. (For an explanation of the term *hectare* and other metric measurements used in the natural sciences, see **Table 1.1**.) In the twentieth century, government action in the United Kingdom produced laws such as the National Parks and Access to the Countryside Act, passed in 1949 for the "protection and public enjoyment of the wider countryside," and the Wildlife and Countryside Act, passed in 1981, for the protection of endangered species, their habitats, and the marine environment.

Many other developed countries also have strong traditions of nature conservation and land protection, most notably Austria, the Netherlands, Germany, Switzerland, Australia, and Japan. In these countries as well, conservation is both enacted by the government and supported by private conservation organizations. Over the last two decades, regional initiatives to protect species, habitats, and ecosystem processes have been coordinated by the European Union and, increasingly, by international organizations such as the World Wildlife Fund.

Awareness of the value of biodiversity greatly increased following the international Earth Summit held in Rio de Janeiro, Brazil, in 1992. At this meeting, representatives of 178 countries formulated and eventually signed the Convention on Biological Diversity (CBD), which obligates countries to protect their biodiversity but also allows them to obtain a share in the profits of new products developed from that diversity. Tropical countries such as Brazil, Costa Rica, and Indonesia have responded to the CBD and related international initiatives by expanding the numbers and areas of their

TABLE 1.1 Some Useful Units of Measurement

Length	
1 meter (m)	1 m = 39.4 inches = ~3.3 feet
1 kilometer (km)	1 km = 1000 m = 0.62 mile
1 centimeter (cm)	1 cm = 1/100 m = 0.39 inch
1 millimeter (mm)	1 mm = 1/1000 m = 0.039 inch
Area	
square meter (m^2)	Area encompassed by a square, each side of which is 1 meter
1 hectare (ha)	1 ha = 10,000 m^2 = 2.47 acres 100 ha = 1 square kilometer (km^2)
Mass	
1 kilogram (kg)	1 kg = 2.2 pounds
1 gram (g)	1 g = 1/1000 kg = 0.035 ounce
1 milligram (mg)	1 mg = 1/1000 g = 0.000035 ounce
Temperature	
degree Celsius (°C)	
°C = 5/9(°F – 32)	0°C = 32° Fahrenheit (the freezing point of water)
	100°C = 212° Fahrenheit (the boiling point of water)
	20°C = 68° Fahrenheit ("room temperature")

national parks. The economic value of these protected areas is constantly increasing because of their importance for tourism and for the valuable ecosystem services they provide, such as purifying water and absorbing carbon dioxide (see Chapter 3). Many tropical countries have established agencies to regulate the exploration and use of their biodiversity. A related movement works to provide benefits to the indigenous peoples who have developed the knowledge of which species are valuable and how to use them.

The interdisciplinary approach: A case study with sea turtles

Interdisciplinary approaches, the involvement of local people, and the restoration of important environments and species all attest to progress in the science of conservation biology.

Throughout the world, scientists are using the approaches of conservation biology to address challenging problems, as illustrated by a Brazilian program for the conservation of highly endangered sea turtles. Many sea turtle populations have shrunk to less than 1% of their original sizes, devastated by a combination of factors that includes coastal development and damage to the beaches they use as nesting habitat, hunting of adult turtles and collecting of turtle eggs for food, and high mortality due to entanglement in fishing gear. Brazil's comprehensive approach to saving these fascinating creatures illustrates the interdisciplinary nature of conservation biology.

Sea turtles spend their lives at sea, with only the females returning to land to lay eggs on sandy beaches. When the Brazilian government set out to design a conservation program, planners discovered that no one knew exactly which species of sea turtles were found in Brazil, how many turtles there were, where they laid their eggs, and how local people were affecting them. To overcome this lack of basic information, in 1980 the Brazilian government established the National Marine Turtle Conservation Program, called Projeto TAMAR* (Marcovaldi and Chaloupka 2007; www.tamar.org.br). The project began with a two-year survey of Brazil's 6000 kilometers of coastline, using boats, horses, and foot patrols, combined with hundreds of interviews with villagers. TAMAR divers aided in these efforts by tagging and monitoring sea turtle populations in the water (**Figure 1.4**). This data-gathering phase is an important initial step in many conservation projects.

The TAMAR survey found that turtle nesting beaches fell into three main zones along 1100 km of the coastline between Rio de Janeiro and Recife, with loggerhead turtles (*Caretta caretta*) the most abundant species and four other species also present. The green turtle (*Chelonia mydas*) was the only species nesting on Brazil's offshore islands.

Interviews with villagers and observations of beaches revealed that adult turtles and turtle eggs were being harvested intensively, with people often collecting virtually every turtle egg laid. In many areas, the construction of resorts, houses, commercial developments, and beach roads had damaged and reduced the available nesting area on beaches. Additionally, the light from the buildings at night disoriented emerging hatchlings: instead of heading straight to the ocean, they often wandered in wrong directions

*TAMAR is an acronym for "TArtarugas MARinhas," which is Portuguese for "marine turtles."

FIGURE 1.4 Researchers gather data and tag endangered sea turtles off the coast of Brazil. (Photograph courtesy of Projeto TAMAR Image Bank.)

and became exhausted. Of the young turtles that did make it to sea, many were caught in the nets of fishermen, where they suffocated and died.

Information from the TAMAR survey was critical to legislation passed in 1986 in Brazil that led to the complete protection of sea turtles and the establishment of two new biological reserves and a marine national park to protect important nesting beaches. To manage these areas, Projeto TAMAR established conservation stations at each of 21 main nesting beaches to control and regulate activities that can affect turtles. Each station has a manager, several university interns, and local employees, many of whom are former fishermen who bring their knowledge of sea turtles to bear on conservation.

The stations' personnel regularly patrol the conservation areas, measuring turtles for size and permanently flipper-tagging all adults observed on the beach. In places where predators are abundant, some nests are covered with wire mesh fitted with small gaps to protect the eggs and then allow movement of the baby turtles after they hatch. Alternatively, the eggs are collected and brought to nearby hatchery areas, where they are reburied (Almeida and Mendes 2007). These measures allow baby turtles emerging from protected nests or hatcheries to enter the ocean just as if they had emerged from natural nests. TAMAR protects over 4000 turtle nests each year and has protected around 100,000 nests and approximately 7 million hatchlings in the years since its inception. On average, the number of turtle nests on the beaches has also been increasing by an impressive 20% a year (Marcovaldi and Chaloupka 2007).

TAMAR also provides local fishermen with information about the importance of turtles and about fishing gear designed to prevent turtle capture. Fishermen are taught techniques for reviving turtles caught in their nets so the turtles will not suffocate. Their increasing appreciation of turtles and their awareness of the new laws lead most fishermen to cooperate with these policies. However, accidental capture remains a leading cause of turtle mortality.

Projeto TAMAR plays a positive role in the communities where it operates. In many areas, TAMAR is the primary source of income for the local people, often providing child care facilities and small medical and dental clinics. Local people are employed in making turtle-themed crafts to sell to tourists. To increase awareness of the program at the local level, TAMAR personnel give talks about marine conservation in village schools and organize hatchling release ceremonies (**Figure 1.5**). Employees and local people have become strong advocates for turtles because they can see that their income and livelihood are linked to the presence of these animals.

The project reaches a wide audience in Brazil through coverage in popular articles and on television programs. In addition, TAMAR operates sea turtle educational centers where hundreds of thousands of tourists, most of whom are from Brazil, visit each year. The tourists get to see conservation in action and receive a large dose of conservation education; in turn, they support the project through their purchases at the centers' gift shops.

FIGURE 1.5 Projeto TAMAR generates publicity for sea turtle conservation by staging festive events involving tourists, school groups, and local people, such as this release of hatchlings that were incubated in the safety of a protected hatchery. (Photograph courtesy of Projeto TAMAR Image Bank.)

As a result of Projeto TAMAR's efforts, sea turtle numbers in Brazil have stabilized and even show signs of increasing. The project has changed people's attitudes, both in coastal villages and in the wider Brazilian society. By integrating conservation goals with community education and development, Projeto TAMAR has improved the future for sea turtles and for local people involved with their conservation.

Conservation Biology's Ethical Principles

Earlier in the chapter, we mentioned that conservation biology is a normative discipline in which certain value judgments are inherent. The field rests on an underlying set of principles that is generally agreed on by practitioners of the discipline (Soulé 1985) and can be summarized as follows:

1. *The diversity of species and biological communities should be preserved.* In general, most people agree with this principle simply because they appreciate biodiversity. The hundreds of millions of visitors each year to zoos, national parks, botanical gardens, and aquariums testify to the general public's interest in observing different species and biological communities. It has even been suggested that humans may have a genetic predisposition, called **biophilia**, to love biodiversity (**Figure 1.6**) (Corral-Verdugo et al. 2009). In addition, many people acknowledge the economic value of biodiversity (see Chapter 3).

> There are ethical reasons why people want to conserve biodiversity, such as belief that species have intrinsic value. Also, people may be naturally disposed to appreciate and value biodiversity.

2. *The untimely extinction of populations and species should be prevented.* The ordinary extinction of species and populations as a result of natural processes is an ethically neutral event. In the past, the local loss of a

FIGURE 1.6 People enjoy seeing the diversity of life, as shown by the growing popularity of butterfly gardens. (Photograph by Richard B. Primack.)

population was usually offset by the establishment of a new population through dispersal. However, as a result of human activity, the loss of populations and the extinction of species has increased by more than a hundredfold with no simultaneous increase in the generation of new populations and species (MEA 2005) (see Chapter 5).

3. *Ecological complexity should be maintained*. Many of the most valuable properties of biodiversity are expressed only in natural environments. For example, plants with unusual flowers are pollinated by specialized insects. These relationships would no longer exist if the animals and plants were housed separately and in isolation at zoos and botanical gardens. Although the biodiversity of species may be partially preserved in zoos and gardens, the ecological complexity that exists in natural communities will be lost without the preservation of natural areas.
4. *Evolution should continue*. Evolutionary adaptation eventually leads to new species and increased biodiversity. Therefore, continued evolution of populations in nature should be supported, in part by preserving genetic diversity and allowing dispersal and exchange of genetic material among populations. Although preserving endangered species in captivity is important, these species are cut off from the natural evolutionary processes and may not be able to survive if returned to the wild.
5. *Biological diversity has intrinsic value*. Species and the biological communities in which they live possess value of their own, regardless of their economic, scientific, or aesthetic value to human society. This value is conferred not just by their evolutionary history and unique ecological role but by their very existence. (See Chapter 3 for a more complete discussion of this topic.)

Not every conservation biologist accepts every one of these principles, and there is no hard-and-fast requirement to do so. Individuals or organizations who agree with even two or three of these principles, such as religious groups and hunters, are often willing to support conservation efforts.

Achievements and Challenges

In many ways, conservation biology is a crisis discipline. Decisions about park design, species management, and other aspects of conservation are made every day under severe time pressure. Conservation biologists and scientists in related fields are well suited to provide the advice that governments, businesses, and the general public need in order to make crucial decisions, but because of time constraints, scientists are often compelled to make recommendations without thorough investigation. Decisions must be made, with or without scientific input, and conservation biologists must be willing to express opinions and take action based on the best available evidence and informed judgment (Black 2011). They must also articulate a long-term conservation vision that extends beyond the immediate crisis (Nelson and Vucetich 2009; Wilhere 2012).

> Over the last several decades, the field of conservation biology has continued to grow in scope and influence. The United Nations even designated 2010 the International Year of Biodiversity and 2011 the Year of Forests.

The field of conservation biology has set itself some imposing—and absolutely critical—tasks: to describe the Earth's biological diversity, to protect what is remaining, and to restore what is degraded. The field is growing in strength, as indicated by increased governmental participation in conservation activities, increased funding of conservation organizations and projects, and an expanding professional society, the Society for Conservation Biology (**Figure 1.7**). The prominence of environmental concerns was highlighted by the award of the 2007 Nobel Peace Prize to former U.S. vice president Al Gore and the Intergovernmental Panel on Climate Change for bringing the issue of global climate change to public attention. The value of biodiversity is further highlighted by the United Nations designation of 2010 as the International Year of Biodiversity and 2011 as the Year of Forests.

FIGURE 1.7 The Society for Conservation Biology has a simple, yet powerful, logo showing the circle of life, within which we live. The ocean waves in the center symbolize the changes that lie ahead. The logo can also be viewed as a bird, which provides us with beauty; on closer look, we see that its wings are really rustling leaves. (Courtesy of the Society for Conservation Biology.)

Despite the threats to biological diversity, we can detect many positive signs that allow conservation biologists to remain cautiously hopeful (Sodhi et al. 2011). The proportion of people living in extreme poverty has been in decline since the Industrial Revolution, and the rate of human population growth has slowed (Sachs 2008). The number of protected areas around the globe continues to increase, with a dramatic expansion in the number of marine protected areas. In 2006, the South Pacific country of Kiribati established the world's largest marine sanctuary. Moreover, our ability to protect biological diversity has been strengthened by a wide range of local, national, and international efforts. Certain endangered species are now recovering as a result of conservation measures (Lotze et al. 2011). We can point to an expansion of our knowledge base and the science of conservation biology, the developing linkages with rural development and social sciences, and our increased ability to restore degraded environments. All of these suggest that progress is being made, despite the enormous tasks still ahead.

Summary

- Biological diversity includes the full range of species, genetic variation, and biological communities and their ecosystem-level interactions with the chemical and physical environment.
- Conservation biology describes biodiversity, identifies the threats biodiversity faces from human activities, and develops methods to protect and restore biodiversity.
- Conservation biology draws on scientific, social, economic, and philosophical ideas and traditions to accomplish its goals. Most conservation biologists accept a set of ethical principles that help to guide their practice.
- The current paradigm of sustainable development advocates an integrated approach to development and the use of natural resources that meets present and future human needs without damaging either the environment or biological diversity.
- The conservation of biodiversity has become an international undertaking. There are many successful projects already, such as the conservation of Brazilian sea turtles, that indicate that progress can be made.

For Discussion

1. How is conservation biology fundamentally different from other branches of biology, such as physiology, genetics, or cell biology? How is it different from environmentalism?
2. What do you think are the major conservation and environmental problems facing the world today? What are the major problems facing your local community? What ideas for solving these problems can you suggest? (Try answering this question now, and once again when you have completed this book.)
3. How would you characterize your own viewpoint about the conservation of biodiversity and the environment? Which of the religious or philosophical viewpoints of conservation biology stated here do you agree or disagree with? How do you, or could you, put your viewpoint into practice?

Suggested Readings

Barnosky, A. D. and 11 others. 2011. Has Earth's sixth mass extinction already arrived? *Nature* 471: 51–57. Evidence from the fossil records and modern extinction rates suggest that we are on the verge of a major extinction event.

Caro, T., J. Darwin, T. Forrester, C. Ledoux-Bloom, and C. Wells. 2012. Conservation in the Anthropocene. *Conservation Biology* 26: 185–188. Even though human activities dominate large areas of the Earth, it is important to remember and plan for the many places and ecosystems where human influence is still minimal.

Dudley, N., L. Higgins-Zogib, and S. Mansourian. 2009. The links between protected areas, faiths, and sacred natural sites. *Conservation Biology* 23: 568–577. In many places, local people are already protecting biodiversity.

Hall, J. A. and E. Fleishman. 2010. Demonstration as a means to translate conservation science into practice. *Conservation Biology* 24: 120–127. Conservation biologists can work with conservation organizations and the government to show the practical application of their ideas.

Leopold, A. 1949. *A Sand County Almanac*. Oxford University Press, New York. Leopold's evocative essays articulate his "land ethic," defining human duty to conserve the land and the living things that thrive upon it.

Lindenmayer, D. and M. Hunter. 2010. Some guiding concepts for conservation biology. *Conservation Biology* 24: 1459–1468. Ten general concepts are developed, with a web-based forum established for discussion.

Nelson, M. P. and J. A. Vucetich. 2009. On advocacy by environmental scientists: What, whether, why, and how. *Conservation Biology* 23: 1090–1101. Scientists have a responsibility to be advocates as well as researchers.

Sodhi, S. N., R. Butler, W. F. Laurance, and L. Gibson. 2011. Conservation successes at micro-, meso- and macroscales. *Trends in Ecology & Evolution* 26: 585–594. There are many examples of successful conservation that can be used to guide future actions.

Stearns, B. P. and S. C. Stearns. 2010. Still watching, from the edge of extinction. *BioScience* 60: 141–146. Many endangered species increasingly rely on human actions to prevent extinction.

Wilhere, G. F. 2012. Inadvertent advocacy. *Conservation Biology* 26: 39–46. Conservation biologists need to be aware of their two separate roles, as scientists providing information about the threats to biodiversity, and as advocates for threatened biodiversity.

KEY JOURNALS IN THE FIELD *Biodiversity and Conservation, Biological Conservation, BioScience, Conservation Biology, Conservation Letters, Ecological Applications, National Geographic, Trends in Ecology and Evolution*

A fire-tailed sunbird (*Aethopyga ignicauda*) visits flowers of the cinnabar rhododendron (*Rhododendron cinnabarinum*) in the eastern Himalayas of northeastern India.

Chapter 2
What Is Biodiversity?

The protection of biological diversity is central to conservation biology. Conservation biologists use the term biological diversity, or simply biodiversity, to mean the complete range of species, genetic variation within species, and ecosystems. By this definition, biodiversity must be considered on three levels:

1. *Species diversity*. All the species on Earth, including single-celled bacteria and protists as well as the species of the multicellular kingdoms (plants, fungi, and animals)
2. *Genetic diversity*. The genetic variation within species, both among geographically separate populations and among individuals within single populations
3. *Ecosystem diversity*. The range of different ecosystems, that is, biological communities and their associations with the chemical and physical environment (**Figure 2.1**)

All three levels of biological diversity are necessary for the continued survival of life as we know it, and all are important to people (Levin 2001; MEA 2005). **Species diversity** reflects the entire range of evolutionary and ecological adaptations of species to particular environments. It provides people with resources and resource alternatives—for example, a tropical rain forest or a temperate swamp with many species produces a wide variety of plant and animal

FIGURE 2.1 Biological diversity includes genetic diversity (the genetic variation found within each species), species diversity (the range of species in a given ecosystem), and community/ecosystem diversity (the variety of habitat types and ecosystem processes extending over a given region). (After Palumbi 2009.)

products that can be used as food, shelter, and medicine. **Genetic diversity** is necessary for any species to maintain reproductive vitality, resistance to disease, and the ability to adapt to changing conditions (Laikre et al. 2010). In domestic plants and animals, genetic diversity is of particular value in the breeding programs necessary to sustain and improve modern agricultural species and their disease resistance. **Ecosystem diversity** results from the collective response of species to different environmental conditions. Biological communities found in deserts, grasslands, wetlands, and forests support the continuity of proper ecosystem functioning, which provides crucial services to people, such as water for drinking and agriculture, flood control, protection from soil erosion, and filtering of air and water. We will now examine each level of biodiversity in turn.

Species Diversity

Species diversity includes the entire range of species found on Earth. Recognizing and classifying species is one of the major goals of conservation biology. How do biologists identify individual species among the mass of living organisms on Earth, many of them small in size and with few distinguishing features? And what is the origin of new species? Identifying the process whereby one species evolves into one or more new species is one of the ongoing accomplishments of modern biology. The origin of new species is normally a slow process, taking place over hundreds, if not thousands, of generations. The evolution of higher taxa, such as new genera and families, is an even slower process, typically lasting hundreds of thousands or even millions of years. In contrast, human activities are destroying in only a few decades the unique species built up by these slow natural processes.

What is a species?

A species is generally defined in one of two ways:

1. A group of individuals that is morphologically,* physiologically, or biochemically distinct from other groups in some important characteristic is the **morphological definition of species.**
2. A group of individuals that can potentially breed among themselves in the wild and that do not breed with individuals of other groups is the **biological definition of species.**

Because the methods and assumptions used are different, these two approaches to distinguishing species sometimes do not give the same results. Increasingly, differences in DNA sequences and other molecular markers distinguish species that look almost identical, such as types of bacteria (Francis et al. 2010).

The morphological definition of species is the one most commonly used by **taxonomists**, biologists who specialize in the identification of unknown specimens and the classification of species (**Figure 2.2**). In practice, the biological definition of species is difficult to use, because it requires a knowledge of which individuals actually have the potential to breed with one another and their relationships to each other—information that is rarely available. As a result, practicing field biologists learn to recognize one or more

> Using morphological and genetic information to identify species is a major activity for taxonomists; accurate identification of a species is a necessary first step in conservation.

*An individual's morphology is its form and structure—or, to put it more simply (if not totally accurately), its appearance.

(A)

(B)

FIGURE 2.2 (A) A plant ecologist prepares a museum specimen using a plant press. The flattened and dried plant will later be mounted on heavy paper with a label giving detailed collection information. (B) An ornithologist at the Museum of Comparative Zoology, Harvard University, classifying collections of orioles: black-cowled orioles (*Icterus prosthemelas*) from Mexico, and Baltimore orioles (*Icterus galbula*) that occur throughout eastern North America. (A, photograph by Richard Primack; B, photograph courtesy of Jeremiah Trimble, Museum of Comparative Zoology, Harvard University © President and Fellows of Harvard College.)

individuals that look different from other individuals and might represent a different species, sometimes referring to them as **morphospecies** or another such term until taxonomists can give them official scientific names (Norden et al. 2009).

Taxonomists collect specimens in the field and store them in places such as the world's 6500 natural history museums. These permanent collections form the basis of species descriptions and systems of classification. Each species is given a unique two-part name (a **binomial**), such as *Canis lupus* for the gray wolf. The first part of the name, *Canis*, identifies the genus (the canids, or dogs). The second part of the name, *lupus*, identifies the smaller group within the genus, the species that is the gray wolf. This naming system both separates a species and connects it to similar species—such as *Canis latrans*, the coyote, and *Canis rufus*, the red wolf.

Problems in distinguishing and identifying species are more common than many people realize (Bickford et al. 2007; Haig et al. 2006). A single species may have several varieties that have observable morphological differences, yet the individuals may be similar enough to interbreed and thus are considered a single biological species. The many breeds of domestic dogs, for instance, all belong to one species (*Canis familiaris*) and readily interbreed despite the conspicuous morphological differences among them. Alternatively, closely related sibling species appear very similar in morphology and physiology yet are biologically separate and may not interbreed.

To further complicate matters, individuals of related but distinct species may occasionally mate and produce **hybrids**, intermediate forms that blur the distinction between species. Sometimes hybrids are better suited to their environment than either parent species and go on to form new species. Hybridization is particularly common among plant species in disturbed habitats. Hybridization in both plants and animals frequently occurs when a few individuals of a rare species are surrounded by large numbers of a closely related species. A case in point is the endangered California tiger salamander (*Ambystoma californiense*) and the introduced, and now common, barred tiger salamander (*A. mavortium*). They are thought to be separate species that evolved from a common ancestor that lived 5 million years ago, yet they readily mate in California (**Figure 2.3**). Their hybrid offspring have a higher fitness than the native species, further complicating the conservation of this endangered species (Fitzpatrick et al. 2010).

The inability to clearly distinguish one species from another, whether due to similarity of characteristics or to confusion over correct scientific

FIGURE 2.3 The hybrid tiger salamander (left) is larger than its parent species, the California tiger salamander (right), and is increasing in abundance. Note the much larger head of the hybrid salamander. (Photograph courtesy of H. Bradley Shaffer.)

names, often slows efforts toward species protection (Gerson et al. 2008). It is difficult to write precise, effective laws to protect a species if the name of that species has not been defined. Tens of thousands of new species are being described each year, but even this rate is not fast enough. The key to solving this problem is to train more taxonomists, especially for work in the species-rich tropics (Wilson 2003).

Measuring species diversity

Conservation biologists often want to identify locations of high species diversity. Quantitative definitions of species diversity have been developed by ecologists as a means of comparing the overall diversity of different communities at varying geographical scales (Flohre et al. 2011).

Three quantitative indexes are used to denote species diversity at three different geographical scales:

- At its simplest level, species diversity has been defined as the number of species found in a given community, such as a lake or a meadow, a measure called **species richness** or **alpha diversity** (Gabriel et al. 2006).
- **Gamma diversity** applies to larger geographical scales and refers to the number of species found across a large region with a number of ecosystems, such as a mountain range or a continent.
- **Beta diversity** links alpha and gamma diversity and represents the *rate of change of species composition as one moves across a large region*. For example, if every lake in a region contained a similar array of fish species, then beta diversity would be low; on the other hand, if the bird species found in one forest were entirely different from the bird species in separate but nearby forests, then beta diversity would be high. A measure of beta diversity can be obtained by dividing gamma diversity by alpha diversity.

We can illustrate the three types of diversity with the theoretical example in **Figure 2.4**. Region 1 has the highest alpha diversity, with more species per mountain on average (six species) than the other two regions; Region 2 has the highest gamma diversity, with a total of 10 species. Dividing gamma by alpha informs us that Region 3 has a higher beta diversity than Region 1 or 2 because each of its species is found on only one mountain.

Identifying patterns of species diversity helps conservation biologists establish which locations are most in need of protection.

In practice, indexes of diversity are often highly correlated. The plant communities of the eastern foothills of the Andes Mountains, for instance, show high levels of diversity in terms of the alpha, beta, and gamma scales. These quantitative definitions of diversity capture only part of the broad definition of biodiversity used by conservation biologists. However, they are useful for comparing regions of the world and highlighting areas that have large numbers of native species requiring conservation protection.

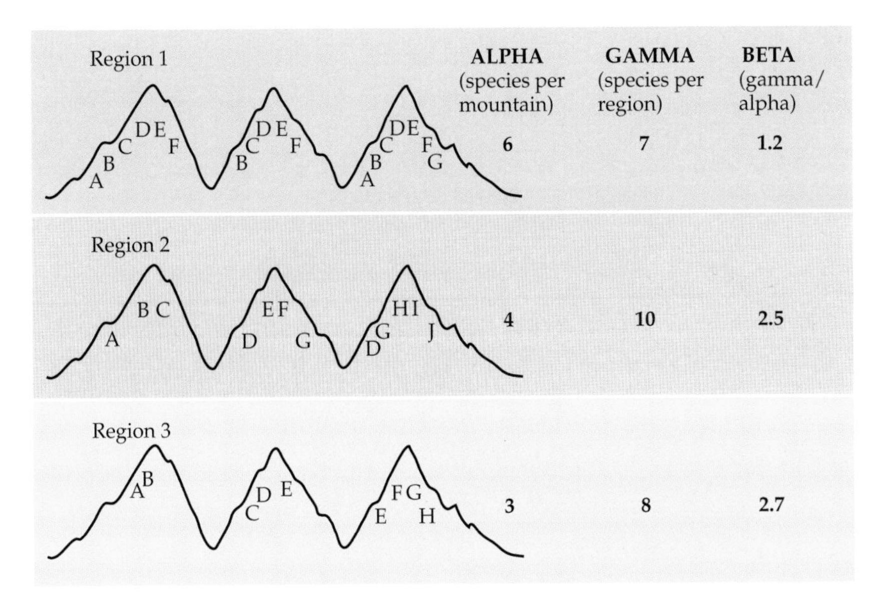

FIGURE 2.4 Biodiversity indexes for three regions, each consisting of three separate mountains. Each letter represents a population of a species; some species are found on only one mountain, while other species are found on two or three mountains. Alpha, gamma, and beta diversity values are shown for each region. If funds were available to protect only one *mountain range*, Region 2 should be selected because it has the greatest gamma (total) diversity. However, if only one *mountain* could be protected, a mountain in Region 1 should be selected because these mountains have the highest alpha (local) diversity, that is, the greatest average number of species per mountain. Each mountain in Region 3 has a more distinct assemblage of species than the mountains in the other two regions, as shown by the higher beta diversity. If Region 3 were selected for protection, the relative priority of the individual mountains should then be judged based on how many unique species are found on each mountain.

Genetic Diversity

A species consists of one or more populations, and the individuals in those populations may be genetically different from one another. A **population** is a group of individuals at a certain place that can potentially mate with one another and produce offspring. A population may consist of only a few individuals or millions of individuals, provided that some of the individuals actually produce offspring together.

Individuals within a population may be genetically different from one another to varying degrees. This genetic diversity, more properly called **genetic variation**, arises because individuals have slightly different forms of the DNA sequences that make up **genes**—the functional units that are the blueprints for the proteins of life (such as the enzymes in a mammal's

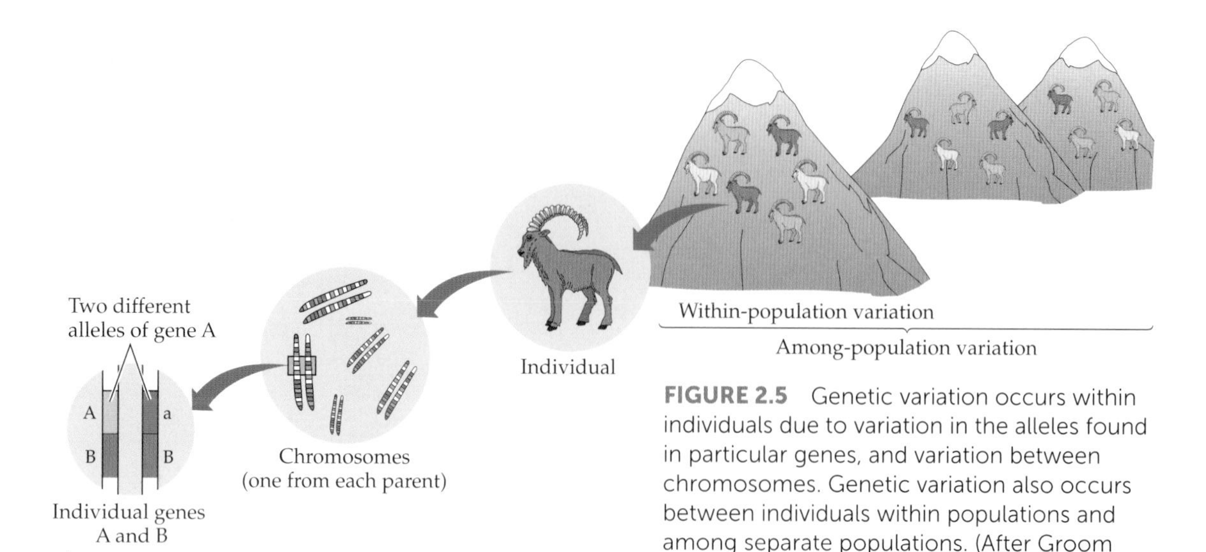

FIGURE 2.5 Genetic variation occurs within individuals due to variation in the alleles found in particular genes, and variation between chromosomes. Genetic variation also occurs between individuals within populations and among separate populations. (After Groom et al. 2006.)

digestive system or the hemoglobin that carries oxygen in the blood). Different forms of a gene are known as **alleles**, and the differences originally arise through **mutations** that change the DNA sequence. The various alleles of a gene may affect the development and physiology of an individual organism and may affect its **fitness**—the relative ability of an individual to survive and reproduce. In addition to genetic differences among individuals within populations, there may be genetic differences among populations of the same species in the types of alleles and the frequency of alleles (**Figure 2.5**).

Genetic variation increases as a result of sexual reproduction because of the **recombination** of genes. New genetic combinations are created when sets of chromosomes from two parents combine to form a genetically unique offspring. Although mutations provide the basic material, the random rearrangement of alleles in different combinations that characterizes sexually reproducing species dramatically increases the potential for genetic variation.

The total array of genes and alleles in a population or a species is its **gene pool**, while the particular combination of alleles that any individual possesses is its **genotype**. The **phenotype** of an individual represents the morphological, physiological, anatomical, and biochemical characteristics of the individual that result from the expression of its genotype in a particular environment. Examples of phenotypes include eye color, blood type, and forms of certain enzymes. These are physical qualities that are determined predominantly by an individual's genotype.

The amount of genetic variation in a species is determined by both the number of **polymorphic genes**—genes that have more than one allele—and the number of alleles that exist for each of these genes. The existence of a polymorphic gene also means that some individuals in a population will be **heterozygous** for the gene; that is, they will receive a different allele of the same gene from each parent. On the other hand, some individuals will be **homozygous**: they will receive the same allele from each parent. All these levels of genetic variation contribute to a population's ability to adapt to a changing environment.

Many rare species have less genetic variation than widespread species and, consequently, less flexibility to adapt when environmental conditions change, which in turn leaves them vulnerable to extinction (Frankham et al. 2009). Genetic variation is also important to the continued improvement of the crop plants and domestic animals on which we depend for food, as humans artificially select and hybridize species to obtain crop varieties with qualities that make them valuable for agriculture.

Genetic variation within a species can allow the species to adapt to environmental change; genetic variation can also increase the value of domestic species to people.

Ecosystem Diversity

Ecosystems are diverse, and this diversity is apparent even across a particular landscape. As we climb a certain mountain, for example, the kinds of plants and animals present gradually change from those found in a tall forest to those found in a low, moss-filled forest to those in an alpine meadow to those of a cold, rocky mountain peak. As we move across the landscape, physical conditions (soil type, temperature, precipitation, and so forth) change, and one by one the species present at the original location drop out, and we encounter new species that were not found at the starting point. The landscape as a whole is dynamic and changes in response to physical and biological components of the environment.

A **biological community** is defined as all the species that occupy a particular locality and the interactions among those species. A biological community together with its associated physical and chemical environment is termed an **ecosystem**. Many characteristics of an ecosystem result from ongoing ecosystem processes, including water cycles, nutrient cycles, and energy capture. Water evaporates from leaves, the ground, and other surfaces, to fall again elsewhere as rain or snow and replenish terrestrial and aquatic environments. Soil is built up from parent rock material and decaying organic matter. Photosynthetic plants absorb energy from sunlight, which is used in the plants' growth. This energy is then captured by animals that eat the plants, and also by the animals that eat the animals that eat the plants. The energy is also released as heat—both during the animals' lives and after the plants and animals die and decompose. Plants absorb carbon dioxide and release oxygen during photosynthesis, whereas animals and fungi absorb oxygen and release carbon dioxide during respiration. Mineral nutrients such as nitrogen and phosphorus cycle between the living and the nonliving components of the ecosystem. Ecosystem processes occur at geographical scales ranging from square meters to hectares to square kilometers, all the way to regional scales involving tens of thousands of square kilometers (MEA 2005).

The physical environment, especially annual cycles of temperature and precipitation and the characteristics of the land surface, affects the structure and characteristics of a biological community and profoundly influences whether a site will be a forest, grassland, desert, or wetland. In aquatic ecosystems, physical characteristics of the water, such as turbulence, clarity, chemistry, temperature, and depth, affect the characteristics of the associated **biota** (the full complement of living things in a region). The biological community can also alter the physical characteristics of an environment. For example, the vegetation present in a given location of a terrestrial ecosystem can affect wind speed, humidity, and temperature. Likewise, marine communities such as kelp forests and coral reefs can affect the temperature and motion of the water in their locations.

Within a community, each species has its own requirements for food, temperature, water, and other resources, any of which may limit its population size and its distribution.

Within a biological community, species play different roles and have varying requirements for survival (Marquard et al. 2009). A given plant species may grow best in one type of soil under certain conditions of sunlight and moisture, be pollinated by only certain types of insects, and have its seeds dispersed by certain bird species. Animal species differ in their requirements, such as the types of food they eat and the types of resting places they prefer. Any of these requirements may become a **limiting resource** that restricts the population size and distribution of the species.

As a result of its particular requirements, behaviors, or preferences, a given species often ends up appearing at a particular time during the process of ecological succession (Swanson et al. 2011). **Succession** is the gradual process of change in species composition, community structure, soil chemistry, and microclimatic characteristics that occurs following natural or human-caused disturbance in a biological community. For example, sun-loving butterflies and annual plants most commonly are found early in succession, in the months or few years immediately following a hurricane or after a logging operation has cleared an old-growth forest. At that time, with the tree canopy disrupted, the ground receives high levels of sunlight, with high temperatures and low humidity during the day. Over the course of decades, the forest canopy is gradually reestablished. Different species, including shade-tolerant wildflowers of moist soils, and birds that nest in holes in dead trees, thrive in the mid- and late-successional stages. Human management patterns often upset the natural pattern of succession; for example, grasslands overgrazed by cattle, and forests from which all the large trees have been harvested for timber no longer contain late-successional species that would otherwise exist there.

Species interactions

The composition of communities is often affected by competition and predation (Cain et al. 2008). **Predators** are animals that hunt and kill **prey**, which are the animals that are eaten. Predators may dramatically reduce the densities of certain prey species and even eliminate some species from particular habitats. Indeed, predators may indirectly increase the diversity of prey species in a community by keeping the density of each species low enough that severe competition for resources does not occur.

In many communities, predators keep the number of individuals of a particular prey species below the number that the resources of an ecosystem can support—a number termed the habitat's **carrying capacity**. If the predators are removed by hunting, fishing, or some other human activity, the prey population may increase to carrying capacity, or it may increase beyond carrying capacity to a point at which crucial resources are overtaxed and the population crashes. In addition, the population size of a species may be controlled by other species that compete with it for the same resources; for example, the population size of rare terns that nest on a small island may decline if a gull species that uses the same nesting sites becomes abundant, or it may grow if the gull species is eliminated from the community.

Community composition is also affected when two species benefit each other in a **mutualistic relationship** (**Figure 2.6**). Mutualistic species reach higher densities when they occur together than when only one of the species is present. One example of mutualism is the relationship between fruit-eating birds and plants with fleshy fruit containing seeds that are dispersed by birds. Another example is flower-pollinating insects and flowering plants. At the extreme of mutualism, two species that are always found together and apparently cannot survive without each other form a **symbiotic relationship**. For example, the death of certain types of coral-inhabiting algae—a result of unusually high seawater temperatures—is followed by the weakening and

FIGURE 2.6 In Peru, a white-necked Jacobin hummingbird (*Florisuga mellivora*) feeds on nectar from a brightly colored legume flower and, in the process, pollinates the plant. (Photograph courtesy of L. Mazariegos.)

subsequent death of their associated coral species. In such cases, conservation efforts must target both species.

Biological communities can be organized into **trophic levels** that represent the different ways in which species obtain energy from the environment (**Figure 2.7**). **Primary producers** comprise the *first trophic level*. These organisms obtain their energy directly from the sun via photosynthesis. In terrestrial environments, the flowering plants, gymnosperms (e.g., conifers), and ferns are responsible for most photosynthesis, while in aquatic environments, seaweeds, single-celled algae, and cyanobacteria ("blue-green algae") are the most important primary producers. All of the species use solar energy to turn simple chemicals like carbon and nitrogen into the organic molecules they need to live and grow. As a consequence of less energy being transferred to each successive trophic level, the greatest biomass (living weight) in a terrestrial ecosystem is usually that of the plants.

The *second trophic level* contains the **herbivores**, which eat photosynthetic species and are thus known as **primary consumers**. The intensity of grazing by herbivores often determines the relative abundance of plant species and even the amount of plant material present.

Carnivores are in the *third and higher trophic levels*. **Carnivores** obtain energy by eating other animals. At the third trophic level are **secondary consumers** (e.g., foxes), predators that eat herbivores (e.g., rabbits). At the *fourth trophic level* are tertiary consumers (e.g., bass), predators that eat other predators (e.g., frogs). Some secondary and higher consumers combine direct predation with scavenging behavior, and others, known as **omnivores**, include both animal and plant foods in their diets. In general, predators occur in lower densities than their prey, and populations at higher trophic levels contain fewer individuals than those at lower trophic levels.

Parasites and disease-causing organisms form an important subclass of predators. Parasites of animals, including mosquitoes, ticks, intestinal worms, and protozoans, as well as microscopic disease-causing organisms such as some bacteria and viruses, do not kill their hosts immediately, if ever. Plants can also be attacked by bacteria, viruses, and a variety of parasites that include fungi, other plants (such as mistletoe), nematode worms, and insects. The effects of parasites range from imperceptibly weakening their hosts to totally debilitating or killing them over time. The spread of parasites and disease from captive or domestic species, such as domestic dogs, to wild species, such as lions, is a major threat to many rare species (see Chapter 4).

Decomposers and **detritivores** feed on dead plant and animal tissues and wastes (detritus), breaking down complex tissues and organic molecules into the simple chemicals that are the building blocks of primary production. Decomposers release minerals such as nitrates and phosphates back into the soil and water, where they can be taken up again by plants and algae. Decomposers are usually much less conspicuous than herbivores and carnivores, but their role in the ecological community is vital. The most

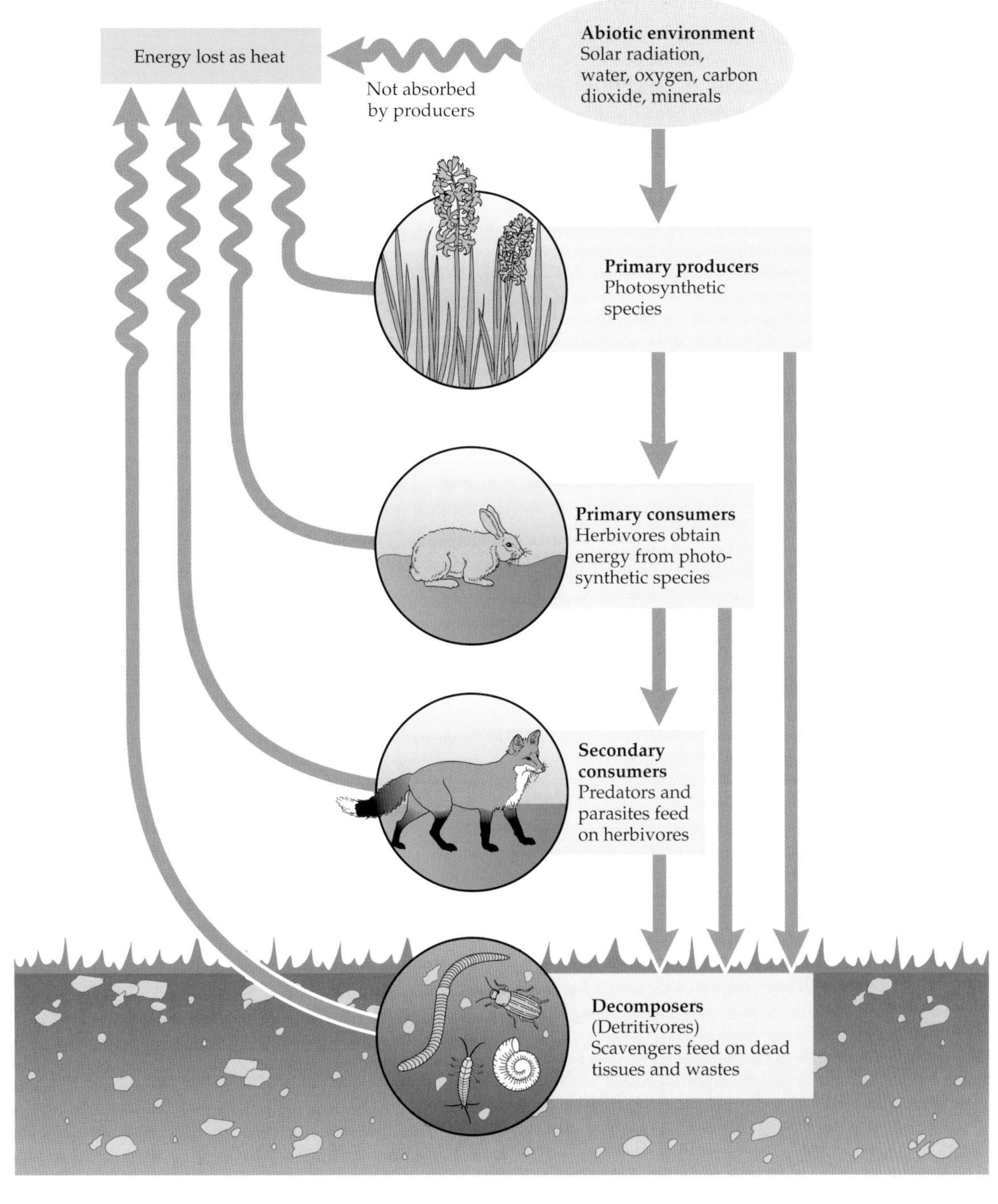

FIGURE 2.7 A model of a field ecosystem, showing its trophic levels and simplified energy pathways.

important decomposers are fungi and bacteria, but a wide range of other species play a role in breaking down organic materials. For example, vultures and other scavengers tear apart and feed on dead animals, dung beetles feed on and bury animal dung, and worms break down fallen leaves and other organic matter. Crabs, worms, mollusks, fish, and numerous other organisms eat detritus in aquatic environments. If decomposers were to die off, organic material would accumulate and plant growth would decline greatly (Gessner et al. 2010).

Food chains and food webs

Although species can be organized into general trophic levels, their actual requirements or feeding habits within the trophic levels may be quite restricted. For example, a certain lady beetle species may feed on only one type of aphid, and a certain aphid species may feed on only one type of plant. These specific feeding relationships are termed **food chains**. The more common situation in many biological communities, however, is for one species to feed on several other species at the lower trophic level, to compete for food with several species at its own trophic level, and to be preyed upon by several species at the next higher trophic level. Consequently, a more accurate description of the organization of biological communities is a **food web**, in which species are linked together through complex feeding relationships. Species at the same trophic level that use approximately the same environmental resources are considered a **guild** of competing species. Humans can substantially alter the relationships in food webs. For example, in urban settings, bird populations may increase because of reduced predation, reducing insect abundance in the process (Faeth et al. 2005).

Keystone species and resources

Within biological communities, certain species or guilds of species with similar ecological features may determine the ability of many other species to persist in the community (**Figure 2.8**). These **keystone species** affect the organization of the community to a far greater degree than one would predict if considering only the number of individuals or the biomass of the keystone species (Estes et al. 2011). Protecting keystone species is a priority for conservation efforts because loss of a keystone species or guild may lead to loss of numerous other species as well.

Keystone species strongly affect the abundance and distribution of other species in an ecosystem. Protecting and restoring keystone species is a conservation priority.

Top predators may be keystone species, because predators often have marked influence on herbivore populations (Wallach et al. 2009). The elimination of even a small number of individual predators, though representing only a minute amount of the community biomass, may result in dramatic changes in the vegetation and a great loss in biodiversity

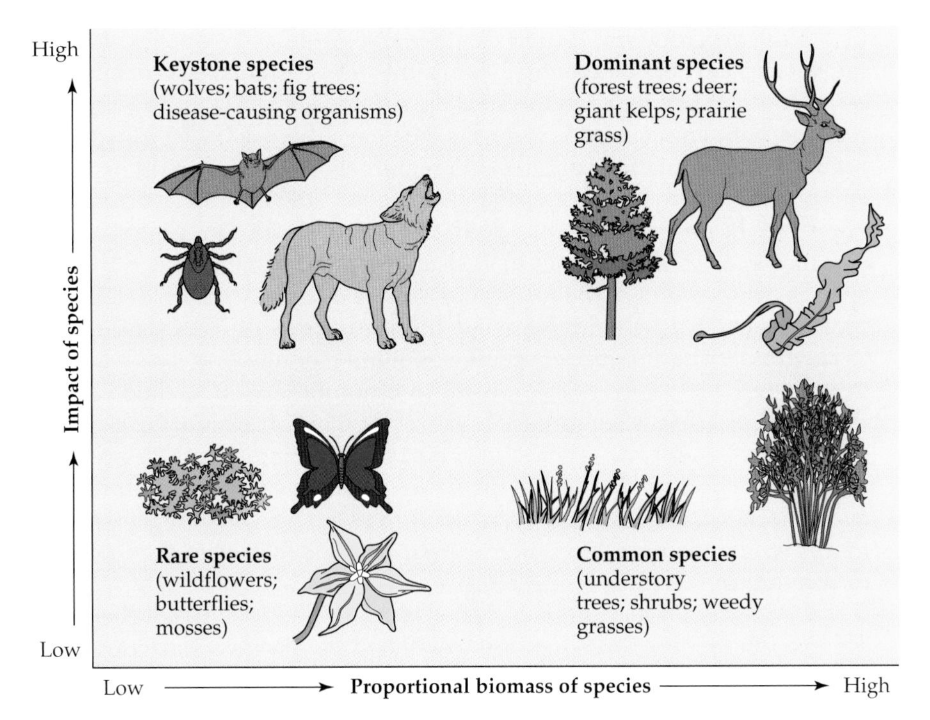

FIGURE 2.8 Keystone species determine the ability of large numbers of other species to persist within a biological community. Although keystone species make up only a small percentage of the total biomass, a community's composition would change radically if one of them were to disappear. Rare species have minimal biomass and seldom have significant impact on the community. Dominant species constitute a large percentage of the biomass and affect many other species in proportion to this large biomass. Some species, however, have a relatively low impact on the community organization despite being both common and heavy in biomass. (After Power et al. 1996.)

that is sometimes called a **trophic cascade** (Ripple and Beschta 2012). For example, in some places where gray wolves have been hunted to extinction by humans, deer populations have exploded. The deer severely overgraze the habitat, eliminating many herb and shrub species and tree seedlings. The loss of these plants, in turn, is detrimental to the deer and to other herbivores, including insects. The reduced plant cover may lead to soil erosion, also contributing to the loss of species that inhabit the soil. When wolves are restored to ecosystems, trophic relationships can sometimes be reestablished (Beyer et al. 2007).

Species that extensively modify the physical environment through their activities, often termed "ecosystem engineers," are considered keystone species as well (Byers et al. 2006). Beavers, for example, build dams that flood temperate forests, creating new wetland habitat for many species.

Losing keystone species can create a series of linked extinction events, known as an **extinction cascade**, that results in a degraded ecosystem with much lower diversity at all trophic levels (Letnic et al. 2009). This may already be happening in tropical forests where overharvesting has drastically reduced the populations of birds and mammals that act as predators, seed dispersers, and herbivores (Nunez-Iturri et al. 2008). While such a forest appears to be green and healthy at first glance, it is really an "empty forest" in which ecological processes have been irreversibly altered such that the species composition of the forest will change over succeeding decades or centuries. In the marine environment, the loss of key structural species such as sea grasses and seaweeds because of pollution or trawling can lead to the loss of specialized species that inhabit communities based on the structure they provide, such as delicate sea horses and other fish (Hughes et al. 2009).

If the few keystone species in a community being affected by human activity can be identified, sometimes they can be carefully protected or even encouraged. Most notably, when wolves and beavers are returned to places where they used to occur, many other native species also increase in abundance.

Often nature reserves are compared and valued in terms of their size because, on average, larger reserves contain more species and habitats than smaller reserves. However, area alone does not ensure that a nature reserve contains the full range of crucial habitats and resources. Particular habitats may contain critical **keystone resources**, often physical or structural, that occupy only a small area yet are crucial to many species in the ecosystem (Kelm et al. 2008). For example, deep pools in streams, springs, and ponds may be the only refuge for fish and other aquatic species during the dry season, when water levels drop. For terrestrial animals, these water sources may provide the only available drinking water for a considerable distance. Hollow tree trunks and tree holes are keystone resources as breeding sites for many bird and mammal species and may limit their population sizes (Cockle et al. 2011). Protecting old hollow trees as a keystone resource is a priority during certain logging activities.

Ecosystem dynamics

An ecosystem in which the chemical and physical processes are functioning normally, whether or not there are human influences, is referred to as a **healthy ecosystem**. The health of an ecosystem might be measured by the weight of plant material it produces per year, or the numbers of individual animals it contains. In many cases, ecosystems that have lost some of their species will remain healthy because there is often some redundancy in the roles performed by ecologically similar species. Ecosystems that are able to remain in the same state are referred to as **stable ecosystems**. These systems remain stable either because of lack of disturbance or because they have special features that allow them to remain stable in the face of disturbance. Such stability despite disturbance could result from one or both of two features: resistance and resilience. **Resistance** is the ability to maintain the same state even with ongoing disturbance; that would be the case if after an oil spill, a river ecosystem retained its major ecosystem processes. **Resilience** is the property of being able to return to the original state quickly after disturbance has occurred; that would be true if following contamination by an oil spill and the deaths of many animals and plants, a river ecosystem returned to its original condition. For example, when nonnative fish are introduced in previously fish-free ponds, the number of native aquatic insect species declines, indicating low resistance; but when the fish are removed as part of a conservation program, the number of native species soon recovers, indicating high resilience (Knapp et al. 2005).

Biodiversity Worldwide

Developing a strategy for conserving biodiversity requires a firm grasp of how many species exist on Earth and how those species are distributed across the planet. The answers to both questions can be complex.

How many species exist?

At present, about 1.5 million species have been described. At least twice this number of species, primarily insects and other arthropods in the tropics, remain undescribed (**Figure 2.9**). Our knowledge of species numbers is imprecise because inconspicuous species have not received their proper share of taxonomic attention. For example, spiders, nematodes, and fungi living in the soil, insects living in the tropical forest canopy, and microorganisms in seawater are small and difficult to study. These poorly known groups could number in the hundreds of thousands, or even millions, of species. Our best estimate, based on DNA analysis of well-known groups and on mathematical patterns, is that there are between 5 million and 10 million species (Strain 2011). This number could be many times higher, or even up to 100 million species, if it turns out that each species or genus of animal and plant has many unique species of bacteria, protists, and fungi living on or inside it.

Many scientists are working to determine the number of species in the world. The best estimate is that there are about 5 million to 10 million species, with about half of them being insects.

Amazingly, around 20,000 new species are described each year. While certain groups of birds, mammals, and temperate flowering plants are relatively well known, several new species in these groups are discovered annually (Joppa et al. 2011). Even among a group as well studied as the primates, new monkey species have been discovered in Brazil and dozens of new lemur species have been found in Madagascar—all since 1990. Between 500 and 600 new species of amphibians are described each decade.

New species may be discovered in unexpected places, as members of an international research team found when they noticed an unusual entrée on the grill in a Laotian food market. Natives called it "kha-nyou"; although

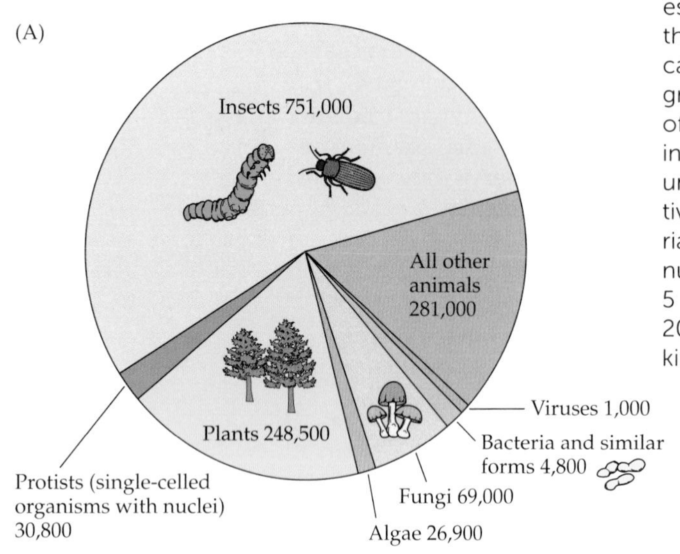

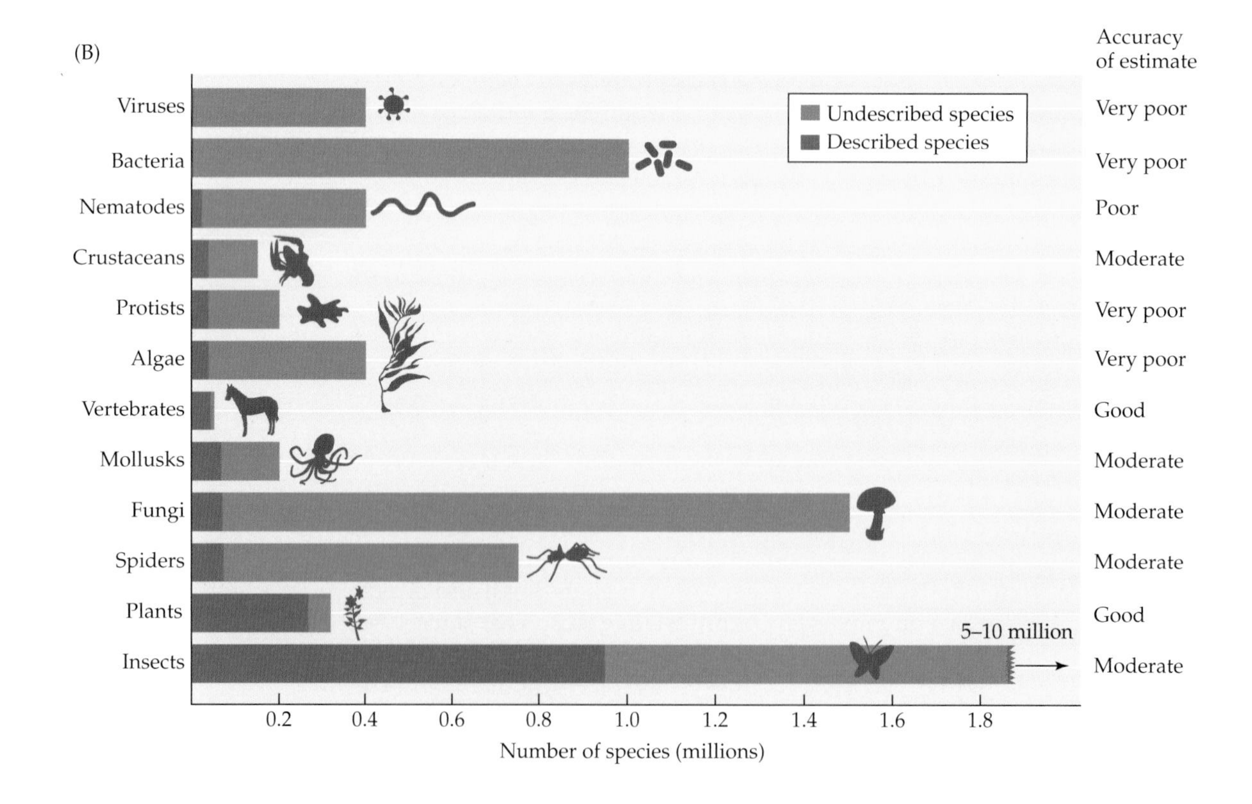

FIGURE 2.9 (A) Approximately 1.5 million species have been identified and described by scientists; the majority of these are insects and plants. (B) For several groups estimated to contain over 100,000 species, the numbers of described species are indicated by the blue portions of the bars; the green portions are estimates of the number of undescribed species. The vertebrates are included for comparison. The number of undescribed species is particularly speculative for the microorganisms (viruses, bacteria, protists). Most estimates of the possible number of identifiable species range from 5 million to 10 million. (A, data from Wilson 2010; B, data from Groombridge and Jenkins 2010.)

FIGURE 2.10 Researchers first encountered *Laonastes aenigmamus* being sold as a delicacy in Laotian food markets. This recently discovered species belongs to a group of rodents previously believed to have been extinct. In 2006, David Redfield of Florida State University led an expedition that was able to obtain the first photographs of a living *L. aenigmamus*. (Photograph by Uthai Treesucon, courtesy of *Research in Review*, FSU.)

clearly a rodent, it was not an animal known to any of the researchers. In 2006, after several years of studying skeletons and dead specimens, taxonomists deemed kha-nyou to be a heretofore unknown species belonging to a rodent family thought to have been extinct for 11 million years (Dawson et al. 2006). The newly discovered species was given the scientific name *Laonastes aenigmamus*, or "rock-dwelling, enigmatic mouse." More commonly called the Laotian rock rat or rock squirrel, this rodent is neither a mouse nor a rat nor a squirrel, but a unique species (**Figure 2.10**).

In addition to new species, entire biological communities continue to be discovered, usually in remote and inaccessible localities. These communities often consist of inconspicuous species of bacteria, protists, and small invertebrates that have escaped the attention of earlier taxonomists. Specialized exploration techniques have aided in these discoveries, particularly in the deep sea and in the forest canopy. Some recently discovered communities include the following:

- Diverse communities of animals, particularly insects, are adapted to living in the canopies of tropical trees and rarely, if ever, descend to the ground (Lowman et al. 2006). The use of technical climbing equipment, canopy towers and walkways, and tall cranes is opening this habitat to exploration.
- Using DNA technology to investigate the interior of leaves of healthy tropical trees has revealed an extraordinarily rich group of fungi, consisting of thousands of undescribed species (Arnold and Lutzoni 2007). These fungi appear to aid the plants in excluding harmful bacteria and fungi in exchange for receiving a place to live and perhaps some carbohydrates.

- Investigations of bacterial communities using new sampling techniques have revealed a great diversity of species. For example, the floor of the deep sea has unique communities of bacteria and animals that grow around geothermal vents and in marine sediments (Scheckenback et al. 2010). Drilling projects have shown that diverse bacterial communities exist 2.8 km deep in the Earth's crust, at densities of up to 100 million bacteria per gram of solid rock (Fisk et al. 1998). These bacterial communities are being actively investigated as a source of novel chemicals, for their potential usefulness in degrading toxic chemicals, and for insight into whether life could exist on other planets.

Although the marine environment appears to be a great frontier of biodiversity, the difficulty of collecting there has particularly hampered our knowledge of the number of marine species (Koslow 2007). Huge numbers of marine species remain to be discovered as the methods of investigation improve. A recent expedition to the oceans around Antarctica collected 674 different species of isopods, a group of small arthropods that includes the common pillbug. In that collection, 585 were new species (Brandt et al. 2007).

DNA analyses suggest that many thousands of species of bacteria have yet to be described. The marine environment also contains large numbers of species unknown to science.

Bacteria are especially poorly known and thus underrepresented in estimates of the total number of species on Earth (Azam and Worden 2004). Only about 5000 species of bacteria are currently recognized by microbiologists. However, work analyzing bacterial DNA indicates that there may be 6400 to 38,000 species in a single gram of soil and 160 species in a liter of sea water (Nee 2003). Such high diversity in such small samples suggests that there could be hundreds of thousands or even millions of undescribed bacteria species. And closer to home, the Human Microbiome Project has demonstrated that the human body is occupied by 10 times more bacterial cells than human cells, with many still undescribed species occupying specific places both inside our bodies and on our skin. While some of these bacteria species are involved in disease, many of them are probably important in maintaining our health. This is a very active area of current research.

If the task of describing the world's species continues at its present rate, the job will not be completed for over 250 years. This fact underlines the critical need for taxonomists trained to use the latest molecular technology and for web-based information sharing. International databases such as those of the Global Biodiversity Information Facility (http://data.gbif.org) and the Encyclopedia of Life project (www.eol.org) will make species names and descriptions more widely and readily available.

Where is the world's biodiversity found?

The most species-rich environments appear to be tropical rain forests and deciduous forests, coral reefs, the deep sea, large tropical lakes and river systems, and regions with Mediterranean climates (Vadeboncoeur et al. 2011; Anderson et al. 2011; Forzza et al. 2012).

> Species diversity is greatest in the tropics, particularly in tropical forests and coral reefs.

TROPICAL FORESTS Even though the world's tropical forests occupy only 7% of the land area, they appear to contain over half the world's species, most of which are insects (Gaston and Spicer 2004; Corlett and Primack 2010). Tropical forests also have many species of birds, mammals, amphibians, and plants. For flowering plants, gymnosperms, and ferns, about 40% of the world's 275,000 species occur in the world's tropical forest areas. Each of the rain forest areas in the Americas, Africa, Madagascar, Southeast Asia, New Guinea, Australia, and various tropical islands has a different biogeographic history, resulting in unique assemblages of species (see Figure 4.4). For example, lemurs are found only in Madagascar, and hummingbirds are found only in the Americas.

CORAL REEFS Colonies of tiny coral animals build the large coral reef ecosystems—the marine equivalent of tropical rain forests in both species richness and complexity (**Figure 2.11**) (Knowlton and Jackson 2008). These

FIGURE 2.11 Coral reefs are built up from the skeletons of billions of tiny individual animals. The intricate coral landscapes create a habitat for a diversity of other marine species, including many different kinds of fish. This reef is in the Maldives, an island nation in the equatorial Indian Ocean. (Photograph © Wolfgang Amri/istock.)

coral reefs provide homes for huge numbers of species of fish, mollusks, and other marine mammals. One explanation for this richness is the high primary productivity of coral reefs, which produce 2500 grams of biomass per square meter per year, in comparison with 125 $g/m^2/y$ produced in the open ocean. The clarity of the water in the reef ecosystem allows sunlight to penetrate deeply so that high levels of photosynthesis occur in the algae that live mutualistically inside the coral. Extensive niche specialization among coral species and adaptations to varying levels of disturbance may also account for the high species richness found in coral reefs.

The world's largest coral reef is Australia's Great Barrier Reef, with an area of 349,000 km^2. The Great Barrier Reef contains over 400 species of coral, 1500 species of fish, 4000 species of mollusks, and 6 species of turtles, and it provides breeding sites for some 252 species of birds. Although the Great Barrier Reef occupies only 0.1% of the ocean surface area, it contains about 8% of the world's fish species. Scientists are also now beginning to learn about deep sea corals that live in deep, cold environments without light (Roark et al. 2006). These deep sea coral communities are still poorly known, but they appear to be rapidly declining due to destructive trawling practices.

OCEANIC DIVERSITY In coral reefs and the deep sea, diversity is spread over a much broader range of phyla and classes than in terrestrial ecosystems. These marine systems contain representatives of 28 of the 35 animal phyla that exist today; one-third of these phyla exist only in the marine environment (Grassle 2001). In contrast, only one phylum is found exclusively in the terrestrial environment. Diversity in the ocean may be due to great age, enormous water volume, degree of isolation of certain seas by intervening landmasses, the stability of the environment, and specialization on particular sediment types and water depths (adding a third dimension to the space occupied). High freshwater diversity is also found in complex river systems and tropical lakes, with individual species having restricted distribution.

MEDITERRANEAN-TYPE COMMUNITIES Great diversity is found among plant species in southwestern Australia, the Cape Region of South Africa, California, central Chile, and the Mediterranean basin, all of which are characterized by a Mediterranean climate of moist winters and hot, dry summers (see Figure 7.5). The Mediterranean basin is the largest in area (2.1 million km^2) and has the most plant species (22,500) (Conservation International and Caley 2008); the Cape Floristic Region of South Africa has an extraordinary concentration of unique plant species (9000) in a relatively small area (78,555 km^2). The shrub and herb communities in these areas are rich in species apparently because of their combination of considerable geological age, complex site characteristics (such as topography and soils), and severe environmental conditions. The frequency of fire in these areas also may favor rapid speciation and prevent the domination of just a few species.

The distribution of species

At various spatial scales, there are concentrations of species in particular places, and there is often a correspondence in the distribution of species richness between different groups of organisms (Schuldt and Assman 2010). For example, in South America, concentrations of amphibians, birds, mammals, and plants are greatest in the western Amazon, with secondary concentrations in the highlands of northeastern South America and the Atlantic forests of southeastern Brazil (**Figure 2.12**). In North America, large-scale patterns of species richness for amphibians, birds, butterflies, mammals, reptiles, snails, trees, and tiger beetles are highly correlated; that is, a region with numerous species of one group will tend to have numerous species of the other groups. On a local scale, this relationship may break down; for example, amphibians may be most diverse in wet, shady habitats, whereas reptiles may be most diverse in drier, open habitats. At a global scale, each group of living organisms may reach its greatest species richness in a different part of the world because of historical circumstances or the suitability of the site to its needs.

Places with large concentrations of species often have a high percentage of **endemic species**, that is, species that occur there and no place else. The countries with the largest numbers of endemic mammals, which represent important targets for conservation efforts, are Indonesia (259), Australia (241), Brazil (183), Madagascar (187), and Mexico (158) (www.iucnredlist.org). Geographically isolated countries and islands also have a high percentage of endemic species. For example, most of the species in Madagascar are endemic because it is a large, ancient island on which many unique species have evolved in isolation. Species-rich countries such as South Africa, have

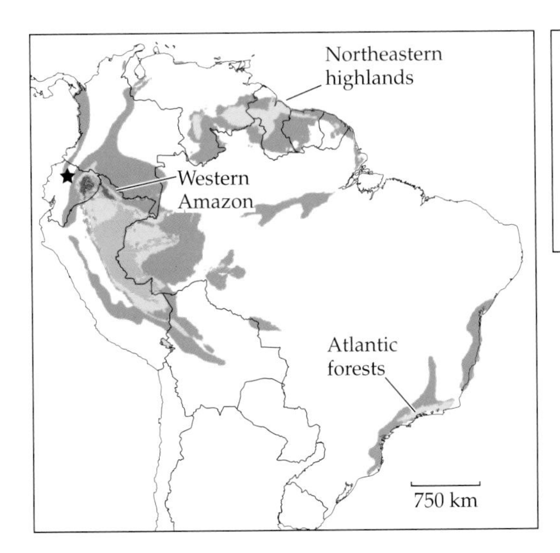

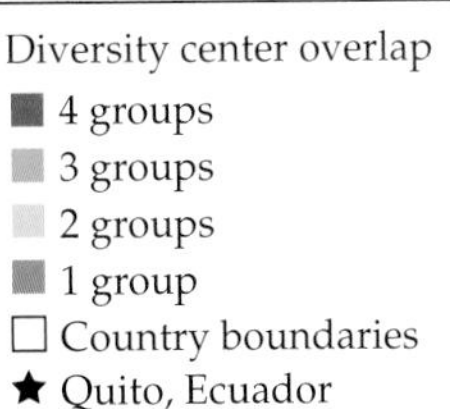

FIGURE 2.12 Certain areas of South America are characterized by high species concentrations of amphibians, birds, mammals, and plants, with overlap among the four groups in the western Amazon basin along the border of Peru and Ecuador. (From Bass et al. 2010.)

TABLE 2.1 Number of Native Mammal Species in Selected Tropical and Temperate Countries Paired for Comparable Size

Tropical country	Area (1000 km²)	Number of mammal species	Temperate country	Area (1000 km²)	Number of mammal species
Brazil	8456	604	Canada	9220	207
DRC[a]	2268	425	Argentina	2737	378
Mexico	1909	529	Algeria	2382	84
Indonesia	1812	471	Iran	1636	150
Colombia	1039	443	South Africa	1221	278
Venezuela	882	363	Chile	748	147
Thailand	511	241	France	550	104
Philippines	298	180	United Kingdom	242	75
Rwanda	25	111	Belgium	30	71

Source: Data from IUCN Red List 2009.
[a]DRC = Democratic Republic of the Congo.

comparatively fewer endemic species because they share many species with neighboring countries.

Almost all groups of organisms show an increase in species diversity toward the tropics. For example, Thailand has 241 species of mammals, while France has only 104, despite the fact that the countries have roughly the same land area (**Table 2.1**). The contrast is particularly striking for trees and other flowering plants: 10 ha of forest in Amazonian Peru or Brazil might have 300 or more tree species, whereas an equivalent forest area in temperate Europe or the United States would probably contain 30 species or less. Within a given continent, the number of species increases toward the equator.

For many groups of marine species, the greatest diversity of coastal species is found in the tropical areas, with a particular richness of species in the western Pacific. For open ocean species, the greatest diversity is found in midlatitudes. Temperature is the most important variable explaining these patterns (Tittensor et al. 2010).

Local variation in climate, sunlight and rainfall, topography, and geological age are also factors that affect patterns of species richness (Gaston 2000). In terrestrial communities, species richness tends to be greatest in sunny, lowland locations with abundant rainfall. Species richness can be greater where complex topography (such as mountains and valleys) and great geological age provide more environmental variation, which in turn allows genetic isolation, local adaptation, and speciation to occur. Geologically complex areas can produce a variety of soil conditions with very sharp boundaries between them, leading to multiple communities as plant species are adapted to one specific soil type or another.

With better methods of exploration and investigation, we are now able to appreciate the great diversity of the living world. This is truly a golden age of biological exploration. Amateur naturalists and other interested people can contribute to this effort by helping scientists carry out breeding-bird surveys and recording the presence of species on special "biodiversity days." Yet with this knowledge comes both the awareness of the damaging impacts of human activity, which are diminishing that diversity right before our eyes, and the responsibility to protect and restore that biodiversity which still remains.

Summary

- Taxonomists use morphological and genetic information to describe and identify the world's species. Places vary in their species richness, the number of species found in a particular location.
- There is genetic variation among individuals within a species. Genetic variation allows species to adapt to a changing environment, and it is valuable for the continuing improvement of crop plants and domestic animals.
- Within an ecosystem, species play different roles and have varying requirements for survival. Certain keystone species are important in determining the ability of other species to persist in an ecosystem.
- It is estimated that there are 5 million to 10 million species, most of which are insects. The majority of the world's species have still not been described and named. Further work is needed to describe microorganisms such as bacteria.
- The greatest biological diversity is found in tropical regions, with particular concentrations of species in rain forests and coral reefs. The ocean may also have great species diversity but needs further exploration.

For Discussion

1. How many species of birds, trees, and insects can you identify in your neighborhood? How could you learn to identify more? Is it important to be able to identify species in the wild?
2. What are the factors promoting species richness? Why is biological diversity diminished in particular environments? Why aren't species able to overcome these limitations and undergo the process of speciation?

Suggested Readings

Corlett, R. and R. B. Primack. 2010. *Tropical Rainforests: An Ecological and Biogeographical Comparison*, 2nd ed. Wiley-Blackwell Publishing, Malden, MA. Rain forests on different continents have distinctive assemblages of animal and plant species.

Estes, J. A. and 23 others. 2011. Trophic downgrading of planet earth. *Science* 333: 301–306. Complex species interactions can change over decades, affecting multiple trophic levels.

Groombridge, B. and M. D. Jenkins. 2010. *World Atlas of Biodiversity: Earth's living resources in the 21st century*. University of California Press, Berkeley. Great resource, with numerous figures; available on-line.

Joppa, L. N., D. L. Roberts, and S. L. Pimm. 2011. The population ecology and social behavior of taxonomists. *Trends in Ecology & Evolution* 26: 551–553. The number of taxonomists and the number of species described per year are steadily increasing.

Kelm, D. H., K. R. Wiesner, and O. von Helversen. 2008. Effects of artificial roosts for frugivorous bats on seed dispersal in a Neotropical forest pasture mosaic. *Conservation Biology* 22: 733–741. Putting up artificial roosts can dramatically increase the number of seeds being dispersed by bats.

Laikre, L. and 19 others. 2010. Neglect of genetic diversity in implementation of the Convention on Biological Diversity. *Conservation Biology* 24: 86–88. A greater emphasis on genetic diversity needs to be part of conservation efforts.

Lowman, M. D., E. Burgess, and J. Burgess. 2006. *It's a Jungle Up There: More Tales from the Treetops*. Yale University Press, New Haven, CT. Anecdotes and adventures while exploring the diversity of the tropical forest canopy.

Marquard, E. and 8 others. 2009. Plant species richness and functional composition drive overyielding in a six-year grassland experiment. *Ecology* 90: 3290–3302. Plant species richness and functional diversity increase biomass in a grassland ecosystem.

Forzza, R. C., J. F. A. Baumgratz, C. E. M. Bicudo, D. A. L. Canhos, A. A. Carvalho Jr., and 21 others. 2012. New Brazilian floristic list highlights conservation challenges. *BioScience* 62: 39–45. The new list of Brazilian plants shows a huge number of endemic species.

Ripple, W. J. and R. L. Beschta. 2012. Trophic cascades in Yellowstone: The first 15 years after wolf reintroduction. *Biological Conservation* 145: 205–213. Restoring a keystone species has resulted in large changes to this ecosystem.

Strain, D. 2011. 8.7 million: A new estimate for all the complex species on earth. *Science* 333: 1083. A variety of methods have been developed for estimating the total numbers of species on Earth.

Tittensor, D. P. and 6 others. 2010. Global patterns and predictors of marine biodiversity across taxa. *Nature* 466: 1098–1101. Temperature is the most important factor affecting marine diversity.

Ecotourists viewing a tiger in Ranthambhore National Park, Rajasthan, India.

Chapter 3
The Value of Biodiversity

Conserving biodiversity will require a significant shift in current political and social thinking. Governments and communities throughout the world must realize that biodiversity is extremely valuable—indeed, it is essential to human existence. Ultimately, change will occur only if people believe that by continuing to destroy biodiversity we are truly losing something of value. But what exactly are we losing? Why should anyone care if a species becomes extinct or an ecosystem is destroyed? What factors induce humans to act in an unsustainable and ultimately destructive manner?

Most environmental degradation and species loss occur as thoughtless or accidental by-products of human activities. Species are hunted to extinction. Sewage is released into rivers. Poor-quality land is cleared for short-term cultivation. An understanding of a few fundamental economic principles will clarify the reasons why people treat the environment in what appears to be a shortsighted, wasteful manner.

One of the most universally accepted tenets of modern economic thought centers on the idea of "voluntary transaction"—that a monetary transaction takes place only when it is beneficial to both of the parties involved. However, there is a notable exception to this principle that directly applies to environmental issues. It is assumed that

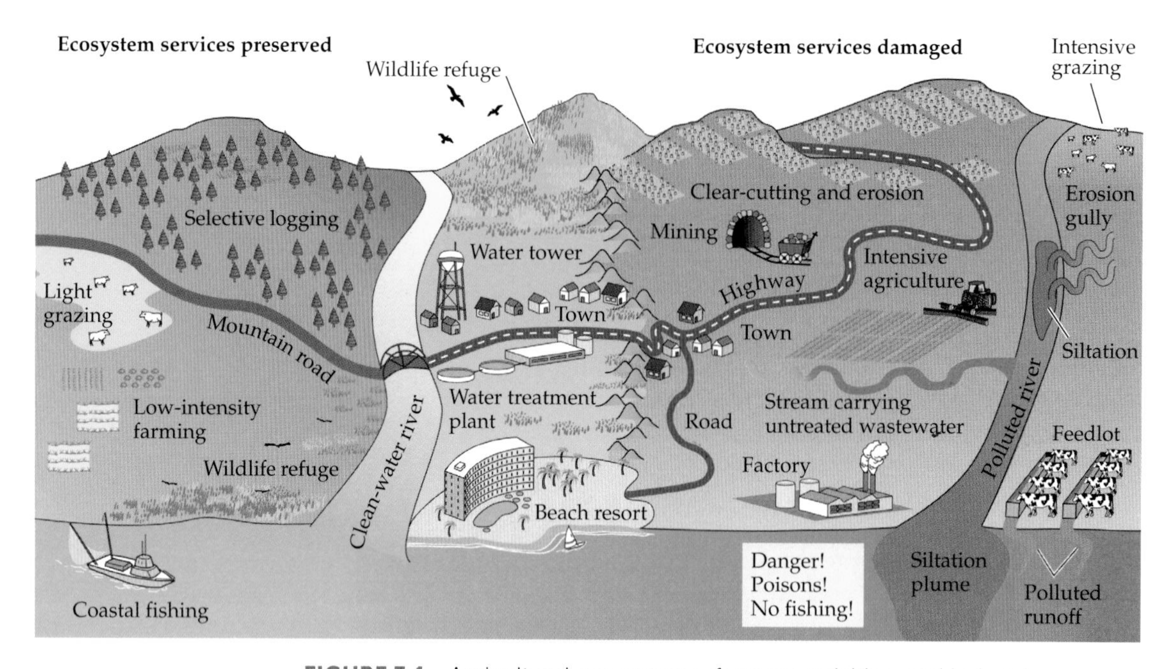

FIGURE 3.1 Agricultural ecosystems, forestry activities, and industries are usually valued by the products that they produce. In many cases these activities have negative externalities in that they erode soil, degrade water quality, and harm aquatic life (right side of figure). Farms, forests, and other human activities could also be valued on the basis of their public benefits, such as soil retention and maintaining water quality and fish populations, and their owners might receive subsidies for these benefits (left side of figure).

the costs and benefits of free exchange are accepted and borne by the participants in the transaction. During many of these transactions, individuals not directly involved in the exchange, or society at large, are forced to accept some of the costs. These hidden costs (or sometimes benefits) are known as **externalities** (**Figure 3.1**) (Abson and Termansen 2011). Where externalities exist, the market fails to maximize the net benefits to society as a whole. **Market failure** occurs when resources are misallocated, allowing a few individuals or businesses to benefit at the expense of the larger society. As a result of such misallocation, the society as a whole becomes *less* prosperous, not more prosperous, from certain activities.

Perhaps the most important and frequently overlooked market failure is the environmental damage that occurs to **open-access resources**, such as water, air, and sometimes soil, as a consequence of human economic activity. Open-access resources are collectively owned and used by society at large or are owned by no one, with availability to everyone who is part of that society. When there are no property rights or when property rights are ill-defined or lack proper enforcement, then people, industries, and governments have incentives to overuse and damage these resources without paying more than a minimal private cost, or sometimes they pay nothing at all. In these situations—also described as **the tragedy of the commons**—the value of the open-access resource is gradually lost to all of society (Lant et al. 2008). For example, consider the dumping of chemical wastes into a river as a by-product of mining. Without formal regulations establishing property rights, the river can be treated as an open-access resource for waste disposal purposes. The externalities associated with this activity can include degraded drinking water and an increase in disease, loss of opportunity to bathe and swim in the water, fewer fish that are safe to eat, and the loss of many species unable to survive in the polluted river. The external cost of "free" waste disposal is borne by society rather than the mining company.

Market failure can also occur when there is a lack of enforcement of property rights relating to the direct overexploitation of open-access natural resources such as water, forests, and fisheries. For example, harvesting fish in an area may be managed according to a sustainable system. However, if this system is not adequately enforced, overharvesting may occur, leading to a collapse of the fishing stock and damage to the ecosystem.

Ecological Economics and Environmental Economics

The fundamental challenge facing conservation biologists is to ensure that *all* the costs and benefits of economic behavior are understood and taken into account when decisions are made that will affect biological diversity (Hoeinghaus et al. 2009). In an effort to account for all costs of economic transactions, including environmental costs, two closely related subdisciplines of economics—**ecological economics** and **environmental economics**—have evolved that integrate economics, environmental science, ecology, and public policy and that include valuations of biodiversity in economic analyses (Common and Stagl 2005). Conservation biologists are now using the concepts and vocabulary of ecological and environmental economics because government officials, bankers, and corporate leaders often can be more readily convinced of the need to protect biodiversity if they are provided with an economic justification for doing so.

> Arguments for the protection of biodiversity are often strengthened by evidence provided by ecological economics.

Some people would argue that any attempt to place a strictly monetary value on biological diversity is inappropriate and potentially corrupting, since many aspects of the natural world are unique and thus truly priceless (Redford and Adams 2009). Supporters of this position point out that there is no way to assign monetary value to the wonder that people experience when they see an animal in the wild or a beautiful natural landscape; nor can economic value realistically be assigned to the human lives that have been and will be saved through the medicinal compounds derived from wild species. In fact, it is to the advantage of conservationists to develop economic models—both to improve such models' accuracy and to appreciate their limitations—since these models often provide surprisingly strong support for the crucial role of biological diversity in local economies and for the need to protect ecosystems.

Economic methods are now being used to review development projects and evaluate their potential environmental effects *before* the projects proceed. **Environmental impact assessments**, in particular, consider the present and future effects of the projects on the environment. The environment is often broadly defined to include not only harvestable natural resources but also air and water quality, the quality of life for local people, and the preservation of endangered species. Frequently included in an environmental impact assessment is a **cost-benefit analysis**, which compares the values gained against the costs of the project or resource use. In practice, though, cost-benefit analyses are notoriously difficult to calculate accurately because benefits and costs change over time and are difficult to measure. Today there is an increasing tendency by governments, conservation groups, and economists to apply the **precautionary principle**. That is, it may be better not to approve a project that has risk associated with it and to err on the side of doing no harm to the environment than to do harm unintentionally or unexpectedly (Prato 2005).

It would be highly beneficial to apply cost-benefit analysis to many of the industries and practices of modern society. Many economic activities that appear to be profitable are actually losing money, because governments subsidize some industries that are involved in environmentally-damaging activities with tax breaks, direct payments or price supports, cheap fossil fuel, free water, and road networks—sometimes referred to as **perverse subsidies**. Government subsidies that promote the activities of specific industries, such as agriculture, fishing, and energy production, may amount to as much as 5% of the world economy—several trillion dollars a year (Myers et al. 2007). Without these subsidies, many environmentally damaging activities, such as farming on marginal lands that yield low returns, logging in remote areas, and inefficient and highly polluting energy use, would be reduced or eliminated because the *true* costs exceed the benefits.

Attempts have been made to include the loss of natural resources in calculations of gross domestic product (GDP) and other indexes of national productivity (Balmford et al. 2005; Dobson 2005). Currently, GDP measures the economic activity in a country without accounting for all the costs of unsustainable activities (such as overfishing of coastal waters and poorly managed strip-mining). This practice results in a positive GDP, even though these activities may be destructive to a country's long-term economic well-being; ironically even activities related to cleaning up environmental damage contribute to calculations of GDP. In actuality, the economic costs associated with environmental damage can

Unsustainable activities such as clear-cut logging, strip-mining, and overfishing may cause a country's apparent productivity to increase for the present moment but are generally destructive to long-term economic well-being.

be considerable, often offsetting the gains attained through agricultural and industrial development.

Another attempt to account for natural resource depletion, pollution, and unequal income distribution in measures of national production is the development of the Index of Sustainable Economic Welfare (ISEW), the updated version of which is called the Genuine Progress Indicator (GPI; www.progress.org). The GPI suggests what conservation biologists have long feared—that many modern economies are achieving their growth only through the unsustainable consumption of natural resources and loss of biodiversity. As these resources run out, the economies on which they are based may be seriously disrupted.

A third measure is the Environmental Sustainability Index (ESI), which uses 21 environmental indicators to rank countries according to the health of, and threats to, their ecosystems, the vulnerability of their human population to an adverse environment, the ability of their society to protect the environment, and participation in global environmental protection efforts (van de Kerk et al. 2009; http://sedac.ciesin.columbia.edu/es/epi/). There is a concern among many economists and businesspeople that a country that rigorously protects its environment as shown by a high ESI may not be competitive in the world economy as measured by a competitiveness index that includes worker productivity and a country's ability to grow and prosper. However, **Figure 3.2** shows that environmental sustainability is not linked to a country's economic competitiveness. Countries such as Finland have an economy that is both sustainable and competitive, whereas Belgium is competitive but ranks poorly in sustainability. The rapidly growing economies of China and India are intermediate in competitiveness but rank low in environmental sustainability.

New measures of national productivity take environmental sustainability into account. These include both the benefits and costs that result from human activities.

As yet there is no universally accepted framework for assigning values to biological diversity, but a variety of approaches have been proposed. Among

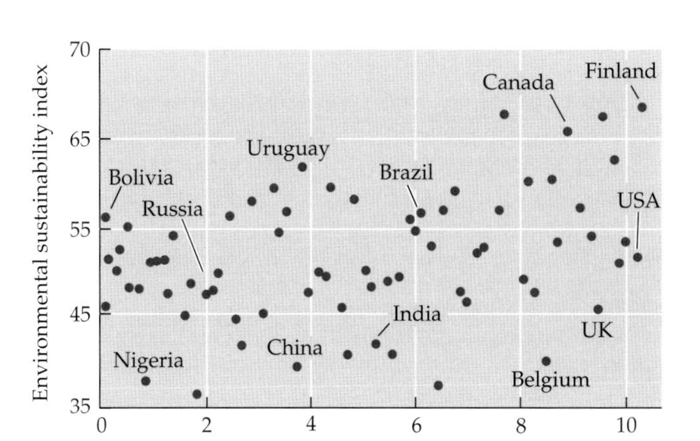

FIGURE 3.2 The economic competitiveness of a country is not closely related to its environmental sustainability, as measured by five sets of indicators: healthy ecosystems, low stress on ecosystems, low human vulnerability to environmental change, ability of the society and institutions to cope with environmental changes, and cooperation in environmental initiatives. (After Esty et al. 2005.)

the most useful is a system in which economic values are first divided into **use values** and **non-use values**. Use values of biodiversity are divided between **direct use values** (also known in other frameworks as **commodity values**, and **private goods**) and **indirect use values**. Direct use values are values assigned to products harvested by people, such as timber, seafood, and medicinal plants from the wild, while indirect use values are benefits provided by biological diversity that do not involve harvesting or destroying the resource. Indirect use values provide current benefits to people, such as recreation, education, scientific research, and scenic amenities, and include the benefits of **ecosystem services** such as water quality, pollution control, natural pollination and pest control, ecosystem productivity, soil protection, and regulation of climate. **Option value** is also part of indirect use value and is determined by the prospect for possible future benefits for human society, such as new medicines, potential future food sources, and future genetic resources. **Existence value** is the non-use value that can be assigned to biodiversity—for example, economists can attempt to measure how much people and their governments are willing to pay to protect a species from going extinct or an ecosystem from being destroyed. A category of existence value is **bequest value**, which is how much people are willing to pay to protect something for their children or future generations. The combination of all these individual values can be used to calculate the total economic value of biodiversity. (See Figure 3.13 for an example of how these different values can be applied to a tropical wetland ecosystem.)

Direct Economic Values

Direct use values can often be calculated by observing the activities of representative groups of people, by monitoring collection points for natural products, and by examining import and export statistics. Direct use values are further

(A)

(B)

FIGURE 3.3 (A) Along a river in India, fishermen catch small fish to eat. (B) A woman in India returns to her village with a load of wood. Fuelwood is one of the most important natural products consumed by local people, particularly in Africa and southern Asia. (A, photograph courtesy of Sandesh Kadur; B, photograph © Borderlands/Alamy.)

divided into **consumptive use value**, for goods that are consumed locally, and **productive use value**, for products that are sold in markets.

Consumptive use value

Goods such as wild game and fuelwood that are locally consumed are assigned consumptive use value (**Figure 3.3**). People living close to the land often derive a considerable proportion of the resources they require for their livelihood from the surrounding environment. These goods do not appear in a country's GDP, because they are consumed or exchanged on a local level and are neither bought nor sold in a national or international marketplace. However, if rural people are unable to obtain these products (as might occur following environmental degradation, overexploitation of natural resources, or even the creation of a protected reserve), their standard of living will decline, possibly to the point where they are forced to relocate.

Studies of traditional societies in the developing world show how extensively these people use their natural environment to supply themselves with fuelwood, vegetables, fruit, meat, medicine, rope and string, and building materials. About 80% of the world's population still relies on traditional medicines derived from thousands of plant and animal species as their primary source of treatment (**Figure 3.4**) (Shanley and Luz 2003).

One of the most crucial requirements of rural people is protein, which they obtain by hunting and collecting wild animals for meat (see Figure 3.3A), Wild meat, known as **bushmeat**, constitutes a significant portion of the

FIGURE 3.4 In much of the world, medical treatment relies on natural products found in the immediate environment. Here Hortense Robinson, a village healer and midwife in Belize, gives a demonstration recreating the application of a medicinal poultice made of leaves from the palm *Chamaedorea tepejilote*. (Photograph courtesy of Michael J. Balick.)

protein in the average person's diet in many developing countries—about 40% in Botswana and about 75% in the Democratic Republic of the Congo (formerly Zaire) (Rao and McGowan 2002). Extraction rates for Africa are undeniably unsustainable, perhaps by a factor of six. This wild meat includes not only birds, mammals, and fish but spiders, snails, caterpillars, and insects. In certain areas of Africa, because of overharvesting of larger animals, insects may constitute the majority of the dietary protein and supply critical vitamins.

In areas along coasts, rivers, and lakes, wild fish represent an important source of protein. Throughout the world, 130 million tons of fish, crustaceans, and mollusks, mainly wild species, are harvested each year, with 100 million tons constituting marine catch and 30 million tons constituting freshwater catch (Chivian and Bernstein 2008). Much of this catch is consumed locally. In coastal areas fishing is often the most important source of employment, and seafood is the most widely consumed protein. Even though fish farming is increasing rapidly, much of the feed used is fish meal derived from wild-caught fish (Gross 2008a).

Consumptive use value can be assigned to a product by considering how much people would have to pay if they had to buy an equivalent product when their local source was no longer available. This is sometimes referred to as a **replacement cost approach**. In many cases local people do not have the money to buy products in the market.

Consumptive use value can be calculated by considering how much people would have to pay to buy an equivalent product if their local source were no longer available.

Although dependency on local natural products is primarily associated with the developing world, there are rural areas of the United States, Canada, Europe, and other developed countries where hundreds of thousands of people are dependent on fuelwood for heating, on wild game and seafood for their protein needs, and on intact ecosystems for clean drinking water and sewage treatment. Many of these people would be unable to survive in these locations if they had to pay for these necessities.

Productive use value

Resources that are harvested from the wild and sold in both national and international commercial markets are assigned productive use value. In standard economics these products are valued at the price paid at the first point of sale minus the costs incurred up to that point, whereas other methods value the resource at the final retail price of the products. For example, the bark and leaves from wild shrubs and trees of the common witch-hazel (*Hamamelis virginiana* and related species) are used to make a variety of astringent herbal products, including after-shave lotions, insect bite creams, and hemorrhoid preparations. The final retail price of the medicine, which includes the values of all inputs (labor, energy, other materials, transportation, marketing, and witch-hazel bark and leaves), is vastly greater than the purchase price of the witch-hazel raw materials.

The productive use value of natural resources is significant, even in industrial nations. Approximately 4.5% of the U.S. GDP ($15 trillion)—about $675 billion for the year 2011—depends in some way on wild species (Prescott-Allen and Prescott-Allen 1986). The percentage is far higher for developing countries that have less industry and a higher percentage of the population living in rural areas. The international trade in wildlife, fisheries, and timber products harvested from the wild has been estimated recently to be $332 billion (Engler 2008 cited in Barber-Meyer 2010).

The range of products obtained from the natural environment and sold in the marketplace is enormous: fuelwood, construction timber, fish and shellfish, medicinal plants, wild fruits and vegetables, wild meat and skins, fibers, rattan (a vine used to make furniture and other household articles), honey, beeswax, natural dyes, seaweed, animal fodder, natural perfumes, and plant gums and resins (Chivian and Bernstein 2008). Additionally, there are large international industries associated with collecting tropical cacti, orchids, and other plants for the horticultural industry, and birds, mammals, amphibians, and reptiles for zoos and private collections. The value of ornamental fish in the aquarium trade is estimated at $1 billion per year, with wild-caught fish representing 15% to 20% of the total. A surprisingly large area within 23 countries in sub-Saharan Africa is managed for trophy hunting by foreigners (Lindsey et al. 2007).

A wide variety of natural resources are sold commercially and have enormous total market value. Their value can be considered the productive value of biodiversity.

FOREST PRODUCTS Wood is one of the most significant products obtained from natural environments, with an export value of about $135 billion per year (WRI 2003). The total value of timber and other wood products is far greater, perhaps about $400 billion per year, because most wood is used locally and is not exported. Wood products from the forests of tropical countries, including timber, plywood, and wood pulp, are being exported at a rapid rate to earn foreign currency, to provide capital for industrialization, to pay foreign debt, and to provide employment. In tropical countries such as Indonesia, Brazil, and Malaysia, timber products earn billions of dollars per year (Corlett and Primack 2010).

Nonwood products from forests, including bushmeat, fruits, gums and resins, rattan, and medicinal plants, also have a large productive use value. These nonwood products are sometimes erroneously called "minor forest products"; in reality they are often very important economically and may even rival the value of wood.

THE NATURAL PHARMACY Effective drugs are needed to keep people healthy, and they represent an enormous industry, with worldwide sales of around $300 billion per year (Chivian and Bernstein 2008). The natural world is an important source of medicines currently in use and those that may be used in the future. All 20 of the most frequently used pharmaceutical products in the United States are based on chemicals that were first identified in natural organisms. More than 25% of the prescriptions filled in the United States contain active ingredients derived from plants, and many of the most important antibiotics, including penicillin and tetracycline, are derived from fungi or microorganisms.

Even in the case of medicines that are now produced synthetically by chemical processes, many were first discovered in wild species used in traditional medicine. For example, the use of coca (*Erythroxylum coca*) by natives of the Andean highlands eventually led to the development of synthetic derivatives such as novocaine and lidocaine, commonly used as anesthetics in dentistry and surgery. Another species with great medicinal use is the rose periwinkle (*Catharanthus roseus*) from Madagascar. Two potent drugs derived from this plant are effective in treating Hodgkin's disease and leukemia and other blood cancers. Treatment using these drugs has increased the survival rate for childhood leukemia from 10% to 90%. Venomous animals such as rattlesnakes and bees have been especially rich sources of chemicals with valuable medical and biological applications. To take another example, an enzyme derived from a heat-tolerant bacteria (*Thermus aquaticus*) collected from hot springs at Yellowstone National Park forms a key component in the polymerase chain reaction used to amplify DNA in the biotechnology industry and in biological research. This enzyme is also used in the medical field to detect human diseases. The industries using this enzyme have generated hundreds of billions of dollars of value and employ tens of thousands of people. How many more such valuable organisms will be discovered in the years ahead—and how many will go extinct before they are discovered?

Indirect Economic Values

Indirect use values can be assigned to aspects of biodiversity—such as ecosystem services and recreation—that provide economic benefits without being harvested or destroyed during use (**Figure 3.5**). Because these benefits are not goods or services in the usual economic sense, they do not typically appear in the statistics of national economies, such as GDP. However, they may be crucial to the continued availability of the natural products on which the economies depend. If natural ecosystems are not available to provide these benefits, substitute sources must be found—often at great expense—or local and even regional economies may face collapse (Granek et al. 2010).

The value of the great variety of environmental services provided by biodiversity can be classified as a particular type of indirect use value, known as **nonconsumptive use value** (because the services are not consumed). Economists are just beginning to calculate the value of ecosystem services at regional and global levels (Peterson et al. 2010). The point can be made that human societies are totally dependent on the free services that we obtain from natural ecosystems, because we could not pay to replace these ecosystems if they were permanently degraded or destroyed. Especially important in this regard are wetland ecosystems for their role in water purification and nutrient recycling, as well as their enormous importance in flood control (Horwitz and Finlayson 2011) (see the section "Water and soil protection," p. 59).

Many economists are in sharp disagreement about how to calculate the value of ecosystem services, or if these calculations should be attempted at all. One such calculation suggests that the value of ecosystem services is enormous, actually exceeding the current $62 trillion annual value of the world's economy (Costanza et al. 1997). Using different approaches, other ecological economists have come up with much lower estimates, but the estimates have still amounted to trillions of dollars a year. The disparity in these various estimates indicates that much more work needs to be done on this topic.

The great variety of environmental services that ecosystems provide can be separated into particular types of indirect use value. The following sections discuss some of the specific indirect use values derived from conserving biodiversity. Later in the chapter, we will consider two other ways of valuing biodiversity: option value, or the value that biodiversity may have in the future, and existence value, which is the amount that people are willing to pay to protect biodiversity for its own sake rather than its use by humans.

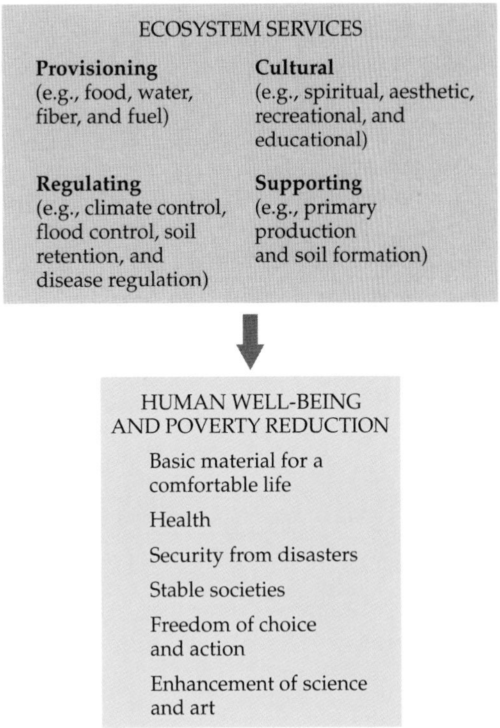

ECOSYSTEM SERVICES

Provisioning
(e.g., food, water, fiber, and fuel)

Cultural
(e.g., spiritual, aesthetic, recreational, and educational)

Regulating
(e.g., climate control, flood control, soil retention, and disease regulation)

Supporting
(e.g., primary production and soil formation)

HUMAN WELL-BEING AND POVERTY REDUCTION

Basic material for a comfortable life

Health

Security from disasters

Stable societies

Freedom of choice and action

Enhancement of science and art

FIGURE 3.5 Natural ecosystems provide many important products and services that are essential for human well-being. (After Millennium Ecosystem Assessment 2005.)

Ecosystem productivity

The photosynthetic capacity of plants and algae allows the energy of sunlight to be captured in living tissue. Some of the energy stored in plants is harvested by humans for use as food, fuelwood, and animal feed. This plant material is also the starting point for innumerable food chains, from which people harvest many animal products. Human needs for natural resources dominate approximately half of the productivity of the terrestrial environment (MEA 2005). The destruction of the vegetation in an area due to overgrazing by domestic animals, overharvesting of timber, or frequent fires will destroy the system's ability to make use of solar energy. Eventually it will lead to the loss of production of plant biomass, loss of animals that live at that site, and loss of a place where people can make a living.

Likewise, estuaries are areas of rapid growth of plants and algae that provide the starting points for food chains that lead to commercial stocks of fish and shellfish. Damage to coastal estuaries results in lost productive value of commercial fish and shellfish and in lost recreational value of fish caught for sport. Even when degraded or damaged aquatic ecosystems are

> Ecosystems with reduced species diversity are less able to adapt to the altered conditions associated with rising carbon dioxide levels and global climate change.

rebuilt or restored at great expense, they often do not function as well as before and almost certainly do not contain their original species compositions.

Scientists are actively investigating how the loss of species from biological communities affects ecosystem processes such as the ability of plants to grow, absorb atmospheric carbon dioxide, and adapt to global climate change (Isbell et al. 2011). Many studies of natural and experimental grassland communities confirm that as species are lost, overall productivity declines and the community is less flexible in responding to environmental disturbances, such as drought (**Figure 3.6**). We know that species diversity is being reduced in major ecosystems as a result of human activities. At what

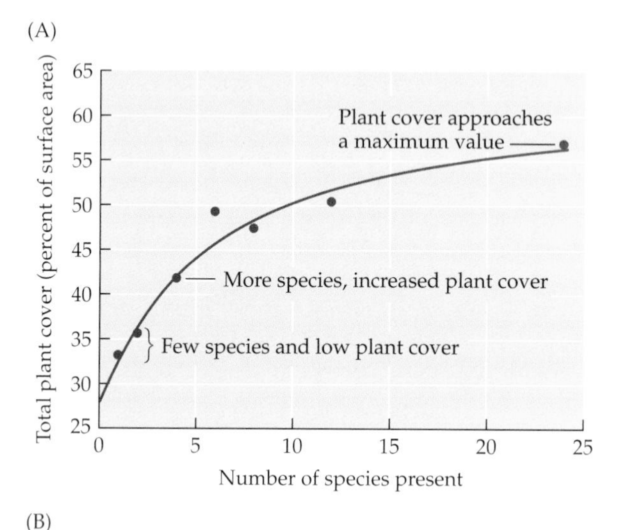

(B)

FIGURE 3.6 (A) Varying numbers of grassland species were grown in experimental plots. The plots containing the most species had the greatest overall amount of growth, as measured by the total plant cover (the percentage of the total surface area occupied by plants). (B) View of grassland research plots in Germany. Plots vary by color and shading depending on the species present and the density of planting. The experimental field is 300 m across the top, with individual squares being 20 m on a side. (A, after Tilman 1999; B, photograph by Dr. Alexandra Weigelt.)

point will the productivity of these ecosystems decline as well? We need to know the answer to this question before the world's forestry, ranching, agriculture, and fishing industries become irreparably damaged by the consequences of species decline. The value of intact and restored forests in retaining carbon and absorbing atmospheric carbon dioxide is now being recognized by ecological economists. As countries and corporations reduce their carbon dioxide emissions as part of the worldwide effort to address global climate change, they are paying to protect and restore forests and other ecosystems (Venter et al. 2010).

Water and soil protection

Biological communities are of vital importance in protecting watersheds, buffering ecosystems against extremes of flood and drought, and maintaining water quality (Thorp et al. 2010). Plant foliage and dead leaves intercept the rain and reduce its impact on the soil. Roots and soil organisms aerate the soil, increasing its capacity to absorb water. This increased water-holding capacity reduces the flooding that would otherwise occur after heavy rains and allows for a slow release of water for days and weeks after the rains have ceased.

When vegetation is disturbed by logging, farming, and other human activities, the rates of soil erosion, and even occurrences of landslides, increase rapidly, reducing the value of the land for human activities. Damage to the soil limits the ability of plant life to recover from disturbance and can render the land useless for agriculture. Erosion and flooding can destroy the plants and animals living in nearby and downstream aquatic communities and can make water supplies undrinkable, leading to an increase in human health problems. Soil erosion increases sediment loads into the reservoirs behind dams, causing a loss of electrical output, and creates sandbars and islands that reduce the navigability of rivers and ports.

Throughout the world, people have made wetlands protection a priority in order to prevent flooding of developed areas. The development of floodplains and their use for agriculture along the Mississippi River in the midwestern United States and along the Rhine River in Europe are considered major factors in the massive, damaging floods that have occurred in past years (**Figure 3.7**). This includes the devastating flooding of the city of New Orleans in 2005 after Hurricane Katrina struck the surrounding Mississippi delta. A century of human activities had removed marshes and swamps south of New Orleans, greatly increasing its vulnerability to storm damage. The risk of flooding in these regions would be substantially reduced if some of the wetlands along these rivers were restored to a natural condition.

The economic value of ecosystems in protecting watersheds is enormous. In the late 1980s, the government of New York City paid $1.5 billion to county and town governments in rural New York State to maintain forests on the watersheds that surround its reservoirs and to improve agricultural practices. Water filtration plants doing the same job would have cost $8 to

FIGURE 3.7 A flooded area with only the rooftops showing along the Mississippi River, Missouri in 1993. The river channel can be seen in the upper left. (Photograph by Andrea Booher/FEMA photo.)

$9 billion (nyc.gov/watershed). The value of just the U.S. national forests in contributing to the water supply has been estimated at $4 billion per year.

When forests' ability to absorb atmospheric carbon dioxide and their ecosystem value as sources of drinking water, protectors of soil, and preventers of floods are considered along with the value of nonwood products, then maintaining and utilizing intact forests may prove to be more productive than intensive logging, converting the forest into commercial plantations, or establishing cattle ranches (Kirkby et al. 2010). Careful selective tree harvesting that minimizes damage to the forest and the associated ecosystem services it provides, combined with gathering nontimber products, may be the most profitable (Fisher et al. 2011).

Wetland ecosystem services whose value is typically not accounted for in the current market system include waste treatment, water purification, and flood control—all of which are essential to healthy human societies.

Aquatic communities in swamps, lakes, rivers, floodplains, tidal marshes, mangroves, estuaries, coastal shelves, and the open ocean are also important in breaking down and immobilizing toxic pollutants—sewage, industrial wastes, heavy metals, and pesticides—released into the environment by human activities (Balmford et al. 2002). When such aquatic communities, especially the bacteria and other microorganisms they contain, are damaged by a combination of sewage overload and habitat destruction, alternative systems have to be developed. These contrived systems, such as waste treatment facilities and giant landfills, cost tens of billions of dollars. In regions that cannot afford to build such facilities, people's quality of life can be severely harmed.

Climate regulation

Plant communities are important in moderating local, regional, and probably global climate conditions (West et al. 2011). At the local level, trees provide shade and transpire water, which reduces the temperature in hot weather. This cooling effect reduces the need for fans and air conditioners and increases people's comfort and work efficiency. Trees are also locally important because they act as windbreaks, reducing soil erosion from agricultural fields and heat loss from buildings in cold weather.

At the regional level, plants capture water that falls as rain and then transpire it back into the atmosphere, from which it can fall as rain again. At the global level, loss of vegetation from large forested regions such as the Amazon basin and western Africa may result in a reduction of average annual rainfall or greatly altered weather patterns. In both terrestrial and aquatic environments, plant growth is tied to the carbon cycle. A reduction in plant life results in reduced uptake of carbon dioxide, contributing to the rising carbon dioxide levels that lead to global warming (Pan et al. 2011; McKinley et al. 2011). International programs are being developed to assign economic value to forests for their ability to absorb and store carbon. And ultimately, plants are the "green lungs" of the planet, producing the oxygen on which all animals, including people, depend for respiration.

Species relationships and environmental monitors

Many of the species harvested for their productive use value depend on other wild species for their continued existence. For example, the wild game and fish harvested by people are dependent on wild insects and plants for their food. A decline in insect and plant populations will result in a decline in animal harvests. Thus, a decline in a wild species of little immediate value to humans may result in a corresponding decline in a harvested species that is economically important.

Crop plants also benefit from wild insects, as well as from birds and bats (Gardiner et al. 2009). Predatory insects such as praying mantises, and many bird and bat species, feed on pest insect species that attack the crops, both increasing crop yields and reducing the need to spray pesticides (Boyles et al. 2011; Kross et al. 2012). Insects, birds, and bats act as pollinators for numerous crop species (**Figure 3.8**). About 150 species of crop plants in the United States require insect pollination of their flowers, often involving a combination of wild insects and honeybees (Kremen and Ostfeld 2005). The value of these pollinators in increasing crop yield has been estimated to be $20 billion to $40 billion per year. The value of wild insect

> Relationships between species are often essential for preserving biodiversity and providing value to people. For example, many insects pollinate the crops on which people depend for food.

FIGURE 3.8 Wild species benefit crop plants. (A) A house wren feeds on white cabbage butterflies whose caterpillars are a major pest of vegetable crops such as cabbage and broccoli. (B) A bumblebee visits apple blossoms; without such pollination, the apple fruits would not develop. (A, photograph © Steve Byland/istockphoto; B, photograph © Sergey Ladanov/shutterstock.)

pollinators will increase in the near future if they take over the pollination role of domestic honeybees and wild bumblebees, whose populations are declining in many places because of disease and pests (Cameron et al. 2011). Many useful wild plant species depend on fruit-eating birds, primates, and other animals to act as seed dispersers. Where these animals have been overharvested, fruits remain uneaten, seeds are not dispersed, and species head toward local extinction (Sethi and Howe 2009). However, it should be noted that there is redundancy in guilds of similar species, and the service of one natural predator, pollinator, or seed disperser may be carried out equally well by another species.

Numerous species of bacteria, fungi, and other microorganisms provide key services to agriculture, forestry, and fisheries, on which people depend (Beattie and Ehrlich 2010). One of the most economically significant relationships in ecosystems is the one between many forest trees (and crop plants) and the soil organisms (especially fungi) that provide them with essential nutrients through the breakdown of dead plant and animal matter (Hart and Trevors 2005). The poor growth and dieback of many trees in certain areas of North America and Europe are attributable in part to the deleterious effects of acid rain and air pollution on soil fungi that help supply trees with mineral nutrients and water.

Species that are particularly sensitive to chemical toxins serve as "early warning indicators" of the health of the environment. Some species can even serve as a substitute for expensive detection equipment. Among the best known indicator species are lichens, which absorb large amounts of chemicals from rainwater and airborne pollution (Jovan and McCune 2005). High levels of toxic materials kill lichens, so the distribution and abundance of lichens can identify areas of contamination around sources of air pollution, such as smelters. Aquatic filter feeders such as clams are also effective in monitoring pollution, because they process large volumes of water and concentrate toxic chemicals such as poisonous metals, PCBs, and pesticides in their tissues.

Amenity value

Ecosystems provide many recreational services for humans; for instance, they furnish a place to enjoy nonconsumptive activities such as hiking, photography, and birdwatching (Buckley 2009). The monetary value of these activities, sometimes called their **amenity value**, can be considerable and can have a major impact on local economies. In the United States, people spend around 7 billion hours per year enjoying nature at national parks, state parks, wildlife refuges, and other protected public lands (Siikamaki 2011). These visitors enjoy nature and in the process spend billions of dollars on fees, travel, lodging, food, and equipment. If we estimate that these nature experiences in the park have a value of $7 per hour (the amount people might spend at a movie or dinner), U.S. protected areas have an estimated value of $49 billion per year.

Even sport hunting and fishing, which in theory are consumptive uses, can be considered nonconsumptive because the food value of the animals caught by fishermen and hunters is insignificant compared with the time and money spent on these activities. In national and international sites known for their conservation value or exceptional scenic beauty, such as Yellowstone National Park, nonconsumptive recreational value often dwarfs the value generated or captured by all other economic enterprises, including ranching, mining, and logging (Power and Barrett 2001).

Ecotourism is a special category of recreational and amenity value that involves people visiting places and spending money wholly or in part to experience unique ecosystems (such as coral reefs, the African savanna, the Galápagos archipelago, and the Everglades) and to view "flagship" species such as the elephants of the African savanna (www.ecotourism.org; Balmford et al. 2009). Tourism is among the world's largest industries (on the same scale as the petroleum and automotive industries), and ecotourism currently represents some 20% of the $940 billion worldwide tourist industry. Ecotourism has traditionally been a key industry in eastern African countries such as Kenya and Tanzania, and it has also become important in Latin America and many other parts of the world. Tourism associated with the Great Barrier Reef in Australia is estimated to be worth $5.5 billion per year and employs more than 50,000 people, which is 36 times more than the commercial fishing industry (McCook et al. 2010). In addition to international tourism, the rapidly growing middle classes in developing countries, such as China and India, are increasingly traveling within their own countries to visit national parks and nature reserves (Karanth and DeFries 2011).

The revenue from ecotourism potentially provides one of the most immediate justifications for protecting biodiversity and restoring degraded lands, particularly when ecotourism activities are integrated into overall management plans (Fennel 2007; Vianna et al. 2012). In integrated conservation and development projects (ICDPs), local communities develop accommodations, expertise in nature guiding, local handicraft outlets, and other sources of income; the revenue income from ecotourism allows the local people to give up unsustainable or destructive hunting, fishing, or grazing practices (see Chapter 8). The key to successful ecotourism is making sure that enough tourist money is paid to the local people, to maintain and improve their way of life, and to park authorities for park protection (**Figure 3.9**). Travel companies are increasingly promoting measures to minimize their impacts and provide greater benefits to local people. Ecotourism activities can even become certified as sustainable through programs such as Green Globe 21.

The rapidly developing ecotourism industry can provide income to protect biodiversity, but possible costs must be weighed along with benefits.

A danger of ecotourism is that tourists themselves will unwittingly damage the sites they visit—by trampling wildflowers, breaking coral, or disrupting nesting bird colonies, for instance—thereby contributing to the degradation and disturbance of sensitive areas (Lamb and Willis 2011). Tourists might also indirectly damage sites through their presence, for example by creating a demand for fuelwood for heating and cooking, thus contributing to deforestation. In addition, the presence, affluence, and demands of tourists can transform traditional human societies in tourist areas by changing employment opportunities and often serving as a magnet for outside people looking for work (Dahles 2005). As local people increasingly enter a cash-based economy, their values, customs, and relationship to nature may be lost along the way. A final potential danger of this industry is that ecotourist facilities may provide a sanitized fantasy experience rather than help visitors understand the serious social and environmental problems that endanger biological diversity.

Educational and scientific value

Many books, television programs, movies, and websites produced for educational and entertainment purposes are based on nature themes (Osterlind 2005). These natural history materials are incorporated continually into school curricula and are worth billions of dollars per year. To take one example, recent movies with penguins as main characters or themes have had revenues estimated at around $1.6 billion. They represent a nonconsumptive use value of biodiversity because they use nature only as intellectual content. A considerable number of professional scientists, as well as highly motivated amateurs, are engaged in making ecological observations and preparing educational materials. In rural areas, their activities often take

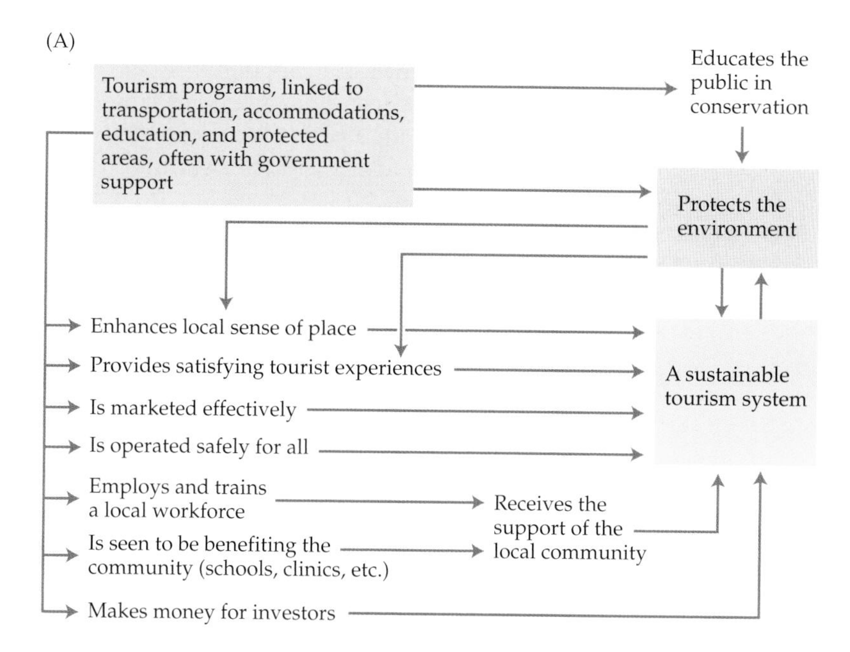

FIGURE 3.9 Ecotourism can provide an economic justification for protecting biodiversity and also can provide benefits to people living nearby. (A) The diagram illustrates some of the main elements in a successful ecotourism program. (B) Ecotourists watch African wild dogs (*Lycaon pictus*) in Hluhluwe-iMfolozi Park, South Africa. (A, after Braithwaite 2001; B, photograph courtesy of Micaela S. Gunther.)

FIGURE 3.10 Horseshoe crabs (*Limulus polyphemus*) gather in great numbers to spawn in shallow coastal waters. These aggregations have significant value for contending groups of people. (Photograph © Prisma Bildagentur AG/Alamy.)

place in scientific field stations, which can become sources of training and employment for local people. While these scientific activities provide economic benefits to the areas surrounding field stations, their real value lies in their ability to increase human knowledge, enhance education, and enrich the human experience.

Multiple uses of a single resource: A case study

Horseshoe crabs provide an example of the diverse values that can be provided by just one species. They are usually noticed only as clumsy creatures that seem to move with difficulty in shallow seawater (**Figure 3.10**). In the United States, commercial fishermen harvest these animals in large quantities for use as cheap fishing bait. In recent years, however, biologists have realized that horseshoe crab eggs and juveniles are extremely important as a food source for shorebirds and coastal fish, which have a major role in local tourism related to birdwatching and sportfishing. When horseshoe crabs have been overharvested, bird populations and sport fish have also declined in abundance (Niles 2009). Additionally, the blood of horseshoe crabs is collected to make limulus amoebocyte lysate (LAL), a highly valuable chemical used to detect bacterial contamination in medications and vaccines (Odell et al. 2005). This chemical cannot be manufactured synthetically, and horseshoe crabs are its only source. Without this natural source of LAL, our ability to determine the purity of important medicines and vaccines would be compromised.

Currently, commercial and sportfishing interests, environmental groups, birdwatching groups, and the biomedical industry are competing for control of horseshoe crabs along U.S. coastlines and are struggling to find a working compromise. Each group can make a good argument for its own right to use or protect horseshoe crabs. Hopefully, the final result will be a working compromise that allows the crabs a place in a functioning ecosystem and still provides for the needs of people living in the area.

Sometimes a specific species or ecosystem can provide a diversity of goods and services to human society. Compromises are often needed to balance competing uses.

The Long-Term View: Option Value

In addition to the direct and indirect values discussed already, option value is another way of valuing biological diversity. The potential of biodiversity to provide an economic benefit to human society at some point in the future is its **option value**. As the needs of society change, so must the methods of satisfying those needs. Scientists are constantly searching for and finding previously unknown uses of plant or animal species, previously untapped genetic variation of an economically important species, or important and previously unknown ecosystem processes. For example, the continued genetic improvement of cultivated plants is necessary, not only to increase yield but also to guard against pesticide-resistant insects and more virulent strains of fungi, viruses, and bacteria (Sairam et al. 2005).

We are continually searching the ecosystems of the world for new plants, animals, fungi, and microorganisms that can be used to fight human diseases or to provide some other economic value,* an activity referred to as **bioprospecting** (Lawrence et al. 2010). These searches are generally carried out by government research institutes, pharmaceutical and agricultural companies, and university researchers. To facilitate the search for new medicines and to profit financially from new products, the Costa Rican government established the National Biodiversity Institute (INBio) to collect biological products (**Figure 3.11**). Merck, an international pharmaceuticals company, signed an agreement to pay INBio $1 million for the right to screen samples and pay a portion of the royalties on any commercial products that result from the research (Bhatti et al. 2009). The Glaxo Wellcome corporation (now GlaxoSmithKline) and the Brazilian government signed a contract worth $3 million to sample, screen, and investigate approximately 40,000 plants, fungi, and bacteria from Brazil, with part of the funds going to support scientific research and local community-based conservation and development projects. Programs such as these provide financial incentives for countries to protect both

*We discussed the productive use value of natural materials earlier in the chapter; however, their future value as new products also gives them value in the present time, so here we will discuss their option value.

FIGURE 3.11 Taxonomists and technicians at INBio sort and identify Costa Rica's rich array of species. Many of these species are also tested for their effectiveness in treating diseases. (Photograph by Steve Winter.)

their natural resources and the knowledge of biodiversity possessed by their indigenous inhabitants.

The search for valuable natural products is wide-ranging. Entomologists search for insects that can be used as biological control agents, microbiologists search for bacteria that can assist in biochemical manufacturing processes, and agricultural scientists search for species and genetic varieties that can potentially produce more food to feed a growing population (Chivian and Bernstein 2008). The growing biotechnology industry is using the world's biodiversity to find new ways to reduce pollution, to develop alternative industrial processes, and to fight diseases threatening human health. Gene-splicing techniques of molecular biology allow unique genes found in one species to be transferred to other species. If biodiversity is reduced, the ability of scientists to locate and perhaps make use of a broad range of species and novel genetic variation will also be reduced.

A question currently being debated among conservation biologists, governments, environmental economists, lawyers, and corporations is, Who owns the commercial development rights to the world's biological diversity? In the past, species were freely collected from wherever they occurred (often in the developing world) by corporations (almost always headquartered in the developed world). Whatever these corporations found useful in the species they collected during bioprospecting was then processed and sold at a profit, which was entirely kept by the corporations. An excellent example is provided by the immunosuppressant drug cyclosporine. The drug cyclosporine was developed from the fungus *Tolypocladium inflatum* into a family of drugs with sales of $1.2 billion per year by the Swiss company Sandoz, which later merged to become Novartis (Bull 2004). The fungus was found in a sample of soil that was collected in Norway, without permission, by a Sandoz biologist on vacation. As of yet, Norway has not received any payment for the use of this fungus in drug production. Similarly, the U.S. government has not received any payments from the biotechnology industry for the bacteria collected from Yellowstone National Park, mentioned earlier in this chapter. Such past and present unauthorized collecting of biological materials for commercial purposes is now often termed **biopiracy**. Many developing countries have reacted to this situation by passing laws that require permits for the collection of biological material for research and commercial purposes, and they impose criminal penalties and fines for the violation of such laws (Bhatti et al. 2009). People collecting samples without the needed permits have been arrested for violating these laws.

A question currently being debated among conservation biologists, governments, ecological economists, corporations, and local individuals is, Who owns the commercial rights to the world's biodiversity?

Countries in both the developing and developed world now frequently demand a share in the commercial activities derived from the biological diversity contained within their borders, and rightly so. Local people in developing countries who possess knowledge of the species, protect them, and show them to scientists should also share in the profits from any use of them. Writing treaties and developing procedures to guarantee participation in this process will be a major diplomatic challenge in the coming years.

While most species may have little or no direct economic value and little option value, a small proportion may have enormous potential value to supply medical treatments, to support a new industry, or to prevent the collapse of a major agricultural crop. If just one of these species goes extinct before it is discovered, it may be a tremendous loss to the global economy, even if the majority of the world's species are preserved. As Aldo Leopold, the wildlife biologist mentioned in Chapter 1, commented:

> If the biota, in the course of aeons, has built something we like but do not understand, then who but a fool would discard seemingly useless parts? To keep every cog and wheel is the first precaution of intelligent tinkering.

The diversity of the world's species can be compared to a manual on how to keep the Earth running effectively. The loss of a species is like tearing a page out of the manual. Although we may someday need the information from that page to save ourselves and the Earth's other species, the information has been irretrievably lost.

Existence Value

Many people throughout the world care passionately about wildlife and plants and want to protect them. Their concern may be associated with a desire to someday visit an unusual ecosystem or see unique species in the wild; alternatively, concerned individuals may not expect or even desire to see these elements of biodiversity personally. In either case, these individuals recognize an **existence value** in wild nature—the benefit people receive from knowing that a habitat or species exists and quantified as the amount that people are willing to pay to prevent species from being harmed or going extinct, habitats from being destroyed, and genetic variation from being lost (Zander and Garnett 2011). A component of existence value is **beneficiary** or **bequest value**, which is the benefit people receive by preserving a resource or species for their children and descendants or future generations, and quantified as the amount people are willing to pay for this goal.

Particular species—the so-called charismatic megafauna such as pandas, whales, elephants, manatees, and many birds—elicit strong responses in people (**Figure 3.12**). Special groups have been formed to appreciate and protect butterflies and other insects and wildflowers. People place value on wildlife and wild lands in a direct way by joining and contributing to conservation organizations that protect species. In the United States, billions of dollars are donated each year to environmental wildlife organizations, with The Nature Conservancy ($1.8 billion in 2008), the World Wildlife Fund ($110 million), Ducks Unlimited ($180 million), and the Sierra Club ($51 million) high on the list. Citizens also show their concern by directing their

People, governments, and organizations annually contribute large sums of money to ensure the continuing existence of certain species, such as birds and whales, and of ecosystems such as rain forests and lakes.

FIGURE 3.12 Whale watching can make a significant contribution to a local economy, and participants often later contribute money to organizations promoting whale conservation. Here people greet a California gray whale (*Eschrichtius robustus*) that is spyhopping off the Baja Peninsula of Mexico. Spyhopping is an activity in which the whale holds itself in a vertical position with its head out of the water for the purpose of looking around its environment. Such meetings (which usually take place at greater distances, as in a more traditional "whale-watch" setting) can enrich human lives. (Photograph © Specialist Stock/Corbis.)

governments to spend money on conservation programs and to purchase land for habitat and landscape protection. For example, the government of the United States spent millions of dollars to protect a single rare species, the brown pelican (*Pelecanus occidentalis*), once protected under the U.S. Endangered Species Act and now considered recovered.

Existence value can be attached to ecosystems, such as temperate old-growth forests, tropical rain forests, coral reefs, lakes, wetlands such as marshes, and prairie remnants, as well as to areas of scenic beauty. Over the last 20 years and continuing into the present, surveys taken in the United States and the United Kingdom have shown that the public regards environmental protection and the needed government funding as high priorities. Further, people want environmental education included in what their children learn in school (www.neefusa.org; www.epa.gov/enviroed).

In summary, ecological economics has helped to draw attention to the wide range of goods and services provided by biodiversity. This helps scientists to better evaluate projects, because they can now account for environmental impacts that were previously left out of the equation. When analyses of large-scale development projects are finally completed, some projects that initially appear to be successful are seen to be actually

FIGURE 3.13 Evaluating the success of a development project must incorporate the full range of its environmental effects. This figure shows the total economic value of a tropical wetland ecosystem, including direct and indirect use value, option value, and existence value. A development project such as an irrigation project lowers the value of the wetland ecosystem when water is removed for crop irrigation. When that lowered value is taken into account, the irrigation project may represent an economic loss. (After Groom et al. 2006; based on data in Emerton 1999.)

running at an economic loss. For example, to evaluate the success of a development project, such as an irrigation project using water diverted from a tropical wetland ecosystem, the short-term benefits (improved crop yields) must be weighed against the long-term environmental costs (lost ecosystem services). **Figure 3.13** shows the total economic value of a tropical wetland ecosystem, including its use values, option value, and existence value. When the wetland ecosystem is damaged by the removal of water, the ecosystem's ability to provide the services shown in the figure are curtailed, its value greatly diminishes, and the economic success of the project is called into question. It is only by incorporating the wetland's natural value into this equation that an accurate view of the economics of the total project can be gained.

Environmental Ethics

In most modern societies, people attempt to protect environmental quality, biodiversity, and human well-being through regulations, incentives, fines, environmental monitoring, and assessments. A complementary approach is to change the fundamental values of our materialistic society. **Environmental ethics**, a vigorous new discipline within philosophy, articulates the ethical value of the natural world. As a corollary, it challenges the materialistic values that tend to dominate modern societies (Alexander 2009). If contemporary societies de-emphasized the pursuit of wealth and instead focused on furthering genuine human well-being, the preservation of the natural environment and the maintenance of biological diversity would likely become fundamental priorities, rather than occasional afterthoughts. Environmental ethics also has strong linkages to movements for social and economic justice.

Ethical arguments supporting preservation

Ethical arguments advanced to justify the protection of biodiversity have foundations in the value systems of most religions, philosophies, and cultures. Such arguments appeal to the nobler instincts of people and relate to a general respect for life; a reverence for nature; a sense of the beauty, fragility, uniqueness, or antiquity of the living world; or a belief in divine creation (Bhagwat et al. 2011). People will often accept or at least consider these arguments on the basis of their belief systems (Woodhams 2009). In contrast, arguments based on economic grounds, as described earlier in this chapter, are still being developed and may eventually prove to be inadequate, inaccurate, or unconvincing. In some cases, economic arguments by themselves provide a basis for valuing biodiversity, but such arguments can also be used (and misused) to decide that we ought *not* to save a species or ecosystem, or that we should save one species and not another.

Ethical arguments can complement economic and biological arguments for protecting biodiversity. Such ethical arguments are readily understood by many people.

How can we assign rights of existence and legal protection to nonhuman species and to ecosystems when they lack the self-awareness that we associate with the morality of rights and duties? Many advocates for environmental ethics believe that species do assert their will to live through their production of offspring and their continuous evolutionary adaptation to the changing environment. The preventable extinction of a species due to human activities destroys this natural process. Such an extinction can be regarded as "superkilling" because it not only kills living individuals but eliminates future generations of the species, thus limiting the processes of evolution and speciation (Rolston 1989).

The following assertions of environmental ethics are important to conservation biology because they provide the rationale for protecting rare species, species of no obvious economic value, genetic variation within species, and ecosystems:

- Each species has a right to exist. All species represent unique biological solutions to the problem of survival. On this basis, the survival of each species must be guaranteed, regardless of its abundance or its importance to humans. This is true whether the species is large or small, simple or complex, ancient or recently evolved, economically important or unimportant. All species are part of the community of living beings and have just as much right to exist as humans do. Each species has value for its own sake, an **intrinsic value** unrelated to human needs or desires (Sagoff 2008). This argument suggests that we have a moral responsibility to actively protect species from going extinct as the result of our activities, as articulated by the Deep Ecology movement described later in this chapter. The argument also recognizes humans are part of the larger biotic community and reminds us that we are not the center of the universe (**Figure 3.14**).

> An argument can be made that people have a responsibility to protect species and other aspects of biodiversity because of their intrinsic value, not because of human needs.

- Species interact in complex ways in natural communities. The loss of one species may have far-reaching consequences for other members of the community (as described in Chapter 2). Other species may become extinct in response, or the entire ecosystem may become destabilized as the result of cascades of species extinctions. For these reasons, if we value some parts of nature, we should protect all of nature (Leopold 1949). We are obligated to conserve the system as a whole because that is the appropriate survival unit (Diamond 2005).

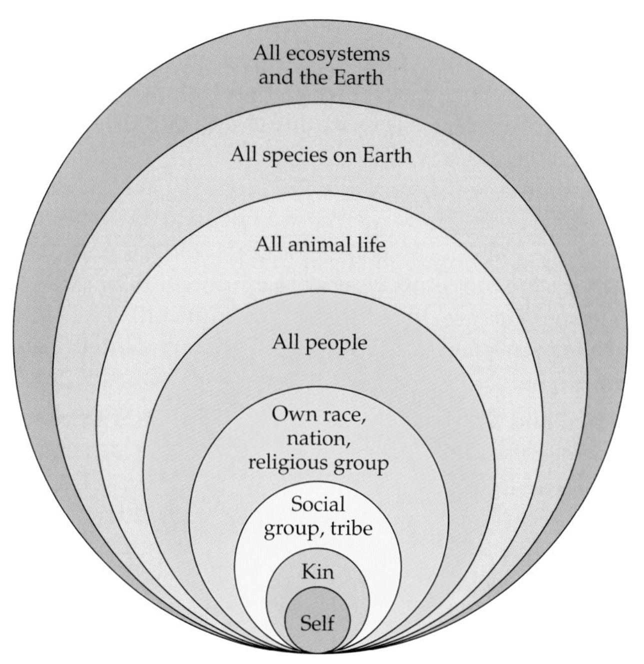

FIGURE 3.14 Environmental ethics holds that an individual has an expanding set of moral obligations that extend outward beyond the self to progressively more inclusive levels. (After Noss 1992.)

All species and ecosystems are interdependent, and so all parts of nature should be protected. It is in the long-term survival interest of people to protect all of biodiversity.

Even if we only value human beings, our instincts toward self-preservation should impel us to preserve biodiversity. When the natural world prospers, people prosper. When the natural world is harmed, people suffer—from widespread health problems such as asthma, food poisoning, waterborne diseases, and cancer and from lower crop yields and fewer recreational opportunities due to or aggravated by environmental pollution.

- People have a responsibility to act as stewards of the Earth. Many religious adherents find it wrong to allow the destruction of species, because they are God's creation (Moseley 2009). If God created the world, then all of the species God created have value. Within the Jewish, Christian, and Islamic traditions, human responsibility for protecting animal species is explicitly described as part of the covenant with God. Other major religions, including Hinduism and Buddhism, strongly support the preservation of nonhuman life.

- If in our daily living we degrade the natural resources of the Earth and cause species to become extinct, future generations will pay the price in terms of a lower standard of living and quality of life (Gardiner et al. 2010). As species are lost and wild lands developed, children are deprived of one of the most exciting experiences in growing up—the wonder of seeing "new" animals and plants in the wild. To remind us to act more responsibly, we might imagine that we are borrowing the Earth from future generations who expect to get it back in good condition.
- Respect for human life and human diversity is compatible with a respect for biological diversity. Some people worry that recognizing an intrinsic value in nature requires taking resources and opportunities away from human beings. But a respect for and protection of biological diversity can be linked to greater opportunities and better health for people (Jacob et al. 2009). Some of the most exciting developments in conservation biology involve supporting the economic development of disadvantaged rural people in ways that are linked to the protection of biological diversity. In both developing and developed countries, the **environmental justice** movement seeks to empower poor and politically weak people, who are often members of minority groups, to protect their own environments; in the process their well-being and the protection of biological diversity are enhanced (Robinson 2011).
- Nature has spiritual and aesthetic value that transcends its economic value. Throughout history, religious thinkers, poets, writers, artists, and musicians of all varieties have drawn inspiration from nature (Thoreau 1854) (**Figure 3.15**). Nearly everyone enjoys wildlife and landscapes aesthetically, and outdoor activities involving nature appreciation are enjoyed by millions of people and are important in creating a sense of well-being (Luck et al. 2011). A loss of biodiversity diminishes this

experience. What if there were no more butterflies? What if there were no more meadows filled with wildflowers?

- Three of the central mysteries in the world of science are how life originated, how the diversity of life evolved and interacts to form complex ecosystems, and how humans evolved. Thousands of biologists are working on these questions and are coming ever closer to the answers. New techniques of molecular biology allow greater insight into the relationships of living species as well as some extinct species, which are known to us only from fossils. However, when species become extinct and ecosystems are damaged, important clues are lost, and these mysteries become harder to solve. For example, if *Homo sapiens'* closest living relatives—chimpanzees, bonobos, gorillas, and orangutans—disappear from the wild, we will lose important clues regarding human physical and social evolution.

FIGURE 3.15 Different people and organizations appreciate nature for different reasons, such as this scientist, priest, and fisherman, but they can still work together to protect nature. (Photograph courtesy of Sudeep Chandra.)

Deep ecology

Throughout the world, environmental organizations such as Greenpeace, Earth First!, and various Green parties in Europe are actively using their knowledge to protect species and ecosystems. One of the most highly developed environmental philosophies that supports this activism is **deep ecology**, which begins with the ethical premise that all species have value in and of themselves and that humans have no right to reduce this richness (Naess 2008). Because ongoing human activities are destroying the Earth's biological diversity, deep ecology states that existing political, economic, technological, and ideological structures must change, and change radically. These changes entail enhancing people's quality of life by emphasizing improvements in environmental quality, aesthetics, culture, and religion rather than economic growth through increased material consumption. The philosophy of deep ecology includes an obligation to work toward implementing the needed changes in society through political involvement and a commitment to personal lifestyle changes. Deep ecology is a philosophy that not only shares many values with conservation biology but also includes the basis of a powerful plan for personal, social, and political change.

Deep ecology is an environmental philosophy that advocates placing greater value on protecting biodiversity through changes in personal attitude, lifestyle, and even societies.

Summary

- Ecological economics is developing methods for valuing biodiversity and, in the process, is providing arguments for its protection. Direct use values are assigned to products harvested from the wild, such as timber, fuelwood, fish, wild animals, edible plants, and medicinal plants. Direct use values can be further divided into consumptive use values, for products that are used locally, and productive use values, for products harvested in the wild and later sold in markets.
- Indirect use values can be assigned to aspects of biodiversity that provide economic benefits to people but are not harvested during their use. Nonconsumptive use values include ecosystem productivity, protection of soil and water resources, positive interactions of wild species with commercial crops, and regulation of climate. Biodiversity also provides value to recreation, education, and ecotourism activities.
- The option value of biodiversity is in terms of its potential to provide future benefits to human society, such as new medicines, industrial products, and crops. Biodiversity also has existence value, based on the amount of money people and their governments are willing to pay to protect species and ecosystems without any plans for direct or indirect use.
- Environmental ethics appeals to religious and secular value systems to justify preserving biodiversity. The most central ethical argument asserts that people must protect species and other aspects of biodiversity because of their intrinsic value, unrelated to human needs. Also biodiversity must be protected because human well-being is linked to a healthy and intact environment.

For Discussion

1. Find a recent large development project in your area, such as a dam, office park, shopping mall, highway, or housing development, and learn all you can about it. Estimate the costs and benefits of this project in terms of biological diversity, economic prosperity, and human health. Who pays the costs and who receives the benefits? Consider other projects carried out in the past, and determine their impact on the surrounding ecosystem and human community.
2. Consider the natural resources that people use near where you live. Can you place an economic value on those resources? If you can't think of any products harvested directly, consider basic ecosystem services such as flood control, freshwater provisioning, and soil retention.
3. Imagine that the only known population of a dragonfly species will be destroyed unless money can be raised to purchase the pond

where it lives and the surrounding land. How could you assign a monetary value to this species?

4. Do living creatures, species, biological communities, and physical entities, such as rivers, lakes, and mountains, have rights? Can we treat them any way we please? Where should we draw the line of moral responsibility?

Suggested Readings

Abson, D. J. and M. Termansen. 2011. Valuing ecosystem services in terms of ecological risks and returns. *Conservation Biology* 25: 250–258. The ecosystem services concept is widely used in conservation strategies but is not without problems.

Chivian, E. and A. Bernstein (eds.). 2008. *Sustaining Life: How Human Health Depends on Biodiversity*. Oxford University Press, New York. Great examples, and beautifully illustrated.

Kross, S. M., J. M. Tylianakis, and X. J. Nelson. 2012. Effects of introducing threatened falcons into vineyards on abundance of passeriformes and bird damage to grapes. *Conservation Biology* 26: 142–149. Introducing threatened falcons into vineyards was 95% effective in preventing introduced birds from eating wine grapes, and saved growers around $300/ha.

Leopold, A. 1949. *A Sand County Almanac and Sketches Here and There*. Oxford University Press, New York. Strong statement for the beauty and value of nature and the many benefits that accrue from protecting it.

Luck, G. W., P. Davidson, D. Boxall, and L. Smallbone. 2011. Relations between urban bird and plant communities and human well-being and connection to nature. *Conservation Biology* 25: 816–826. A greater sense of well-being and satisfaction with the neighborhood is linked to the amount of vegetation cover and species richness.

McCook, L. J. and 20 others. 2010. Adaptive management of Great Barrier Reef: A globally significant demonstration of the benefits of networks of marine reserves. *PNAS* 107: 18,278–18,285. This protected area makes enormous and diverse contributions to the Australian economy.

Naess, A. 2008. *The Ecology of Wisdom: Writings by Arne Naess*. A. Drengson and B. Devall (eds.). Counterpoint, Berkeley, CA. Essays by an influential thinker and a founder of the deep ecology movement.

Siikamaki, J. 2011. Contributions of the US state park system to nature recreation. *PNAS* 108: 14,031–14,036. The recreation value of state parks is unexpectedly large.

Thoreau, H. D. 1854. *Walden; or, Life in the Woods*. Ticknor and Fields, Boston. More than 150 years after its publication, still an eloquent personal statement for protecting nature.

Vianna, G. M. S., M. G. Meekan, D. J. Pannell, S. P. Marsh, and J. J. Meeuwig. 2012. Socio-economic value and community benefits from shark-diving tourism in Palau: A sustainable use of reef shark populations. *Biological Conservation* 145: 267–277. Ecotourism involving just sharks accounts for 8% of the economic activity of this Pacific Island.

An experiment to test how a changing climate affects an ecosystem.

Chapter 4
Threats to Biodiversity

Maintaining a healthy environment means preserving all of its components in good condition—ecosystems, biological communities, species, populations, and genetic variation. If species, ecosystems, and populations are adapted to local environmental conditions, why are they being lost? Why don't they tend to persist in the same place over time? Why can't they adapt to a changing environment? These questions have a single, simple answer: massive disturbances caused by human beings have altered, degraded, and destroyed the landscape on a vast scale, destroying populations, species, and even whole ecosystems.

There are seven major threats to biodiversity: habitat destruction, habitat fragmentation, habitat degradation (including pollution), global climate change, the overexploitation of species for human use, the invasion of exotic species, and the increased spread of disease (**Figure 4.1**). Most threatened species face at least two of these threats, thus speeding their way to extinction and hindering efforts to protect them (MEA 2005; Forister et al. 2010). Typically, these threats develop so rapidly and on such a large scale that species are not able to adapt genetically to the changes or disperse to a more hospitable location. Moreover, multiple threats may interact additively or even synergistically such that their combined impact on a species or an ecosystem is greater than their individual effects.

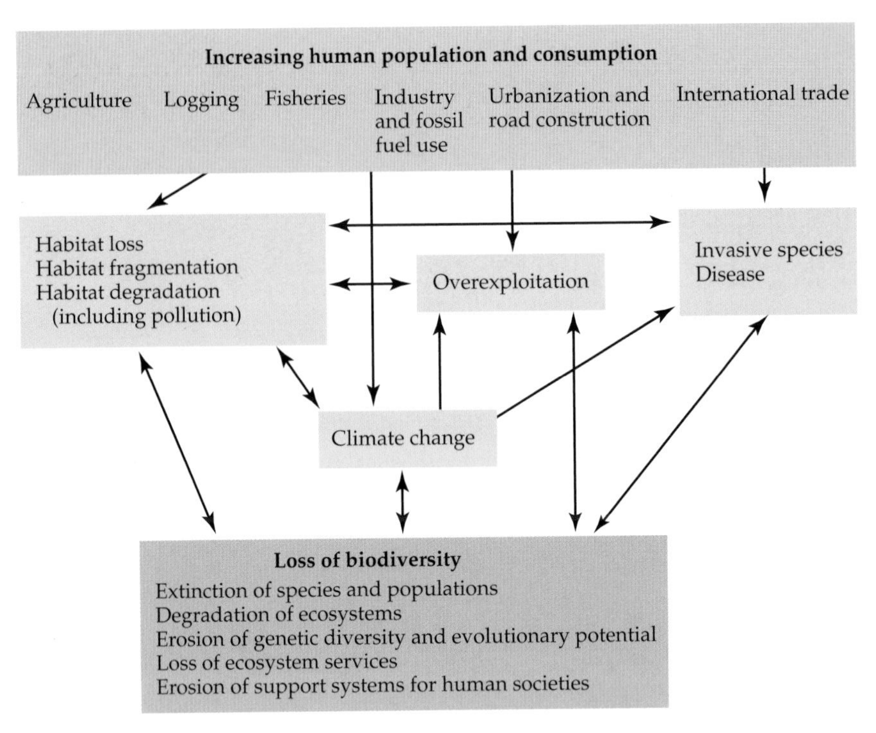

FIGURE 4.1 The major threats to biodiversity (yellow boxes) are the result of human activities. These seven factors can interact synergistically to speed up the loss of biodiversity. (After Groom et al. 2006.)

Human Population Growth and Its Impact

The seven major threats to biological diversity are all caused by an ever-increasing use of the world's natural resources by an expanding human population. Up until the last 300 years, the rate of human population growth had been relatively slow, with the birthrate only slightly exceeding the mortality rate. The greatest destruction of biological communities has occurred over the last 150 years, during which time the human population has exploded from 1 billion in 1850 to 7 billion today. Humans have increased in such numbers because birthrates have remained high while mortality rates have declined—a result of both modern medical achievements (specifically the control of disease) and the presence of more reliable food supplies. Population growth has slowed in the industrialized countries of the world, as well as in some developing countries in Asia and Latin America, but it is still high in other areas,

> The major threats to biodiversity—habitat destruction, habitat fragmentation, pollution, global climate change, overexploitation of resources, invasive exotic species, and the spread of disease—are all rooted in the expanding human population.

TABLE 4.1 Three Ways Humans Dominate the Global Ecosystem

1. Land surface

Human land use, mainly for agriculture, and our need for resources, especially from forestry, have transformed as much as half of the Earth's ice-free land surface.

2. Nitrogen cycle

Each year human activities, such as cultivating nitrogen-fixing crops, using nitrogen fertilizers, and burning fossil fuels, release more nitrogen into terrestrial systems than do natural biological and physical processes.

3. Atmospheric carbon cycle

By the middle of this century, human use of fossil fuels and the cutting down of forests will result in a doubling of the level of carbon dioxide in the Earth's atmosphere.

Sources: Data from MEA 2005 and Kulkarni et al. 2008.

particularly in tropical Africa. If these countries implement immediate and effective programs of population control, human population numbers could possibly peak at "only" 8 billion in 2050 and then gradually decline.

People use natural resources, such as fuelwood, wild meat, and wild plants, and convert vast amounts of natural habitat for agricultural and residential purposes (**Table 4.1**). Agricultural systems now occupy one-fourth of the Earth's land surface. All else being equal, more people equals greater human impact and less biodiversity (Laurance 2007; Clausen and York 2008). More land must be cleared to feed a growing population (Godfrey et al. 2010). Nitrogen pollution is greatest in rivers flowing through landscapes with high human population densities, and rates of deforestation are greatest in countries with the highest rates of human population growth. Therefore, some scientists have argued strongly that controlling the size of the human population is the key to protecting biological diversity (O'Neill et al. 2010).

Healthy ecosystems can persist close to areas with high densities of people as long as human activities are regulated by local custom or government. The sacred groves that are preserved next to villages in Africa, India, and China are examples of locally managed biological communities. When this regulation breaks down during war, political unrest, or other periods of social instability, the result is usually a scramble to collect and sell resources that had been used sustainably for generations. The higher the human population density, and the larger the city, the more closely human activities must be regulated, and the greater the potential for both destruction and conservation (Grimm et al. 2008).

People in industrialized countries (and the wealthy minority in the developing countries) consume a disproportionate share of the world's energy, minerals, wood products, and food and therefore have a disproportionate impact on the environment. Each year, the United States, which has 5% of the world's human population, uses roughly 25% of the world's natural resources. Each year the average U.S. citizen uses 23 times more energy and 79 times more paper products than does the average citizen of India (Randolph and Masters 2008).

The impact (I) of any human population on the environment is captured by the formula $I = P \times A \times T$, where P is the number of people, A is the average income, and T is the level of technology (Ehrlich and Goulder 2007). It is important to recognize that this impact is often felt over a great distance; for example, a citizen of Germany, Canada, the United Kingdom, or Japan affects the environment in other countries through his or her use of foods and other materials produced elsewhere. This increasing interconnectedness of resource and labor markets is termed **globalization**. The fish eaten at home in Washington, DC, perhaps came from Alaskan waters, where its capture contributed to the population decline of sea lions, seals, and sea otters; the chocolate cake and coffee consumed at the end of a meal in Italy or France may have been made with cacao and coffee beans grown in plantations carved out of rain forests in West Africa, Indonesia, or Brazil. This linkage has been captured in the idea of the **ecological footprint**, defined as the influence a group of people has on both the surrounding environment and locations across the globe (**Figure 4.2**). In the western United States, the human footprint is greatest in populated areas and in areas converted to agriculture (Leu et al. 2008).

The enormous consumption of resources in an increasingly globalized world is not sustainable in the long term.

A modern city in a developed country typically has an ecological footprint that is hundreds of times its area. For example, the city of Toronto in Canada occupies an area of only 630 km^2, but each of its citizens requires the environmental services of 7.7 ha (0.077 km^2) to provide food, water, and waste disposal sites; with a population of 2.4 million people, Toronto has an ecological footprint of 185,000 km^2, an area equal to the state of New Jersey or the country of Syria. This excessive consumption of resources is not sustainable in the long term. Unfortunately, this pattern is now being adopted by the expanding middle class in the developing world, including the large, rapidly developing countries of China and India, and this increases the probability of massive environmental disruption (Grumbine 2007). In fact, the developing countries in the world now generate more greenhouse gases and consume more of certain natural resources than the developed countries of the world; China in particular has emerged as a rapidly growing industrial powerhouse that not only exports manufactured goods but also imports resources from around the world. The affluent citizens of developed countries must confront their excessive consumption of resources and reevaluate their lifestyles while at the same time offering aid to curb population growth, protect biological diversity, and assist industries in the developing world to grow in a responsible way.

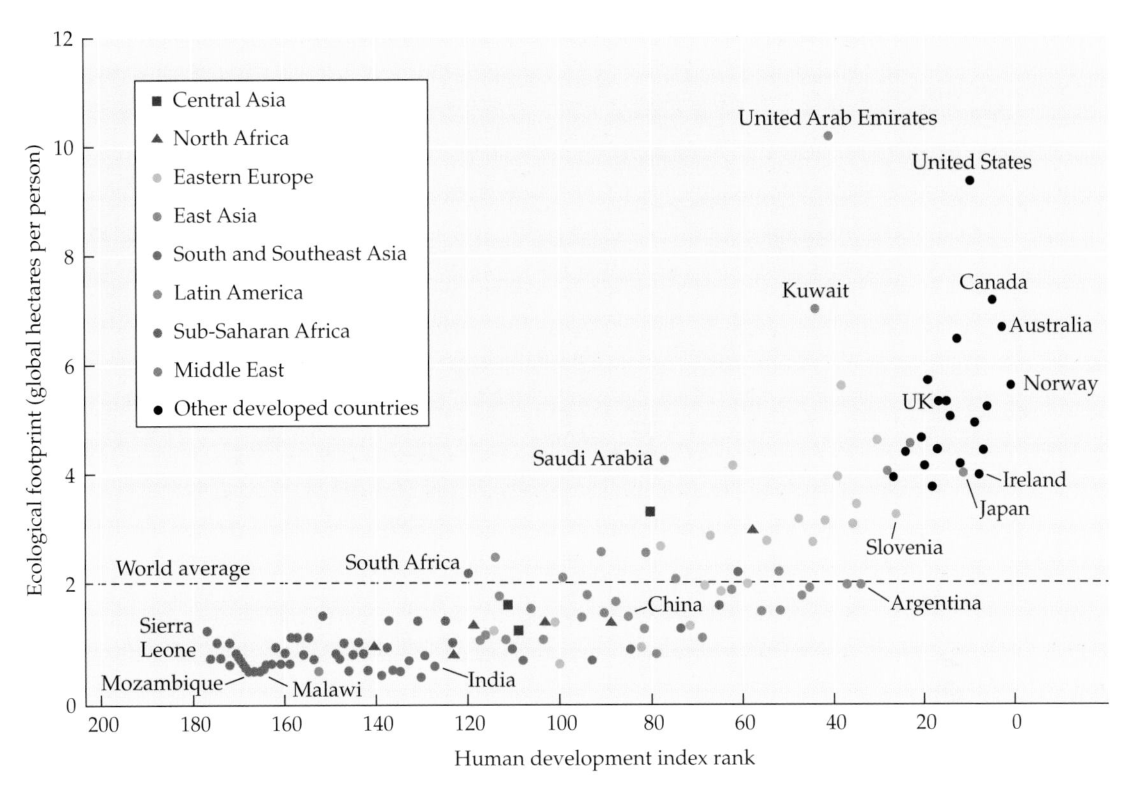

FIGURE 4.2 An ecological footprint for a nation is arrived at by calculations that estimate the number of global hectares needed to support an average citizen of that nation. Although the methods used to arrive at these calculations can be argued, the overall message is clear. When plotted against an economic development index of living standards, global footprints graphically illustrate the disproportionate use of natural resources by people in developed nations. However, the *total* impacts (not shown in this graph) of developing countries such as China, with 1.3 billion people, are also huge because of their large populations. (Data from Global Footprint Network and the United Nations Development Programme 2006.)

Habitat Destruction

The primary cause of the loss of biological diversity, including species, biological communities, and genetic variation, is not direct human exploitation or malevolence but the habitat destruction that inevitably results from the expansion of human populations and human activities (**Figure 4.3**). For the next few decades, land-use change will continue to be the main

> The main threat to biodiversity is habitat destruction.

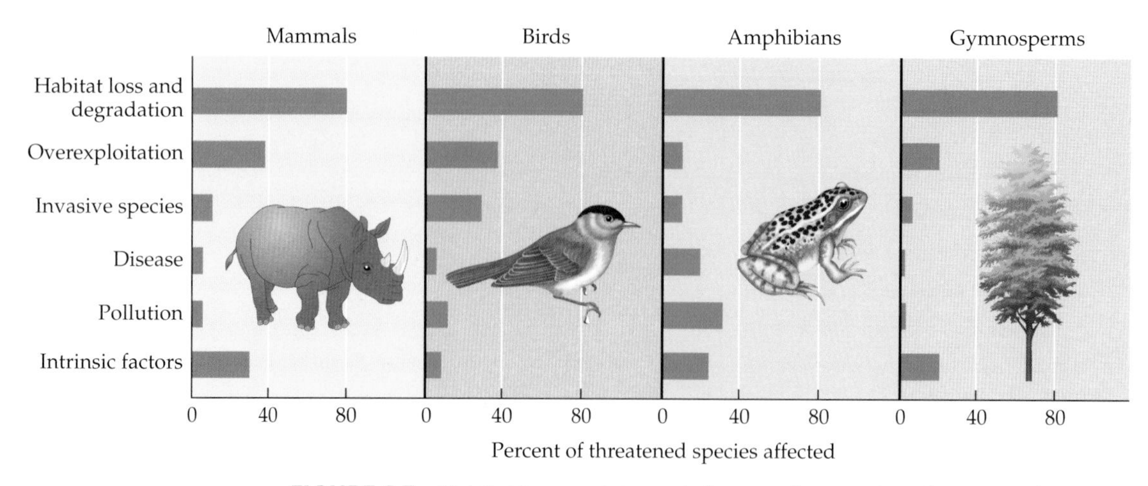

FIGURE 4.3 Habitat loss and degradation are the greatest threats to the world's species, followed by overexploitation and intrinsic factors, which include poor dispersal, low reproductive success, and high juvenile mortality. Groups of species face different threats; birds are more threatened by invasive species, whereas amphibians are more affected by disease and pollution. Percentages add up to more than 100% because species often face multiple threats. (After IUCN 2004.)

factor affecting biodiversity in terrestrial ecosystems, probably followed by overexploitation, climate change, and the introduction of invasive species (IUCN 2004). Consequently, the most important means of protecting biological diversity is habitat preservation. "Habitat loss" includes outright habitat destruction, habitat damage associated with pollution, and habitat fragmentation. When a habitat is degraded and destroyed, the plants, animals, and other organisms living there have nowhere to go and usually die.

In many areas of the world, particularly on islands and in locations where human population density is high, most of the original habitat has been destroyed (MEA 2005; Hambler et al. 2011). Fully 98% of the land suitable for agriculture has already been transformed by human activity (Sanderson et al. 2002). Because the world's population will continue to increase over the next 30 years, the need to protect biological diversity will be forced to compete directly against the need for new agricultural lands.

Habitat disturbance has been particularly severe in many areas of the world: much of Europe; southern and eastern Asia, including the Philippines, China, and Japan; southeastern and southwestern Australia; New Zealand; Madagascar; West Africa; the southeastern and northern coasts of South America; Central America; the Caribbean; and central and eastern North America. In Germany, the United Kingdom, or Japan, one can hardly find any habitat that has not been modified by humans at one time or another. In the United States as a whole, only 42% of the natural vegetation remains, and certain biological communities have declined in area by 98% or more since European settlement (Noss et al. 1995).

The principal habitat threats affecting endangered species in the United States, in decreasing order, are agriculture (affecting 38% of endangered species), commercial developments (35%), water projects (30%), outdoor recreation (27%), livestock grazing (22%), pollution (20%), infrastructure and roads (17%), disruption of fire ecology (13%), and logging (12%).

As a result of habitat fragmentation, farming, logging, and other human activities, very little frontier forest—intact blocks of undisturbed forest large enough to support all aspects of biodiversity—remains in many countries. More than 65% of the wildlife habitat has been destroyed in Old World tropical countries such as Thailand and Ghana (Gallant et al. 2007). In the Mediterranean region, which has been densely populated by humans for thousands of years, only 10% of the original forest cover remains. An important point to remember here is that individual organisms and even entire wild populations are lost in proportion to the amount of habitat that has been lost; even though the Mediterranean forest habitat still exists in places, approximately 90% of the populations of birds, butterflies, wildflowers, frogs, and mosses that once existed there no longer do. However, it is also true that over the centuries, not a single Mediterranean bird species has gone extinct; birds have been able to persist in the habitat that remains and adapt to human-dominated landscapes.

Threatened rain forests

The destruction of tropical rain forests has come to be synonymous with the rapid loss of species. Tropical moist forests occupy 7% of the Earth's land surface, but they are estimated to contain over 50% of its species (**Figure 4.4**) (Corlett and Primack 2010). They are characterized by a complexity of species interactions and specialization unparalleled in any other community. Many rain forest species are important to local economies and have the potential for greater use by the entire world population. Rain forests also have regional importance in protecting watersheds and moderating climate, local significance as home to numerous indigenous cultures, and global importance as reservoirs of carbon that are released into the atmosphere when forests are cut down and burned. About 40% of tropical rain forests are in the Brazilian Amazon (Rodrigues et al. 2009).

Despite the difficulty in obtaining accurate numbers for rain forest deforestation rates, due to varying definitions of forest cover and differing methods, the consensus is that tropical deforestation rates are alarmingly high (Laurance and Luizão 2007). At least one-third of the original extent of tropical rain forest has been destroyed already, and the remainder is being lost at a rate of around 1% per year (Laurance 2007). Strikingly, 55% of all forest losses have been concentrated within only 6% of the total forest area: losses have occurred in an "arc of deforestation" in the south and southeast of the Brazilian Amazon (see Figure 8.6), much of Malaysia, Sumatra, and

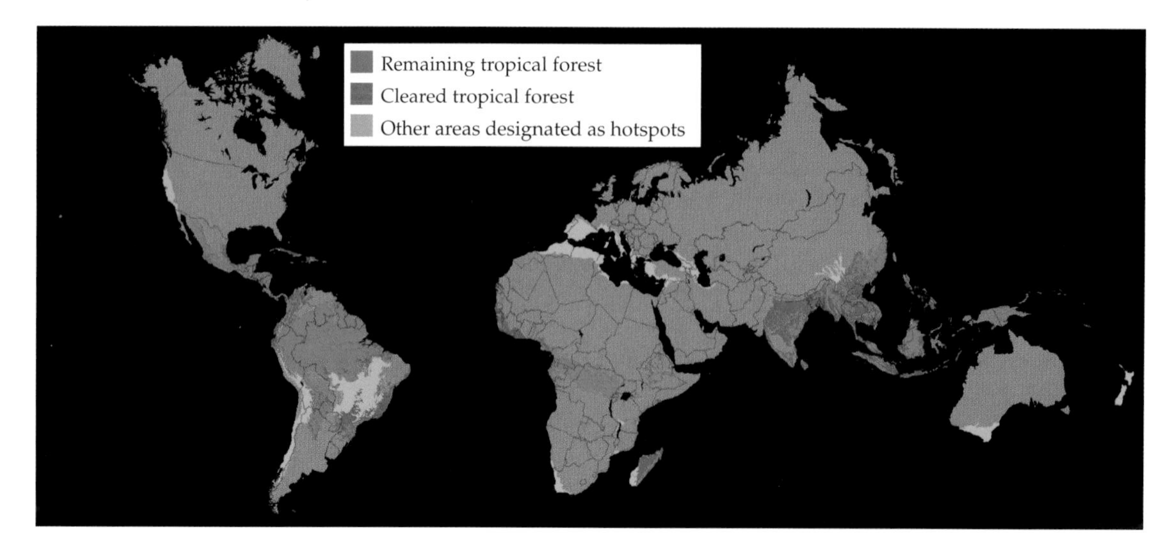

FIGURE 4.4 The current extent of tropical forests, and the areas that have been cleared of tropical forests. Note the extensive amount of land that has been deforested in northern and southeastern South America, India, Southeast Asia, Madagascar, and western Africa. The map also shows hotspots of biodiversity, a subject that will be treated in further detail in Chapter 7. Many of the biodiversity hotspots in the temperate zone have a Mediterranean climate, such as in southwestern Australia, South Africa, California, Chile, and the Mediterranean basin. This map is a Fuller Projection, a type of map that has less distortion of the sizes and shapes of continents than typical maps. (Map created by Clinton Jenkins; originally appeared in Pimm and Jenkins 2005.)

parts of Kalimantan in Indonesia. The loss of tropical forest habitat continues at a rate that guarantees that almost all tropical rain forests will be lost over the next few decades. The only forests that remain will be in protected areas and on rugged or remote terrain.

On a global scale, the majority of rain forest destruction may still result from small-scale cultivation of crops by poor farmers, often forced to remote forest lands by poverty or sometimes moved there by government-sponsored resettlement programs (**Figure 4.5A**). Much of this farming is termed **shifting cultivation**, a kind of subsistence farming, sometimes referred to as slash-and-burn, or swidden, agriculture, in which trees are cut down and then burned away. The cleared patches are farmed for two or three seasons, after which soil fertility usually diminishes to the point where adequate crop production is no longer possible and the land is abandoned (Phua et al. 2008). Shifting cultivation is often practiced because the farmers are unwilling or unable to spend the time and money necessary to develop more permanent forms of agriculture on land that they do not own and may not occupy for very long. Included in this destruction is land degraded each year when trees are cut down for fuelwood, mostly to supply local villagers with wood for cooking fires. More than 2 billion people cook their food with firewood, so their impact is significant. Increasing human population in poor tropical countries will cause further loss of forests in coming decades.

In an increasing proportion of the tropics, however, clearance by peasant farmers to meet subsistence needs is now dwarfed by clearance by large landowners and commercial interests, to create pasture for cattle ranching or to plant cash crops, such as oil palm and soybeans (**Figure 4.5B**) (Butler and Laurance 2008). Cattle ranching and soybean cultivation are particularly important in tropical America, while plantations of tree crops, especially oil palm, are the major cause of deforestation in much of Southeast Asia and are increasing elsewhere. Commercial agriculture displaces poor farmers and justifies the expansion of roads. It is generally

FIGURE 4.5 (A) Tropical forests are cut down for a variety of reasons. In this case, members of the Pemon tribe indigenous to Brazilian Amazonia have cut down trees to build shelters and have burned forest cover in preparation for planting crops. Such settlements are usually abandoned after a few growing seasons as soil fertility decreases—a widespread practice known as shifting cultivation or slash-and-burn agriculture. (B) Large areas of tropical forests have been cleared for oil palm plantations. From the air, these look somewhat like green seas. (A, photograph © David Woodfall/Alamy; B, photograph © jeremy sutton-hibbert/Alamy.)

worse for biodiversity than clearance by peasant farmers, because large areas are maintained under a uniform crop cover. Large areas of rain forest are also damaged during commercial logging operations, most of which are poorly managed selective logging. Also, these logged forests are prone to widespread fires due to the large numbers of branches and dead trees on the ground. Tropical rain forests are easily degraded because the soils are often thin and nutrient poor, and they erode readily in heavy rainfall. In many cases, logging operations precede conversion of land to agriculture and ranching. As these forests are impacted by people, their diversity of species often rapidly declines (Gibson et al. 2011).

The destruction of tropical rain forests is caused frequently by demand in industrialized countries for cheap agricultural products, such as rubber, palm oil, cocoa, soybeans, orange juice, and beef, and for low-cost wood products. During the 1980s, Costa Rica and other Latin American countries had some of the world's highest rates of deforestation as a result of the conversion of rain forests into cattle ranches. Much of the beef produced on these ranches was sold to the United States and other developed countries to produce inexpensive hamburgers. Adverse publicity resulting from this "hamburger connection," followed by consumer boycotts, led major restaurant chains in the United States to stop buying tropical beef from these ranches. The boycott was important in making people aware of the international connections that promote deforestation. Cattle ranching to produce beef for export is still a major contributor to rain forest destruction in Brazil and other Latin American countries. A priority for conservation biology is to help provide the information, programs, and public awareness that will allow the greatest amount of rain forest to persist once the present cycle of destruction ends. At present, most consumers in temperate countries are not aware of how their food choices affect land use.

Other threatened habitats

The plight of the tropical rain forests is perhaps the most widely publicized case of habitat destruction, but many other habitats are also in grave danger. We discuss a few of these threatened habitats below.

TROPICAL DECIDUOUS FORESTS The land occupied by tropical deciduous forests is more suitable for agriculture and cattle ranching than is the land occupied by tropical rain forests. The forests are also easier than rain forests to clear and burn. Moderate rainfall in the range of 250 to 2000 mm per year allows mineral nutrients to be retained in the soil where they can be taken up by plants. Consequently, human population density is five times greater in dry forest areas of Central America than in adjacent rain forests. Today, the Pacific coast of Central America has less than 2% of its original forest remaining (WWF and McGinley 2009), and less than 3% remains in Madagascar, which is home to the lemurs, an endemic group of primates (**Figure 4.6**) (Hogan et al. 2008).

FIGURE 4.6 The Indri (*Indri indri*) are among the largest of the lemurs, a lineage of primates found only on the large island of Madagascar. Virtually all of the numerous lemur species are endangered as a result of the destruction of Madagascar's forests. This species is threatened by the loss of its forest habitat and hunting. Forest fragmentation is particularly harmful to indri populations as these animals rarely descend to the ground. (Photograph © John Warburton-Lee Photography/Alamy.)

GRASSLANDS Temperate grassland is another habitat type that has been almost completely destroyed by human activity. It is relatively easy to convert large areas of grassland to farmland and cattle ranches. Between 1800 and 1950, as much as 98% of North America's tallgrass prairie was converted to farmland (Henwood 2010). Worldwide, only 4% of temperate grasslands are protected. The remaining area of prairie is fragmented and widely scattered across the landscape.

Between 1800 and 1950, as much as 98% of North America's tallgrass prairie was converted to farmland.

WETLANDS The crucial importance of wetlands—habitats in which land and water meet—was discussed in Chapter 3. Healthy wetlands are critical for the well-being of fish, amphibians, aquatic invertebrates, and many birds. Wetlands are often filled in or drained for development, or they are altered by channelization of watercourses, dams, and chemical pollution (Coleman et al. 2008). Over the last 200 years, more than half of the wetlands in the United States have been destroyed, resulting in either extinction or endangerment of 40% to 50% of the freshwater snail species in the southeastern states (Stein et al. 2000). In the United States, 98% of the country's 5.2 million km of streams have been degraded in some way to the point that they are no longer considered wild or scenic. Destruction of wetlands has been equally severe in other parts of the industrialized world, such as Europe and Japan. About 60% to 70% of wetlands in Europe have been lost. Only 2 of Japan's 30,000 rivers can be considered wild, without dams or some other major modification.

In the last few decades, major threats to wetlands in developing countries have included massive development projects involving drainage, irrigation, and dams, organized by governments and often financed by international aid agencies. The Three Gorges Dam on the Yangtze River of China is a recent example. The dam is the largest hydroelectric power plant in the world, generating much-needed clean and renewable energy. However, the dam and reservoir have displaced more than 1 million people, destroyed untold numbers of ecosystems and archeological sites, and altered the river and delta systems, with unknown ecological consequences. The economic benefits of such projects are important, but the rights of local people and the value of the ecosystems are often not adequately considered.

MARINE COASTAL WATERS Human populations are increasingly concentrated in coastal areas. Already more than 20% of marine coastal areas have been degraded or highly modified by human activity. Throughout the world, intensive harvesting of fish, shellfish, seaweed, and other marine products is transforming marine environments. Marine environments are also threatened by pollution, dredging, sedimentation, destructive fishing practices, rising sea temperatures and water levels, the growing acidity of the ocean, and invasive species. Human impacts are less thoroughly studied than in the terrestrial environment but are probably equally severe, especially in shallow coastal areas. Two coastal habitats of special note are mangroves and coral reefs.

MANGROVES Mangrove forests are among the most important wetland communities in tropical areas (Polidoro et al. 2010). Composed of species that are among the few woody plants able to tolerate salt water, mangrove forests occupy shallow coastal areas with saline or brackish water, typically where there are muddy bottoms. Mangroves are extremely important breeding grounds and feeding areas for shrimp and fish. In Australia, for example, two-thirds of the species caught by commercial fishermen depend to some degree on the mangrove ecosystem.

Despite their great economic value and their utility for protecting coastal areas from storms and tsunamis, mangroves are often cleared for rice cultivation and commercial shrimp hatcheries, particularly in Southeast Asian countries, where as much as half of the mangrove area has been removed. Mangroves have also been severely degraded by overcollection of wood for fuel, construction poles, and timber throughout the region. Over 35% of the world's mangrove ecosystems have already been destroyed, and more are being destroyed every year (Martinuzzi et al. 2009).

CORAL REEFS Tropical coral reefs (see Figure 2.11) are particularly significant, as they contain an estimated one-third of the ocean's fish species in

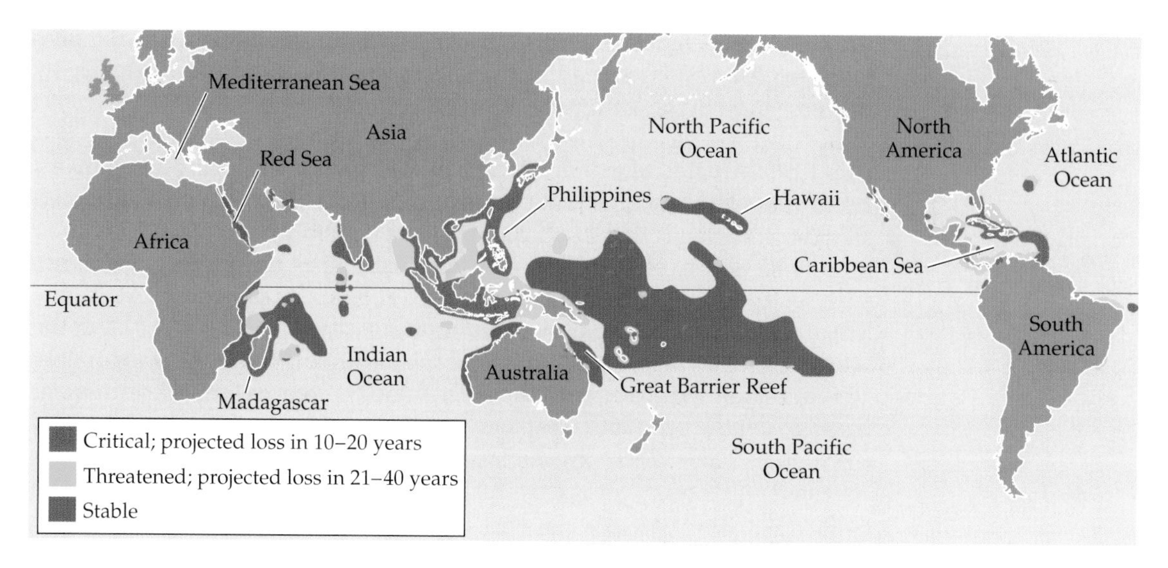

FIGURE 4.7 Extensive areas of coral will be damaged or destroyed by human activity over the next 40 years unless conservation measures can be implemented. (After Bryant et al. 1998, with updates from Burke et al. 2011.)

only 0.2% of its surface area (**Figure 4.7**). Built up over the course of many millennia, 20% of the world's coral reefs have already been destroyed. A further 50% have been degraded (Burke et al. 2011). The main culprits at the local level include pollution associated with coastal development, which either kills the coral directly or allows excessive growth of algae; sedimentation following deforestation; the accidental introduction of invasive species and disease; overharvesting of fish, clams, and other animals; and, finally, blasting with dynamite and releasing cyanide to collect the few remaining living creatures for food and the aquarium trade. The effects of global climate change, including the warming of ocean waters, ocean acidification, and the increasing intensity of tropical storms fueled by warmer surface waters, are also playing a major role in the rapid degradation of coral reefs. Worst of all is the combination of local and global impacts, making it difficult or impossible for coral reefs to bounce back from natural disturbances, such as hurricanes, to which they are otherwise well adapted. The most severe loss of healthy coral reefs is taking place in the Philippines and in the Caribbean region, where a staggering 90% of the reefs are dead or dying.

Over the last 15 years, scientists have discovered extensive reefs of coral living in cold water at depths of 300 m or more, many of which are in the temperate zone of the North Atlantic. These coral reefs are rich in species, with numerous species new to science. Yet at the same time that these communities are first being explored, they are being destroyed by trawlers, which drag nets across the seafloor to catch fish; the trawlers destroy the very coral reefs that protect and provide food for young fish. The damage to these cold-water reefs by careless harvesting is costing the industry its resource base in the long run.

DESERTIFICATION Many biological communities in seasonally dry climates are degraded by human activities into human-made deserts, a process known as **desertification** (Okin et al. 2009). These dryland communities include grassland, scrub, and deciduous forest, as well as temperate shrubland, such as that found in the Mediterranean region, southwestern Australia, South Africa, central Chile, and California. Dry areas cover about 41% of the world's land area and are home to about 1 billion people. Approximately 10% to 20% of these drylands are at least moderately degraded, with more than 25% of the productive capacity of their plant growth having been lost (Neff et al. 2005). These areas may initially support agriculture, but their repeated cultivation, especially during dry and windy years, often leads to soil erosion and loss of water-holding capacity in the soil. Land may also be chronically overgrazed by domestic livestock, and woody plants may be cut down for fuel. Frequent fires during long dry periods often damage the remaining vegetation. The result is the progressive and largely irreversible degradation of the biological community and the loss of soil cover. Ultimately, formerly productive farmland and pastures take on the appearance of a desert.

Worldwide, 9 million km^2 of arid lands have been converted to human-made deserts. These areas are not functional desert ecosystems but wastelands, lacking the flora and fauna characteristic of natural deserts. The process of desertification is most severe in the Sahel region of Africa, just south of the Sahara, where most of the native large mammal species are threatened with extinction. The human dimension of the problem is illustrated by the fact that the Sahel region is estimated to have 2.5 times more people (100 million currently) than the land can sustainably support. Further desertification appears to be almost inevitable, especially when accompanied by the higher temperatures and lower rainfall associated with predictions of future climate change (Verstraete et al. 2009). In such areas, the solution will be programs involving improved and sustainable agricultural practices, the elimination of poverty, the stabilization of civil society, and population control.

Habitat Fragmentation

In addition to being destroyed outright, habitats that formerly occupied wide, unbroken areas are now often divided into pieces by roads, fields, towns, and a broad range of other human constructs. **Habitat fragmentation** is the process whereby a large, continuous area of habitat is both reduced in area and divided into two or more fragments (**Figure 4.8**). When habitat is destroyed, a patchwork of habitat fragments may be left behind. These fragments are often isolated from one another by a highly modified or degraded landscape, and their edges experience an altered set of conditions,

(A)

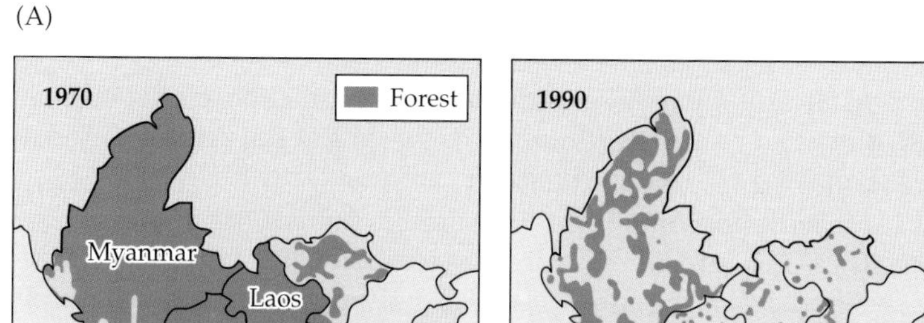

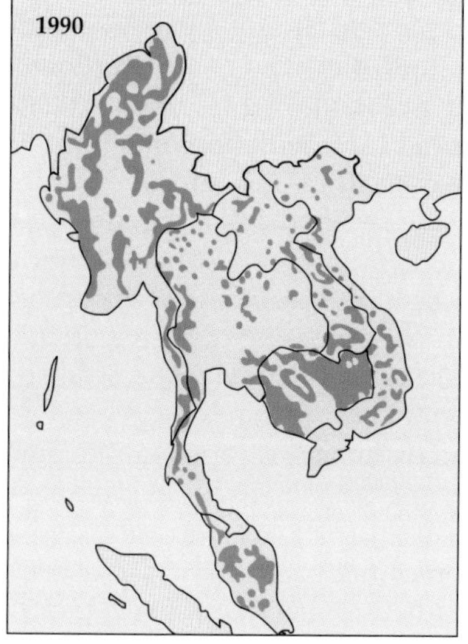

(B)

FIGURE 4.8 The forests of tropical Asia have experienced massive deforestation and fragmentation in recent decades. (A) Two forest maps of Southeast Asia from 1970 and 1990. (B) A wide path (note the car for scale) has been cut through rain forest to allow construction of a gas pipeline in Thailand. Such disturbances often lead to the far-reaching effects of habitat fragmentation. (A, after Bradshaw et al. 2009; B, photograph © Mike Abrahams/Alamy.)

referred to as the **edge effect**. Fragments are often on the least desirable land, with steep slopes, poor soils, or inaccessible areas.

Fragmentation almost always occurs during a severe reduction in habitat area, but it can also occur when area is reduced by only a minor degree if the original habitat is divided by roads, railroads, canals, power lines, fences, oil pipelines, fire lanes, or other barriers to the free movement of species. In many ways, the habitat fragments are islands of original habitats in an inhospitable, human-dominated landscape. Habitat fragmentation is now recognized as a serious threat to biodiversity, as species are often unable to survive under the altered set of conditions.

Habitat fragments differ from the original habitat in three important ways:

1. Fragments have a greater amount of edge per area of habitat (and thus a greater exposure to the edge effect).
2. The center of each habitat fragment is closer to an edge.
3. A formerly continuous habitat hosting large populations is divided into pieces, with smaller populations.

Threats posed by fragmentation

The simple presence of a barrier can threaten the persistence of species in many subtle ways. Edge effects (detailed in the following section) exacerbate the problems of a fragmented habitat.

Habitat fragmentation may limit a species' potential for dispersal and colonization (Laurance et al. 2009; Stouffer et al. 2011). Many bird, mammal, and insect species of the forest interior will not cross even very short stretches of open area, because of the danger of predation or their tendency to avoid sunny, hot, noisy, or dry environments (Ibarra-Macias et al. 2011). Also, animals crossing roads are often killed by motor vehicles (Kociolek et al. 2011). As a result, many species do not recolonize nearby fragments after the original populations disappear. Furthermore, when animal dispersal is reduced by habitat fragmentation, plants with fleshy fruits or sticky seeds that depend on animals for seed dispersal are affected as well. Thus, isolated habitat fragments are not colonized by many of the native species that could potentially live in them. As species become extinct within individual fragments through natural successional and population processes, these dispersal barriers prevent new species from arriving, and the number of native species in the habitat fragment declines over time. Local extinction is most rapid and severe in small habitat fragments.

A second harmful aspect of habitat fragmentation is that it may reduce the foraging ability of native animals. Many animal species, either as individuals or as social groups, need to be able to move freely across the landscape to feed on widely scattered or seasonally available food resources and to find water (Becker et al. 2010). A given resource may be needed for only a few weeks of the year, or even only once in a few years, but when a habitat is fragmented, a species confined to a single habitat fragment may be unable to migrate in search of that scarce resource. For example, fences may prevent the natural migration of large grazing animals such as wildebeest or bison, forcing them to overgraze an unsuitable habitat, eventually leading to starvation of the animals and degradation of the habitat. In river systems, dams may fragment an aquatic habitat (Dudley and Platania 2007).

The barriers that fragment a habitat reduce the ability of animals to forage, find mates, migrate, and colonize new locations. Fragmentation often creates small subpopulations that are vulnerable to local extinction.

Habitat fragmentation may precipitate population decline and extinction by dividing an existing widespread population into two or more subpopulations, each in a restricted area with a limited choice of mates. These smaller populations are more vulnerable to inbreeding depression, genetic drift, and other problems associated with small population size (see Chapter 5). While a large area of habitat may have supported a single large population, it is possible that none of its fragments can support a subpopulation large enough to persist for a long period.

Habitat fragmentation increases the vulnerability of the fragment to invasion by exotic and native pest species (Foxcraft et al. 2011). In the temperate regions of North America, omnivorous native animals such as raccoons, skunks, and blue jays may increase in population size along forest edges, where they can eat foods, including eggs and nestlings of birds, from both undisturbed and disturbed habitats. (Similar increases in nest predation occur on the edges of fragmented tropical forests.) These aggressive feeders seek out the nests of interior forest birds, often preventing successful reproduction for many bird species within hundreds of meters of the nearest forest edge. Nest-parasitizing cowbirds, which live in fields and edge habitats, use habitat edges as invasion points, flying up to 15 km into forest interiors, where they lay their eggs in the nests of forest songbirds (Lloyd et al. 2005). The combination of habitat fragmentation, increased nest predation, and destruction of tropical wintering habitats is probably responsible for the dramatic decline of certain migratory songbird species of North America, such as the cerulean warbler (*Dendroica cerulea*), particularly in the eastern half of the United States (Valiela and Martinetto 2007). (In addition to these local effects, individual bird species in North America and Europe are both increasing and decreasing on regional scales in response to changing land-use patterns, such as those caused by agricultural practices and forest management activities.)

In settled areas with fragmented landscapes, domestic cats may be extremely important predators. In one area of Michigan, 26% of landowners had cats that went outside. Each cat killed an average of one bird per week, including species of conservation concern (Lepczyk et al. 2003).

In many areas of the world, human hunters are the most important predators. When habitat is fragmented by roads, hunters can use the road network to hunt more intensively in the habitat fragments and reach remote areas. Without controls on hunting, there is no refuge for the animals, and their populations decline.

Habitat fragmentation also puts wild populations of animals in closer proximity to domestic animals. Diseases of domestic animals can then spread more readily to wild species, which often have no immunity to them. There is also the potential for diseases to spread from wild species to domestic plants, animals, and even people, once the level of contact increases. In the central United States, fragmented forest habitats often have high densities of white-footed mice, deer, and black-legged ticks and high rates of infection with Lyme disease, along with a corresponding increase in Lyme disease in people living in those areas (Allan et al. 2010).

Edge effects

Habitat fragmentation dramatically increases the amount of edge relative to the amount of interior habitat. A simple example illustrates the problems this can cause. Consider a square conservation reserve that measures 1 km (1000 m) on each side (**Figure 4.9**). The total area of the park is 1 km^2 (100 ha). The perimeter (or edge) of the park totals 4000 m. A point in the middle of the reserve is 500 m from the nearest perimeter. If the principal edge effect on birds in the reserve is predation from domestic cats and introduced

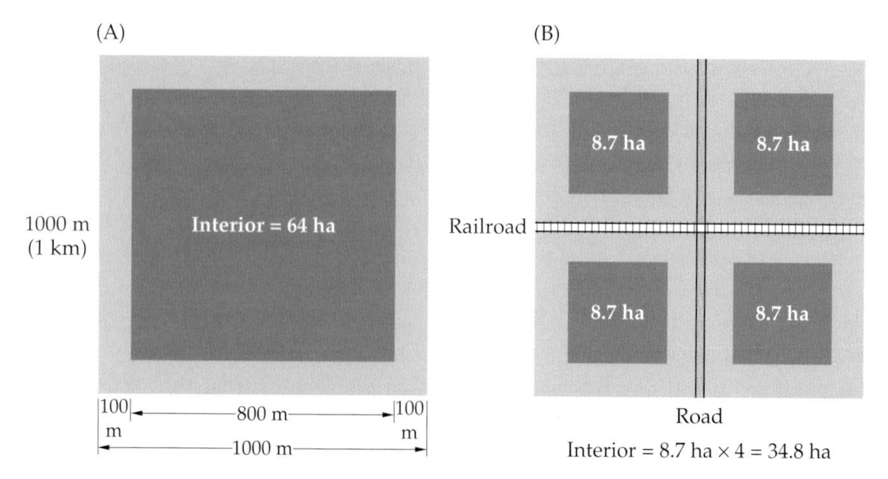

FIGURE 4.9 A hypothetical example shows how habitat area is severely reduced by fragmentation and edge effects. (A) A 1 km^2 protected area. Assuming edge effects (gray) penetrate 100 m into the reserve, approximately 64 ha are available as usable habitat for nesting birds. (B) The bisection of the reserve by a road and a railway, although taking up little in actual area, extends the edge effects so that almost half the breeding habitat is destroyed. Effects are proportionately greater when forest fragments are irregular in shape, as is usually the case.

rats, which forage and hunt 100 m into the forest from the perimeter of the reserve and prevent forest birds from successfully raising their young, then only the reserve's interior—a total of 64 ha—is available to the birds for breeding. Edge habitat, unsuitable for breeding, occupies 36 ha.

Now imagine the park being divided into four equal quarters by a north–south road 10 m wide and by an east–west railroad track, also 10 m wide. The rights-of-way remove a total of 19,900 m^2 (2 ha) from the park. Because only 2% of the park is being removed by the road and railroad, government planners can argue that the effects on the park are negligible. However, the reserve has now been divided into four fragments, each of which is 495 m × 495 m in area. The distance from the center of each fragment to the nearest point on the perimeter has been reduced to 247 m, less than half the former distance. Cats and rats now have access to the forest from along the road and railroad as well as the perimeter, meaning that birds can successfully raise young in only the most interior area of each fragment. Each of these interior areas is now 8.7 ha, for a total of 34.8 ha. Thus, even though the road and railroad removed only 2% of the reserve area, they reduced the habitat available to the birds by about half by increasing edge effects. The implications of this can be seen in the decreased ability of birds to live and breed in small forest fragments, compared with larger blocks of forest.

The microenvironment at a fragment edge is different from that in the forest interior. Some of the more important edge effects are greater fluctuations in levels of light, temperature, humidity, and wind (Laurance et al. 2011). These edge effects are most evident up to 100 m inside the forest, with certain effects detectable up to 400 m from the forest edge (**Figure 4.10**). Because so many plant and animal species are precisely adapted to certain levels of temperature, humidity, and light, changes in those levels eliminate many species from forest fragments. Shade-tolerant wildflower species of temperate forests, late-successional tree species of tropical forests, and humidity-sensitive animals such as amphibians are often rapidly eliminated by habitat fragmentation.

Habitat fragmentation increases edge effects—changes in light, humidity, temperature, and wind that may be less favorable for many species living there.

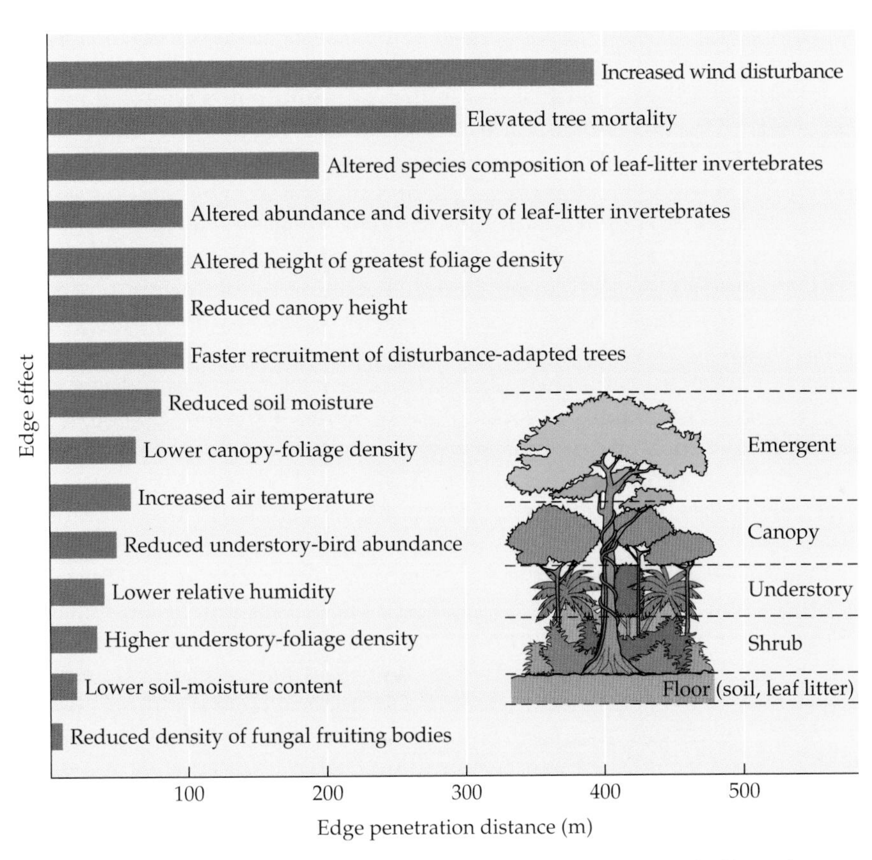

FIGURE 4.10 Edge effects in the Amazon rain forest. The bars indicate how far into the forest fragment the specified effect occurs. For example, trees growing within 300 m of an edge have a higher mortality rate, and the average height of trees in the forest canopy (see drawing) is reduced within 100 m of the edge. (After Laurance et al. 2002.)

When a forest is fragmented, the increased wind, lower humidity, and higher temperatures at the forest edge make fires more likely (Briant et al. 2010). Fires may spread into habitat fragments from nearby agricultural fields that are burned regularly, as in sugarcane harvesting, or from the activities of farmers practicing shifting cultivation. In Borneo and the Brazilian Amazon, millions of hectares of tropical moist forest burned during an unusual dry period in 1997 and 1998; a combination of forest fragmentation due to farming and selective logging, accumulation of brush after logging, and human-caused fires contributed to these environmental disasters (Messina and Cochrane 2007).

Environmental Degradation and Pollution

Even when a habitat is unaffected by overt destruction or fragmentation, the ecosystems and species in that habitat can be profoundly affected by human activities. Biological communities can be damaged and species driven to extinction by external factors that do not change the structure of dominant plants or other features in the community in a way that makes the damage immediately apparent. For example, keeping too many cattle in a grassland community gradually changes it, often eliminating many native plant species and favoring exotic species that can tolerate grazing and trampling. Frequent boating and diving among coral reefs degrade the community, as fragile species are crushed by divers' flippers, boat hulls, and anchors. Out of sight from the public, fishing trawlers drag across an estimated 15 million km^2 of ocean floor each year, an area 150 times greater than the area of forest cleared in the same time period. The trawling destroys delicate creatures such as anemones and sponges and reduces species diversity, biomass, and community structure (**Figure 4.11**) (Hinz et al. 2009). Proposed deep-sea mining operations have the potential to greatly increase the scale of this degradation (Halfar and Fujita 2007).

The most subtle and universal form of environmental degradation is pollution, commonly caused by pesticides, herbicides, sewage, fertilizers from agricultural fields, industrial chemicals and wastes, emissions from factories and automobiles, and sediment deposits from eroded hillsides (Relyea 2005). These types of pollution often are not visually apparent even when they occur all around us, every day, in nearly every part of the world. The general effects of pollution on water quality, air quality, and even the global climate are cause for great concern, not only because of the threats to biological diversity but also because of their effects on human health and agriculture (Barrett et al. 2011). Although environmental pollution is sometimes highly visible and dramatic, it is the subtle, unseen forms of pollution that are probably the most threatening—primarily because they are so insidious.

Pollution of the air, water, and soil by chemicals, wastes, and the by-products of energy production destroys habitats in insidious ways.

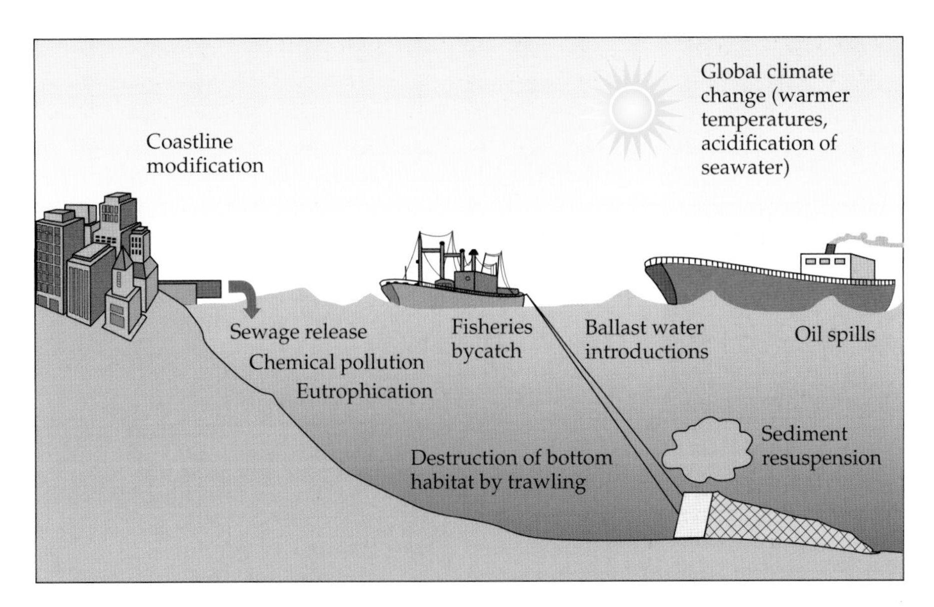

FIGURE 4.11 The aquatic environment faces multiple threats, as shown by this schematic view of damage to the ocean. Trawling is a fishing method in which a boat drags a net along the ocean bottom, harvesting commercial fish indiscriminately with noncommercial species and other sea life ("bycatch") and damaging the structure of the community. (After Snelgrove 2001.)

Pesticide pollution

The dangers of pesticides were brought to the world's attention in 1962 by Rachel Carson's influential book *Silent Spring*. Carson described a process known as **biomagnification**, through which dichlorodiphenyltrichloroethane (DDT) and other pesticides become concentrated as they ascend the food chain (Weis and Cleveland 2008). These pesticides, used on crop plants to kill insects and sprayed on water bodies to kill mosquito larvae, were harming wildlife populations, especially birds that ate large amounts of insects, fish, or other animals exposed to DDT and its by-products. Birds with high levels of concentrated pesticides in their tissues, particularly raptors such as hawks and eagles, became weak and tended to lay eggs with abnormally thin shells that cracked during incubation. As a result of the failure to raise young and the outright death of many adults, populations of these birds showed dramatic declines throughout the world (**Figure 4.12**).

Recognition of this situation in the 1970s led many industrialized countries to ban the use of DDT and other chemically related pesticides. The ban eventually allowed the partial recovery of many bird populations, most notably peregrine falcons (*Falco peregrinus*), ospreys (*Pandion haliaetus*), and bald eagles (*Haliaeetus leucocephalus*). Nevertheless, the continuing massive

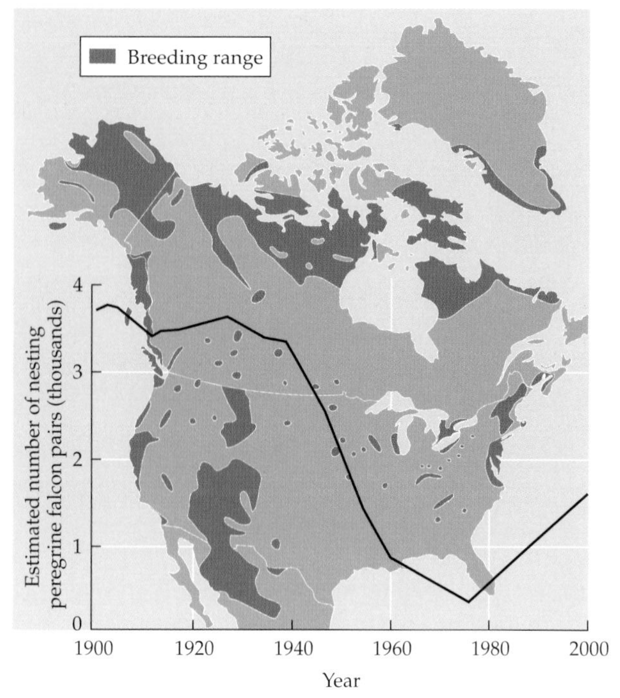

FIGURE 4.12 Peregrine falcons are now breeding in many areas across North America. Their population declined when DDT use began in the 1940s and then recovered after DDT was banned in 1972 (Hoffman and Smith 2003). (After Canadian Wildlife Service and Connecticut Department of Environmental Protection; photograph courtesy of the U.S. Fish and Wildlife Service.)

use of pesticides and even DDT itself in other countries is still cause for concern, not only for endangered animal species, but also for the potential long-term effects on people, particularly the workers who handle these chemicals in the field and the consumers of the agricultural products treated with these chemicals. These chemicals are widely dispersed in the air and water and can harm plants, animals, and people living far from where the chemicals are actually applied (Daly et al. 2007). High concentrations of these toxins have been found in the tissues of polar bears in northern Norway and Russia, where they have a harmful impact on bear health. In addition, even in countries that outlawed these pesticides decades ago, chemicals persist in the environment, where they have a detrimental effect on the reproductive and endocrine systems of aquatic vertebrates (Oehlmann et al. 2009).

Water pollution

Water pollution has negative consequences for people, animals, and all aquatic species: it destroys important food sources and contaminates drinking water with chemicals that can cause immediate and long-term harm to the health of people, fish, and other species that come into contact with the polluted water (Feist et al. 2011). In the broader picture, water pollution often severely damages aquatic ecosystems. Rivers, lakes, and oceans are used as depositories for industrial wastes and residential sewage. And higher densities of people almost always mean greater levels of water pollution. Pesticides, herbicides, oil products, heavy metals (such as mercury, lead, and zinc), detergents, and industrial wastes directly kill organisms, such as insect larvae, fish, amphibians, and even marine mammals living in aquatic environments. Pollution is a threat to 90% of the endangered fishes and freshwater mussels in the United States. An increasing source of pollution in coastal areas is the discharge of nutrients and chemicals from shrimp and salmon farms.

In contrast to a dump in the terrestrial environment, which has primarily local effects, toxic wastes in aquatic environments diffuse over a wide area. Toxic chemicals, even at very low levels in the water, can be lethal to aquatic organisms through the process of biomagnification (**Figure 4.13**). Many aquatic environments are naturally low in essential minerals, such as nitrates and phosphates, and aquatic species have adapted to the natural

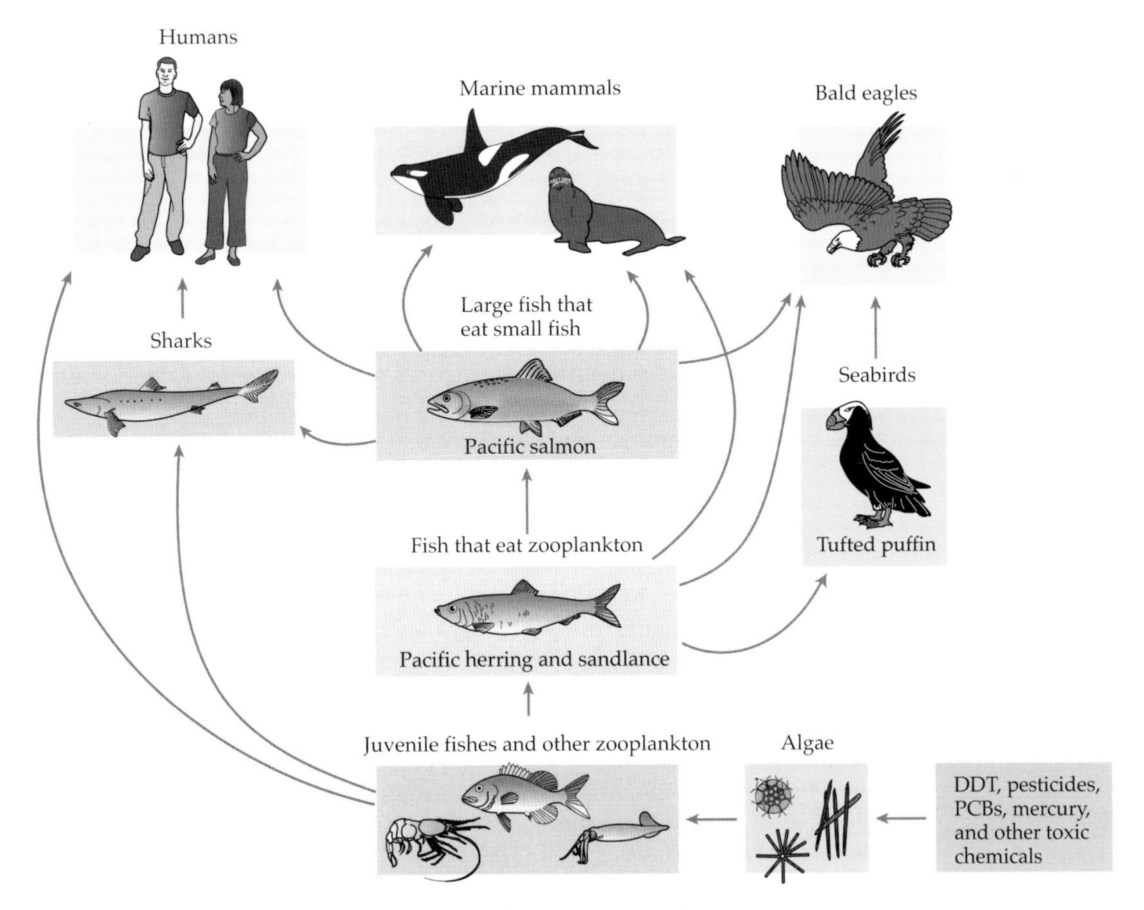

FIGURE 4.13 Toxic chemicals in the water become successively concentrated at higher levels in the food chain, leading to health problems for humans, marine mammals, sea birds, and raptors. (After Groom et al. 2006.)

absence of minerals by developing the ability to process large volumes of water and concentrate these minerals. When these species process polluted water, they concentrate toxic chemicals along with the essential minerals; the toxins eventually poison the plant or animal. Species that feed on these aquatic species ingest these concentrations of toxic chemicals. One of the most serious connections is the accumulation of mercury and other toxins by long-lived predatory fishes, such as swordfish and shark, and its impact on the nervous system of people who eat these types of fish frequently (Jaeger et al. 2009).

Essential minerals that are beneficial to plant and animal life can become harmful pollutants at high levels (Keatley et al. 2011). Human sewage, agricultural fertilizers, detergents, and industrial processes often release large amounts of nitrates and phosphates into aquatic systems, initiating the process of **eutrophication**. Even small amounts of these nutrients can stimulate plant and animal growth, and high concentrations of nutrients released through human activities often result in thick "blooms" of algae at the surface of ponds and lakes. These algal blooms may be so dense that they shade out bottom-dwelling plant species needed as food by many fish and other aquatic animals. As the algal mat becomes thicker, its lower layers sink to the bottom and die. The bacteria and fungi that decompose the dying algae grow in response to this added sustenance and consequently absorb all of the oxygen in the water. Without oxygen, much of the remaining animal life dies off, sometimes visibly in the form of masses of dead fish floating on the water's surface. The result is a greatly impoverished and simplified community, a "dead zone" consisting of only those species tolerant of polluted water and low oxygen levels.

This process of eutrophication can also affect marine systems with large inputs of nutrients from human activities, particularly coastal areas and bodies of water in confined areas, such as the Gulf of Mexico, the Mediterranean, the North and Baltic Seas in Europe, and the enclosed seas of Japan (Greene et al. 2009). Eutrophication increases the growth rate of algae, which then cover and kill the coral reefs and completely change the biological community. The key to stopping eutrophication and its negative effects is to reduce the release of excess nutrients through improved sewage treatment and better farming practices, including reduced applications of fertilizer and establishment of buffer zones between fields and waterways.

Eroding sediments from logged or farmed hillsides can also harm aquatic ecosystems. The sediment covers submerged plant leaves and other green surfaces with a muddy film that reduces light availability and diminishes the rate of photosynthesis. Sediment loads are particularly harmful to many coral species that require crystal-clear waters to survive. Corals have delicate filters that strain tiny food particles out of the clear water. When the water is filled with a high density of soil particles, the filters clog up and the animals cannot feed.

Air pollution

In the past, people assumed that the atmosphere was so vast that materials they released into the air would be widely dispersed and their effects would be minimal. But today several types of air pollution are so widespread that they damage whole ecosystems. These same pollutants also have severe impacts on human health, demonstrating again the common interests shared by people and nature. We discuss specific air pollutants below.

ACID RAIN Acid rain is produced when industries such as smelting operations and coal- and oil-fired power plants release huge quantities of nitrogen oxides and sulfur oxides into the air, where those chemicals combine with moisture in the atmosphere to produce nitric and sulfuric acids. These acids become part of cloud systems and dramatically lower the pH (the standard measure of acidity) of rainwater, leading to the weakening and deaths of trees over wide areas. Acid rain, in turn, lowers the pH of soil moisture and water bodies, such as ponds and lakes, and also increases the concentration of toxic metals such as aluminum.

Increased acidity alone damages many plant and animal species; as the acidity of water bodies increases, many fish either fail to spawn or die outright. Both increased acidity and water pollution are contributing factors to the dramatic decline of many amphibian populations throughout the world (Norris 2007). Most amphibian species depend on bodies of water for at least part of their life cycle, and a decline in water pH causes a corresponding increase in the mortality of eggs and young animals. Acidity also inhibits the microbial process of decomposition, lowering the rate of mineral recycling and ecosystem productivity. Many ponds and lakes in industrialized countries have lost large portions of their animal communities as a result of acid rain. These damaged water bodies are often in supposedly pristine areas hundreds of kilometers from major sources of urban and industrial pollution, because the acid rain is brought to them by the prevailing winds. While the problem of acid rain is decreasing in many areas because of better pollution control, it still remains a serious issue. In developing countries, such as China, the acidity of rain is becoming a greater problem as the country powers its rapid industrial development through the use of fuels high in sulfur.

> Acid rain and other examples of air pollution are increasing rapidly in Asia as countries industrialize. Acid rain is particularly harmful to freshwater species.

OZONE PRODUCTION AND NITROGEN DEPOSITION Automobiles, power plants, and industrial activities release hydrocarbons and nitrogen oxides as waste products. In the presence of sunlight, these chemicals react with the atmosphere to produce ozone and other secondary chemicals, collectively called **photochemical smog**. Although ozone in the upper atmosphere is important in filtering out ultraviolet radiation, high concentrations of ozone at ground level damage plant tissues and make them brittle, harming ecosystems and reducing agricultural productivity. Ozone and smog are detrimental to people and animals when inhaled, so both people and ecosystems benefit from air pollution controls. Smog can be so severe that people may avoid outside activities. When airborne nitrogen compounds are deposited by rain and dust, ecosystems throughout the world are damaged and altered by potentially toxic levels of this nutrient, and many species are unable to survive in the altered conditions (Bobbink et al. 2010).

TOXIC METALS Leaded gasoline (still used in many developing countries, despite its clear danger to human health), mining and smelting operations, coal burned for heat and power, and other industrial activities release large quantities of lead, zinc, mercury, and other toxic metals into the atmosphere (Driscoll et al. 2007). These compounds are poisonous to plant and animal life and can cause permanent injury to children. The effects of these toxic metals are particularly evident in areas surrounding large smelting operations, where life has been destroyed for miles around.

Hope for controlling air pollution in the future depends on building motor vehicles with dramatically lower emissions, increasing the development and use of mass transit systems, developing more-efficient scrubbing processes for industrial smokestacks, and reducing overall energy use through conservation and efficiency measures. Many of these measures are already being actively implemented in European countries and in Japan. The United States lags behind most other industrialized countries, especially in reducing automobile emissions and increasing fuel efficiency.

Global Climate Change

Carbon dioxide, methane, water vapor, and other trace gases in the atmosphere are transparent to sunshine, allowing light energy to pass through the atmosphere and warm the surface of the Earth. These gases and clouds trap some of the energy radiating from the Earth as heat, slowing the rate at which heat leaves the Earth's surface and radiates back into space. These gases are called **greenhouse gases** because they function much like the glass in a greenhouse, which is transparent to sunlight but traps energy inside the greenhouse once it is transformed to heat. The similar warming of Earth by its atmospheric gases is called the **greenhouse effect**. We can imagine that these gases act as "blankets" on the Earth's surface: the denser the concentration of gases, the more heat trapped near the Earth, thus the higher the planet's surface temperature.

The greenhouse effect allows life to flourish on Earth—without it the temperature on the Earth's surface would fall dramatically. Today, however, as a result of human activity, concentrations of greenhouse gases have increased so much that most scientists believe they are already changing the Earth's climate (Gore 2006; IPCC 2007). The term **global warming** is used to describe the increased temperature resulting from the greenhouse effect,

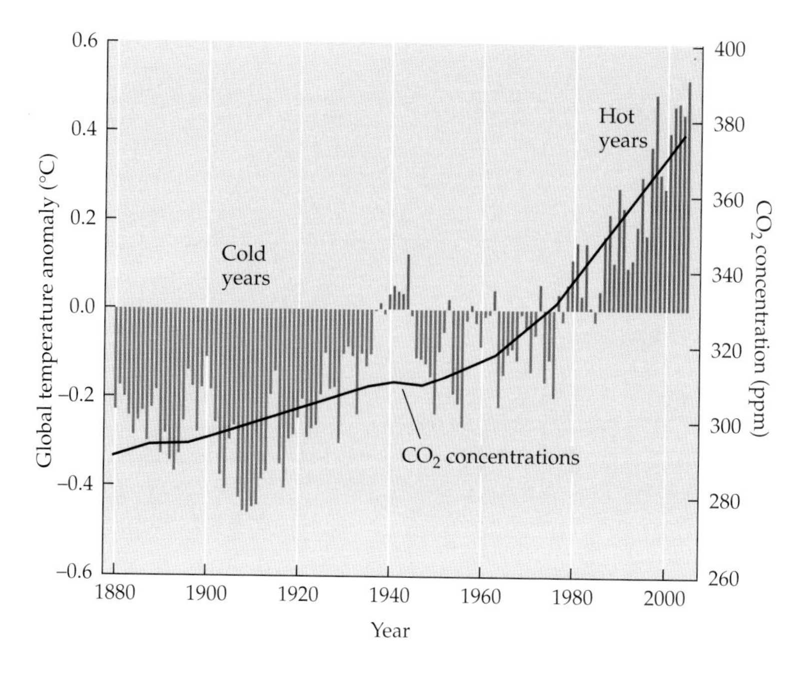

FIGURE 4.14 Over the last 130 years, atmospheric CO_2 concentrations in parts per million (ppm) have increased dramatically as a result of human activities, primarily burning of fossil fuels and clearing of tropical forests. Global annual temperatures were colder than average prior to 1980, when annual temperatures began to be warmer than average. The average annual temperature is based on the period from 1961 to 1990. Results are reported in terms of difference (anomaly) from this average temperature. Most scientists believe the observed increase in global temperature is caused by this increase in carbon dioxide and other greenhouse gases. (After Karl 2006.)

and **global climate change** refers to the complete set of climate characteristics, including patterns of precipitation and wind, that are changing now and will continue to change in the future.

During the past 100 years, global levels of carbon dioxide (CO_2), methane, and other trace gases have been steadily increasing, primarily as a result of burning fossil fuels—coal, oil, and natural gas (IPPC 2007; Kannan and James 2009). Clearing tropical forests during logging and the development of new agricultural lands also contributes to rising concentrations of CO_2. Through these activities, humans currently release about 70 million tons of CO_2 into the atmosphere *every day*. Carbon dioxide concentration in the atmosphere has increased from 290 parts per million (ppm) to 390 ppm over the last 100 years (**Figure 4.14**), and it is projected to double at some point in the latter half of this century.

There is a broad scientific agreement among climate change researchers and the **Intergovernmental Panel on Climate Change (IPCC)**, a study group of leading scientists organized by the United Nations, that the increased levels of greenhouse gases have affected the world's climate and ecosystems already and that these effects will increase in the future (Anderegg et al. 2010). An extensive review of the evidence supports the conclusion that global surface temperatures have increased by 0.6°C during the last century (IPCC 2007; Robinson et al. 2008). In fact, 2010 was the warmest year in Europe for the past 125 years and was probably the warmest year during the past 500 years (Barriopedro et al. 2011). Temperatures at

There is a broad consensus among scientists that increased concentrations of carbon dioxide and other greenhouse gases in the atmosphere, produced as a consequence of human activities, have already resulted in warmer temperatures and will continue to affect Earth's climate in coming decades.

(A)

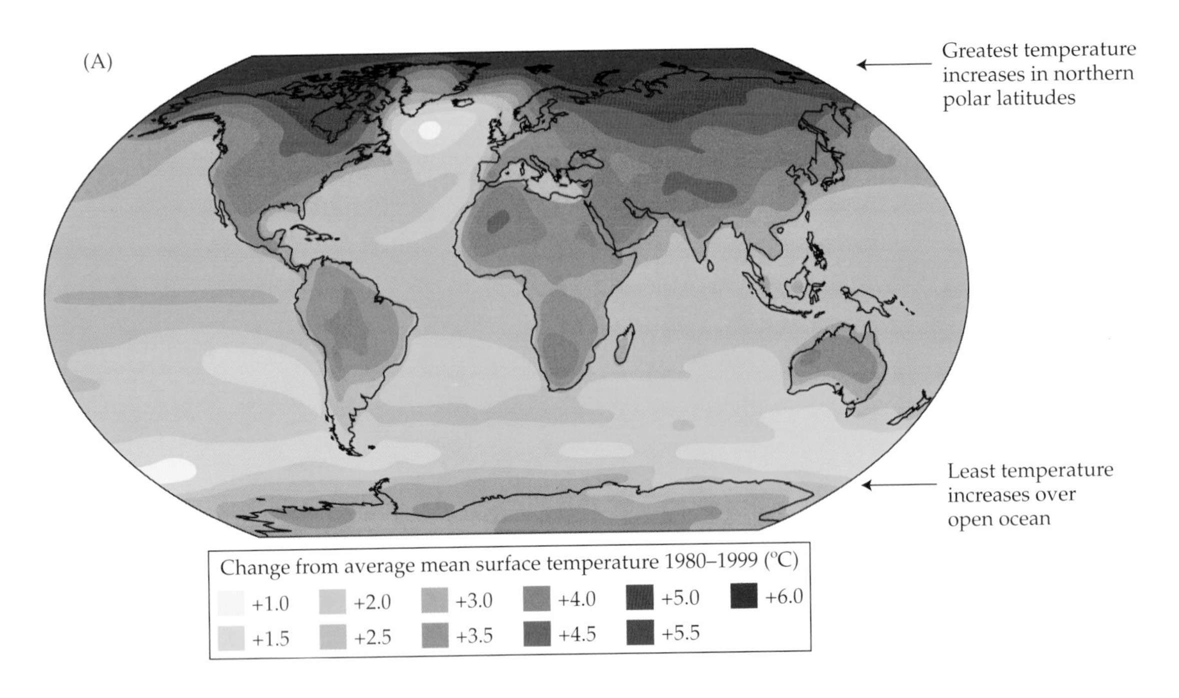

(B)

FIGURE 4.15 (A) Computer models of global climate predict that temperatures will increase significantly when CO_2 levels double in the mid- to late part of this century. Predicted temperature increases for the time frame 2080–2099 are shown, indicated as the amount of deviation (in °C) from mean surface temperatures recorded in 1980–1999. (B) All climate models predict that the greatest warming will take place in the northern polar regions. Polar ice caps are already melting at alarming rates, as these walruses in the Bering Sea off Alaska seem to attest. (A, after IPCC 2007; B, photograph by Budd Christman, courtesy of NOAA.)

high latitudes, such as in Siberia, Alaska, and Canada, have increased more than in other regions.

Climatologists predict an increase in temperature of an additional 2°C to 4°C by 2100, as a result of increased levels of carbon dioxide and other gases (**Figure 4.15**). The increase could be even greater if carbon dioxide levels rise faster than predicted; conversely, it could be slightly less if all countries reduce their emissions of greenhouse gases in the very near future. The increase in temperature will be greatest at high latitudes and over large continents (IPCC 2007; Kannan and James 2009). Rainfall has already started to increase on a global scale and will continue to increase but will vary by region, with some regions showing decreases in rainfall. There will also probably be an increase in extreme weather events, such as hurricanes, flooding, snowstorms, and regional drought, associated with this warming (Bender et al. 2010; Min et al. 2011). In dry forests and savannas, warmer conditions will result in an increased incidence of fire. In coastal areas, storms will cause increased destruction of cities and other human settlements and will severely damage coastal vegetation, including beaches and coral reefs. The series of hurricanes that devastated the southern United States in 2005, including Hurricane Katrina, could be an indication of what the future may bring.

As a result of global climate change, climatic regions in the northern and southern temperate zones will be shifted toward the poles. This change has clearly begun already, with many plants and animals found growing higher on mountains and extending their ranges closer to the poles. It seems likely that many species will be unable to adjust quickly enough to survive this human-caused warming, which will occur far more rapidly than previous, natural climate shifts. Many species will be unable to disperse rapidly enough to track the changing climate and remain within their "climatic envelope" of temperature and precipitation (Jackson et al. 2009; Post et al. 2009). Habitat fragmentation caused by human activities will further slow or prevent many species from migrating to new sites where suitable habitat exists. Species of limited distribution and/or poor dispersal ability, such as snakes, amphibians, and forest birds, will undoubtedly go extinct, while widely distributed, easily dispersed species will be favored in the new communities (Laurance et al. 2011). Extinction rates due to climate change for species of restricted range could be 9% to 13%; over 1 million species are predicted to go extinct by 2050 (Thomas et al. 2004). The loss could be even greater if warmer conditions and elevated carbon dioxide levels favor invasive species and outbreaks of pest species.

As rainfall patterns change and most regions become warmer, many plant and animal species may not be able to adapt quickly enough to survive. Climate change also may have huge impacts on marine ecosystems and coastal areas occupied by people.

The effects of global climate change on temperature and rainfall are also expected to have dramatic effects on tropical ecosystems (IPCC 2007). Many species and ecosystems appear to have narrow tolerances for temperature and rainfall variation, so even small changes in the climate may have major effects on species composition, cycles of plant reproduction, patterns of migration, and susceptibility to fire (Robinson et al. 2008; Primack and Miller-Rushing 2012). Major contractions in the area of rain forest are quite likely (Corlett 2011). Cool-adapted species that live atop tropical mountains may be especially vulnerable to increasing temperatures; as bands of vegetation move higher on mountains, the species at the top will have nowhere to go.

The large areas where temperate agricultural crops, such as wheat, maize (corn), and soybeans, are now grown may show declining yields of 30% or more by the end of the century due to higher temperatures. Such farm areas may have to be moved farther from the equator and perhaps expanded just to maintain production levels (Schlenker and Roberts 2009). Many of the areas potentially suitable for agricultural land in the future are currently protected conservation lands such as national parks. A situation could develop in which the protection of biological diversity will directly compete with supplying the food needs of people.

Warmer waters, acidification, and rising sea levels

Evidence indicates that ocean water temperatures have increased over the last 100 years by an average of 0.1°C. As a consequence, certain marine species are expanding their ranges farther from the equator, and coral reefs and other marine habitats are threatened by rising seawater temperatures (Pandolfi et al. 2011). Episodes of abnormally high water temperatures in the Pacific and Indian Oceans in the past decade led to the death of the symbiotic algae that live inside corals and provide essential carbohydrates. Without their algae, the whitened corals subsequently suffered a massive dieback, with an estimated 70% coral mortality in Indian Ocean reefs. Warmer conditions in the coming decades could be disastrous for coral reefs and the marine species that depend on them. Increased CO_2 levels in the atmosphere are already increasing the acidity of the ocean and reducing the ability of mollusks, crustaceans, and other marine species to secrete calcium carbonate skeletons, with potentially devastating implications for the ecology and chemistry of the marine environment in coming years (Doney 2010).

Warming temperatures in the atmosphere are already causing mountain glaciers to melt and the polar ice caps to shrink, and this process will continue and accelerate. The IPCC predicts that as a result of this release of water plus the thermal expansion of ocean water, over the next 100 years sea levels will rise by 20 to 60 cm (8 to 24 inches) and flood low-lying coastal communities (IPCC 2007). These predictions are considered by many scientists to be too conservative, with values of 130 cm (more than 4 feet) being possible. Such increases in sea level will be catastrophic for low-lying countries such as Bangladesh and for coastal wetlands throughout the world. Many coastal cities, such as Miami and New York, will have to build expensive seawalls, or they will become flooded by the rising waters, especially during the storm surges associated with hurricanes. There is evidence that the process of sea level rise has already begun; sea levels have risen by about 20 cm over the last 100 years, enough to submerge some low-lying islands.

The overall effect of global warming

Global climate change has the potential to radically restructure ecosystems and change the ranges of many species. The pace of this change could overwhelm the natural dispersal abilities of species. There is abundant evidence that this process has already begun (**Table 4.2**), with poleward movements in the distribution of bird and plant species and with reproduction occurring earlier in the spring (Willis et al. 2008; Chen et al. 2011). Increasing temperatures are also associated with summer drought and higher mortality rates of trees in Europe and with devastating outbreaks of tree-killing beetles in the Rocky Mountains (Carnicer et al. 2011). Because the implications of global climate change are so far-reaching, biological communities, ecosystem functions, and climate need to be carefully monitored over the coming decades (Geyer et al. 2011). Global climate change will also have an enormous impact on human populations in coastal areas affected by rising sea levels and increased hurricane impacts. Drought stress and desertification will be exacerbated in many dry areas by higher temperatures and changing rainfall. In many areas of the world, crop yields will decline because of less favorable growing conditions (Lobell et al. 2008). The poor people of the world will be least able to adjust to these changes and will suffer the consequences disproportionately.

TABLE 4.2 Some Evidence for Global Warming

1. Increased temperatures and incidence of heat waves

Examples: 2007 was the warmest year worldwide over the past 125 years; previously the warmest year was 2005. An August 2003 heat wave in France killed over 10,000 people as temperatures reached 40°C (104°F).

2. Melting of glaciers and polar ice

Examples: Arctic Sea summer ice has declined in area by 15% over the past 25 years. Since 1850, glaciers in the European Alps have disappeared from more than 30%–40% of their former range.

3. Rising sea levels

Example: Since 1938, one-third of the coastal marshes in a wildlife refuge in the Chesapeake Bay have been submerged by rising seawater.

4. Earlier spring activity

Example: One-third of English birds are now laying eggs earlier in the year and two-thirds of temperate plant species are now flowering earlier than they did several decades ago.

5. Shifts in species ranges

Example: Two-thirds of European butterfly species studied are now found 35 to 250 km farther north than recorded several decades ago.

6. Population declines

Example: Adélie penguin populations have declined over the past 25 years as their Antarctic sea ice habitat melts away.

Sources: Data from Union of Concerned Scientists (www.ucsusa.org) and NASA.

It is likely that, as the climate changes, many existing protected areas will no longer preserve the rare and endangered species that currently live in them (Mawdsley et al. 2009; Hole et al. 2011). We need to establish new conservation areas now to protect sites that will be suitable for these species in the future, such as sites with large elevational gradients. Potential future migration routes, such as north–south river valleys, need to be identified and established now. If species are in danger of going extinct in the wild because of global climate change, the last remaining individuals may have to be maintained in captivity. Another strategy that we need to consider is to transplant isolated populations of rare and endangered species to new localities at higher elevations and closer to the poles, where they can survive and thrive. This has been termed **assisted colonization**. There is considerable debate within the conservation community about whether assisted migration represents a valid strategy or whether it is too problematic because of the potential for transplanted species to become invasive in their new ranges. Even if global climate change is not as severe as predicted, establishing new protected areas can only help to protect biological diversity.

Although the prospect of global climate change is cause for great concern, it should not divert our attention from the massive habitat destruction that is the principal current cause of species extinction. Preserving intact ecosystems, restoring degraded ones, and increasing the connectivity of existing protected areas are the most important and immediate priorities for conservation, especially in the marine environment. These protected areas will also facilitate the migration of species as they adjust their ranges in response to a changing climate. Over the longer term, we need to reduce our use of fossil fuels and protect and replant forests in order to decrease levels of greenhouse gases. It is widely accepted that local, national, and especially international efforts to reduce greenhouse gas emissions now will result in less severe climate change impacts later.

Overexploitation

People have always hunted and harvested the food and other resources they need to survive. As long as human populations were small and collection methods were unsophisticated, people could sustainably harvest and hunt the plants and animals in their environment. However, as human populations have increased, our use of the environment has escalated and our methods of harvesting have become dramatically more efficient. In many areas, this has led to an almost complete depletion of large animals and the creation of strangely "empty" habitats. Technological advances mean that even in the developing world, guns are used instead of blowpipes, spears, or arrows. Networks of wire snares catch animals of all sizes. Powerful motorized fishing boats and enormous factory ships harvest fish from the world's oceans

> Today's vast human population and improved technology have resulted in unsustainable harvest levels of many biological resources.

FIGURE 4.16 Intensive harvesting has reached crisis levels in many of the world's fisheries. These bluefin tuna are being transferred from a fishing trawler to a factory ship, aboard which huge quantities of fish are efficiently processed for human consumption. Such efficiency can result in massive overfishing. (Photograph © Images & Stories/Alamy.)

and sell them on the global market (**Figure 4.16**). Such overexploitation by humans is estimated to threaten about one-third of endangered mammals and birds (IUCN 2004).

Traditional societies sometimes have imposed restrictions on themselves to prevent overexploitation of jointly owned common property or natural resources (Cinner and Aswani 2007). For example, the rights to specific harvesting territories have been rigidly controlled; hunting and harvesting in certain areas have been banned. There have often been prohibitions against harvesting female, juvenile, and undersized animals. Certain seasons of the year and times of the day have been closed for harvesting. Certain efficient methods of harvesting have not been allowed. (Interestingly enough, these restrictions, which have allowed some traditional societies to exploit communal resources on a long-term, sustainable basis, are almost identical to the fishing restrictions regulators have imposed on or proposed for many fisheries in industrialized nations.)

Such self-imposed restrictions on using common-property resources are often less effective today. In much of the world, resources are exploited opportunistically (de Merode and Cowlishaw 2006). In economic terms, a regulated common-property resource sometimes becomes an open-access resource available to everyone without regulation. The lack of restraint applies to both ends of the economic scale—the poor and hungry as well as the rich and greedy. If a market exists for a product, local people will search their environment to find and sell it. Sometimes traditional groups will sell the rights to a resource, such as a forest or mining area, for cash to buy desired goods. Whole villages are mobilized to systematically remove every usable animal and plant from an area of forest. Where there has been substantial human migration, civil unrest, or war, controls of any type may no longer exist. In countries beset with civil conflict, such as Somalia, Cambodia, the former Yugoslavia, the Democratic Republic of the Congo, and Rwanda, firearms have come into the hands of rural people. The breakdown of legal and social controls leaves the resources of the natural environment vulnerable to whoever can exploit them (Loucks et al. 2009).

Populations of large primates, such as gorillas and chimpanzees, as well as ungulates and other mammals may be reduced by 80% or more by hunting, and certain species may be eliminated altogether, especially those that occur within a few kilometers of a road (Parry et al. 2009; Suárez et al. 2009). In many places, hunters are extracting animals at a rate six or more times greater than the resource base can sustain. The result is a forest with a mostly intact plant community that is lacking its animal community. The decline in animal populations caused by the intensive hunting of animals has been termed the **bushmeat crisis** and is a major concern for wildlife officials and conservation biologists, especially in Africa (www.bushmeat.org; Linder and Oates 2011). Eating primate bushmeat also increases the possibility that new diseases will be transmitted to human populations.

Solutions to the overharvesting of wild animals involve restricting the sale and transport of bushmeat, restricting the sale of firearms and ammunition, closing roads after logging, extending legal protection to key endangered species, establishing protected reserves where hunting is not allowed, and most important, providing alternative protein sources to reduce the demand for bushmeat (Bennett et al. 2007).

International wildlife trade

The legal and illegal trade in wildlife is responsible for the decline of many species (Nijman et al. 2011). Worldwide trade in wildlife is valued at over $10 billion per year, not including edible fish. One of the most pervasive examples of this is the international trade in furs, by which hunted species, such as the chinchilla (*Chinchilla* spp.), vicuña (*Vicugna vicugna*), giant otter (*Pteronura brasiliensis*), and numerous cat species, have been reduced to low numbers. Overharvesting of butterflies by insect collectors; of orchids, cacti, and other plants by horticulturists; of marine mollusks by shell collectors; and of tropical fish by aquarium hobbyists are further examples of targeting of whole groups of species to supply an enormous international demand (**Table 4.3**) (Uthicke et al. 2009). It has been estimated that 350 million tropical

TABLE 4.3 Major Targeted Groups of the Worldwide Trade in Wildlife

Group	Number traded each year[a]	Comments
Primates	70,000	Mostly used for biomedical research; also for pets, zoos, circuses, and private collections.
Birds	4 million	Zoos and pets. Mostly perching birds, but also legal and illegal trade of about 80,000 parrots.
Reptiles	640,000	Zoos and pets. Also 10–15 million raw skins. Reptiles are used in some 50 million manufactured products. Mainly come from the wild, but increasingly from farms.
Ornamental fish	350 million	Most saltwater tropical fish come from wild reefs and may be caught by illegal methods that damage other wildlife and the surrounding coral reef.
Reef corals	1000–2000 tons	Reefs are being destructively mined to provide aquarium decor and coral jewelry.
Orchids	9–10 million	Approximately 10% of the international trade comes from the wild, sometimes deliberately mislabeled to avoid regulation.
Cacti	7–8 million	Approximately 15% of traded cacti come from the wild, with smuggling a major problem.

Sources: Data from WRI 2005, Karesh 2005, and Nijman et al. 2011.
[a]With the exception of reef corals, refers to number of individuals.

fish valued at $1 billion are sold worldwide for the aquarium market, and many times that number are killed during collection or shipping (Karesh et al. 2005). The international trade in other live animals is similarly large: 640,000 reptiles and 4 million birds are sold each year (Karesh et al. 2005). Or consider primates; around 70,000 primates are exported each year, with China being the largest exporter and the United States being the largest importer. In addition, millions of dead primates are sold each year, primarily as bushmeat. In an attempt to regulate and restrict this trade, many declining species are listed as protected under the Convention on International Trade in Endangered Species (CITES; see Chapter 6). Listing species with CITES has often protected species or groups of species from further exploitation.

A striking example of overexploitation is the enormous increase in demand for sea horses (*Hippocampus* spp.) in China (**Figure 4.17**). The Chinese use dried sea horses in their traditional medicine because they resemble dragons and are believed to have a variety of healing powers. About 54 tons of sea horses are consumed in China per year—roughly 19 million animals. Sea horse populations throughout the world are being decimated to supply this ever-increasing demand. The hope is that careful monitoring

FIGURE 4.17 Sea horses have been overfished around the world for traditional medicines, aquarium displays, and curiosities. All exports are now regulated. (Photograph courtesy of ACJ Vincent/Project Seahorse.)

and regulating by international treaty can prevent the extinction of many sea horse populations and species (Foster and Vincent 2005).

Another example is the worldwide trade in frog legs; each year Indonesia exports the legs of roughly 100 million frogs to France and other western European countries for luxury meals. There is no information on how this intensive harvesting affects frog populations, forest ecology, and agriculture. Perhaps not unexpectedly, the names of the frog species on the shipping labels are often wrong, which adds to the difficulty in quantifying the extent of the problem (Warkentin et al. 2009; www.traffic.org).

Besides a surprisingly large legal trade, billions of dollars are involved in the illegal trade of wildlife (Christy 2010). A black market links poor local people, corrupt customs officials, rogue dealers, and wealthy buyers who don't question the sources from whom they buy. This trade has many of the same characteristics, the same practices, and sometimes the same players as the illegal trade in drugs and weapons. Confronting those who perpetuate such illegal activities has become a major and dangerous job for international law enforcement agencies.

Species can often recover when they are protected from overexploitation. Certifying timber, seafood, and other products as sustainable may be a way to prevent overharvesting.

Commercial harvesting

Governments and industries often claim that they can avoid the overharvesting of wild species by applying modern scientific management. As part of this approach, an extensive body of literature has developed in wildlife and fisheries management and in forestry to describe the **maximum sustainable yield**: the greatest amount of a resource, such as Atlantic bluefin tuna (*Thunnus thynnus*), that can be harvested each year and replaced through population growth without detriment to the population. However, in many real-world situations, harvesting a species at the theoretical maximum sustainable yield is not possible, because of unpredictable factors such as weather conditions, disease outbreaks, illegal harvesting, and damage to young individuals during harvesting (Berkes et al. 2006). Also, commercial entities and government officials often lack the key biological information that is needed to make accurate calculations. Not surprisingly, attempts to harvest at high levels can lead to abrupt species declines, particularly for ocean fish (Gutierrez et al. 2011). The Canadian fishing fleet continued to harvest large amounts of cod off Newfoundland during the 1980s, even as populations declined. Eventually, cod stocks dropped to 1% of their original numbers, and in 1992 the government was forced to close the fishery, eliminating 35,000 jobs (MEA 2005). A well-organized industry with defined quotas as well as continuous monitoring of the resource populations are important elements in successful fisheries.

For many marine species, direct exploitation is less important than the indirect effects of commercial fishing (Cox et al. 2007; Zydelis et al. 2009). Many marine vertebrates and invertebrates are caught incidentally as **bycatch** during fishing operations and are killed or injured in the process. Approximately 25% of the harvest in fishing operations is dumped back in the sea to die. The declines of skates, rays, and millions of seabirds have all been linked to their wholesale death as bycatch. The huge number of sea turtles and dolphins killed by commercial fishing boats as bycatch resulted in a massive public outcry in the 1960s and 1970s and is still a cause for concern. The development of improved nets and hooks, as well as other methods to reduce bycatch, is an active area of current fisheries research (Carruthers et al. 2009).

One of the most heated debates over the harvesting of wild marine species has involved the hunting of whales. The debate is due in part to the strong emotional attachment to whales that many people in Western countries have. After recognizing that many whale species had been hunted to dangerously low levels, the International Whaling Commission finally banned all commercial whaling in 1986. Despite that ban, certain species remain at densities far below their original numbers. Those species include the blue whale (*Balaenoptera musculus*) and the northern right whale (*Eubalaena glacialis*), which have been protected since 1967 and 1935, respectively. The densities of other species, such as the gray whale (*Eschrichtius robustus*), appear to have recovered, however (**Table 4.4**). The slow recovery of some species may be due to continued hunting, both legal and illegal. Whale hunting by the Japanese fleet continues under the dubious claim that additional scientific data are needed to assess the status of whale populations.

Finding the methods to protect and manage the remaining individuals of overexploited species is a priority for conservation biologists. As described in Chapter 8, conservation projects linking the conservation of biodiversity and local economic development represent one possible approach. In some cases, this linkage may be made possible by acknowledging the sustainable

TABLE 4.4 Worldwide Populations of Whale Species Harvested by Humans

Species	Numbers prior to whaling[a]	Present numbers	Primary diet items	Status[b]
Baleen whales				
Blue	200,000	10–25,000	Plankton	Endangered
Bowhead	56,000	25,500	Plankton	Least concern
Fin	475,000	60,000	Plankton, fish	Endangered
Gray (Pacific stock)	23,000	15–22,000	Crustaceans	Least concern
Humpback	150,000	60,000	Plankton, fish	Least concern
Minke	140,000	1,000,000	Plankton, fish	Least concern
Northern right	Unknown	1300	Plankton	Endangered
Sei	250,000	54,000	Plankton, fish	Endangered
Southern right	100,000	10,000	Plankton	Least concern
Toothed whales				
Beluga	Unknown	200,000	Fish, crustaceans, squid	Near threatened
Narwhal	Unknown	50,000	Fish, squid, crustaceans	Near threatened
Sperm	1,100,000	360,000	Fish, squid	Vulnerable

Sources: American Cetacean Society (www.acsonline.org); IUCN Red List.

[a]Preexploitation population numbers are highly speculative; genetic evidence suggests the populations might have been even greater (Roman and Palumbi 2003; Alter et al. 2007).

[b]Status is determined by a combination of numbers, threats, and trends. For example, numbers of Southern right whales are low but are increasing.

harvesting of a natural resource through a special certification that allows producers to receive a higher price for their product. Certified timber products and seafood are already entering the market, but it remains to be seen whether they will have a significant positive impact on biodiversity (Butler and Laurance 2008). National parks, nature reserves, marine sanctuaries, and other protected areas can also be established to conserve overharvested species. When harvesting can be reduced or stopped by the enforcement of international regulations, such as CITES, and comparable national regulations, species may be able to recover. Elephants, sea otters, sea turtles, seals, and certain whale species are examples of species that have recovered once overexploitation has been stopped (Lotze and Worm 2009).

Invasive Species

Exotic species are species that occur outside their natural ranges because of human activity. The great majority of exotics do not become established in the places in which they are introduced, because the new environments are not suitable to their needs or because they have not arrived in sufficient

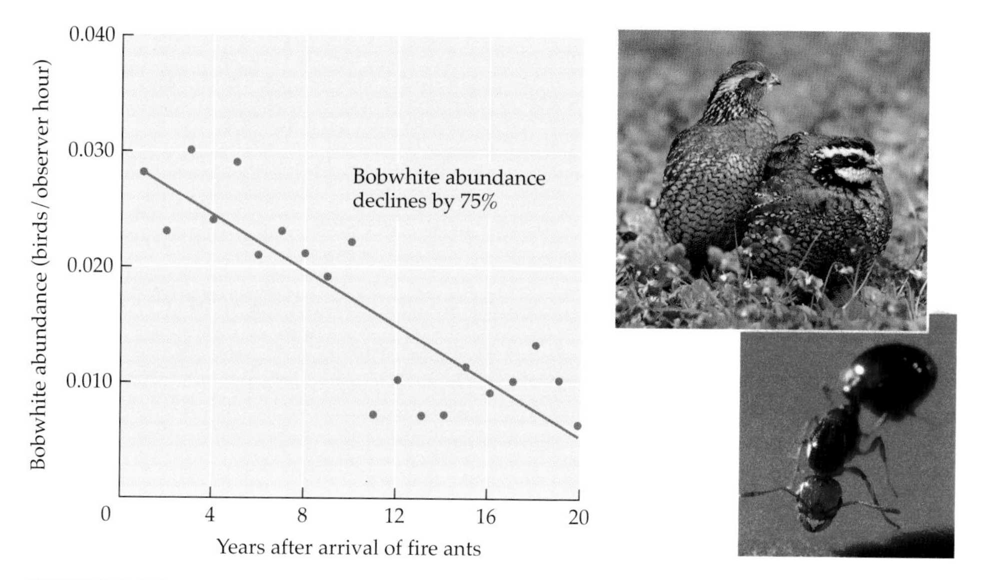

FIGURE 4.18 The abundance of northern bobwhites (*Colinas virginianus*) in Texas has been declining over a 20-year period following the arrival of the exotic red fire ant (*Solenopsis invicta*). The fire ants may directly attack and disturb bobwhites, particularly at the nestling stage, and may compete for food items, such as insects. (After Allen et al. 1995; bobwhite photograph courtesy of Steve Maslowski/U.S. Fish and Wildlife Service; fire ant photograph courtesy of Richard Nowitz/USDA ARS.)

numbers. However, a certain percentage of species do establish themselves in their new homes, and many of these can be considered **invasive species**—that is, they spread and increase in abundance rapidly, sometimes at the expense of native species (Davis 2009; Wilson et al. 2009). These invasive species may displace native species through competition for limiting resources. Introduced animal species may prey upon native species to the point of extinction, or they may alter the habitat so that many natives are no longer able to persist (**Figure 4.18**) (Gooden et al. 2009). Invasive exotic species represent threats to 42% of the endangered species in the United States, with particularly severe impacts on bird and plant species. The thousands of nonnative species in the United States are estimated to cause damages and losses amounting to $120 billion per year, mainly in agriculture, forestry, and fisheries (Pimentel et al. 2000; Aukema et al. 2011). Invasive species are many of the most serious agricultural weeds, costing farmers tens of billions of dollars a year in lost crop yield and extra weed control and herbicide expenses. Globally, over half of all recent animal extinctions are attributable in whole or in part to the effects of invasive species, according to the IUCN database (Clavero and García-Berthou 2005).

Invasive species may displace native species through competition for limiting resources, they may prey upon native species to the point of extinction, or they may alter the habitat so that natives are no longer able to persist.

Many species introductions have occurred by the following means:

- *European colonization.* Settlers arriving at new colonies released hundreds of different species of European birds and mammals into places like New Zealand, Australia, North America, and South Africa to make the countryside seem familiar and to provide game for hunting. Numerous species of fish (trout, bass, carp, etc.) have been widely released to provide food and recreation.
- *Agriculture, horticulture, aquaculture.* Large numbers of plant species have been introduced and grown as ornamentals, agricultural species, pasture grasses, or soil stabilizers. Many of these species have escaped from cultivation or their original habitat and have become established in local communities. As aquaculture develops, there is a constant danger of more plant and animal species escaping and becoming invasive in marine and freshwater environments (Chapman et al. 2003).
- *Accidental transport.* Species are often transported unintentionally. For example, weed seeds are accidentally harvested with commercial seeds and sown in new localities; rats, snakes, and insects stow away aboard ships and airplanes; and disease-causing microbes, parasitic organisms, and insects travel along with their host species, particularly in the leaves and roots of plants and the soil of potted plants. Seeds, insects, and microorganisms on shoes, clothing, and luggage can be transported across the world in a few days by modern jet travelers. Ships frequently carry exotic aquatic species in the water of their ballast tanks, releasing vast numbers of bacteria, viruses, algae, invertebrates, and small fish into new locations. Large ships may hold up to 150,000 tons of ballast water (loaded to add stability during travel). Governments are now developing regulations to reduce the transport of species in ballast water, such as requiring ships to exchange their ballast water 320 km offshore in deep water before approaching a port (Costello et al. 2007).
- *Biological control.* When an exotic species becomes invasive, a common solution is to release an animal species from its original range that will consume the pest and hopefully control its numbers. While biological control can be dramatically successful, there are many cases when a biological control agent does not control its targeted pest or when the introduced species itself becomes invasive, attacking native species along with (or instead of) the intended target species (Elkington et al. 2006). For example, a parasitic fly species (*Compsilura cocinnata*) introduced into North America to control invasive gypsy moths has been found to parasitize more than 200 native moth species, in many cases dramatically reducing population numbers. In order to minimize the probability of such effects, species being considered as biological control agents are tested before release, to determine whether they will restrict their feeding to the species intended as their targets.

Many areas of the world are strongly affected by exotic species. The United States currently has more than 20 species of exotic mammals, 97 species of exotic birds, 138 species of exotic fish, 88 species of exotic mollusks, 5000 species of exotic plants, 53 species of exotic reptiles and amphibians, and 4500 species of exotic insects and other arthropods (Pimentel et al. 2005). Europe has over 1000 nonnative species, of which about 10% have known ecological or economic impacts (Vilá et al. 2010). Exotic perennial plants completely dominate many North American wetlands: purple loosestrife (*Lythrum salicaria*) from Europe dominates marshes in eastern North America. Introduced annual grasses now cover extensive areas of western North American rangelands and increase the probability of ground fires in the summer. Introduced European earthworm species are currently outcompeting native species in soil communities across North America, with negative impacts to certain ground-nesting songbirds and potentially enormous, but as yet unknown, consequences to the rich underground biological communities (Loss and Blair 2011).

Insects introduced both deliberately, such as European honeybees (*Apis mellifera*) and the biocontrol weevil (*Rhinocyllus conicus*), and accidentally, such as fire ants (*Solenopsis invicta*) and gypsy moths (*Lymantria dispar*), can build up huge populations. The effects of such invasive insects on the native insect fauna can be devastating. At some localities in the southern United States, the diversity of insect species declined by 40% following the invasion of exotic fire ants, and there was a similarly large decline in native birds (see Figure 4.18).

Invasive species on islands

The isolation of island habitats encourages the development of a unique assemblage of endemic species, but it also leaves those species particularly vulnerable to depredation by invading species. In many cases, animals introduced onto islands have efficiently preyed upon endemic animal species and have grazed down native plant species to the point of extinction (Garzón-Machado et al. 2010). Introduced plant species, often with tough, unpalatable foliage, are often better able to coexist with the introduced grazers, such as goats and cattle, than are the more palatable native plants. Over time, the exotics begin to dominate the landscape as the native vegetation dwindles. Island animal species that have adapted to a community with few mammalian predators may have limited defenses against introduced predators, such as cats and rats. Moreover, island species often have no natural immunities to mainland diseases; when exotic species such as chicken or ducks are introduced to the island, they frequently carry pathogens or parasites that, though relatively harmless to the carriers, can devastate the native populations.

The introduction of just one exotic species to an island may cause the local extinction of numerous native species. The brown tree snake (*Boiga irregularis*) has been introduced onto a number of Pacific islands, where it devastates endemic bird populations. The snake eats eggs, nestlings, and adult birds. On Guam alone, the brown tree snake has driven 10 of the 11 forest bird species to extinction (Perry and Vice 2009). The government spends $2 million each year on attempts to control the brown tree snake population—so far without success.

Invasive species in aquatic habitats

Freshwater communities are somewhat similar to oceanic islands in that they are isolated habitats surrounded by vast stretches of inhospitable and uninhabitable terrain. Exotic species can have severe effects on vulnerable lake communities and isolated stream systems. There is a long history of the introduction of exotic commercial and sport fish species into lakes; for example, the introduction of exotic tilapia species and the Nile perch (*Lates niloticus*) into Lake Victoria in east Africa was followed by the apparent extinction of hundreds of indigenous species. Although many of these introductions have been deliberate attempts to increase fisheries, others have been the unintentional result of canal building, the transport of ballast water in ships, and escapes from aquaculture. Often these exotic fish are larger and more aggressive than the native fish fauna, and they may eventually drive local fish species to extinction (Maezono et al. 2005). Once these invasive species are removed from aquatic habitats, the native species are sometimes able to recover (Vredenburg 2004).

Aggressive aquatic exotics also include plants and invertebrate animals. One of the most alarming invasions in North America was the arrival in 1988 of the zebra mussel (*Dreissena polymorpha*) in the Great Lakes. This small, striped native of Russian lakes apparently was a stowaway in the ballast tanks of a European tanker that discharged its ballast water into Lake Erie. Within two years, zebra mussels had reached densities of 700,000 individuals per square meter in parts of Lake Erie, encrusting every hard surface (**Figure 4.19A**) and choking out native mussel species in the process. Zebra mussels have now spread south throughout the entire Mississippi

FIGURE 4.19 The zebra mussel (*Dreissena polymorpha*), a native of Russian lakes, was accidentally introduced into Lake Erie in 1988. (A) This current meter was retrieved from Lake Michigan after a crust of thousands of zebra mussels made it inoperable. Such encroachment is typical of the tiny mollusks, which also encrust and destroy native mussel species and other organisms. (B) The map shows distribution of zebra mussel and related quagga mussel infestations throughout North America over the past 20 years and an estimation of the relative risk that these mussels will spread to additional streams and rivers. Each star indicates the location where a boat on a trailer was observed with live mussels on its hull. (A, photograph by M. McCormick, courtesy of NOAA/Great Lakes Environmental Research Laboratory; B, based on data from Whittier et al. 2008 and the U.S. Geological Survey.) ▶

(A)

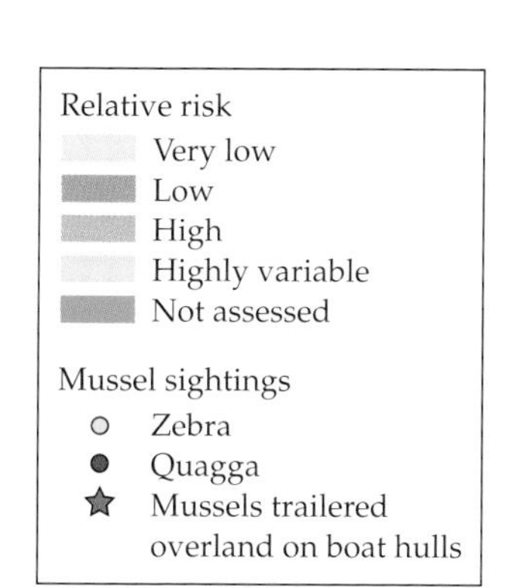
Relative risk
Very low
Low
High
Highly variable
Not assessed
Mussel sightings
Zebra
Quagga
Mussels trailered overland on boat hulls

(B)

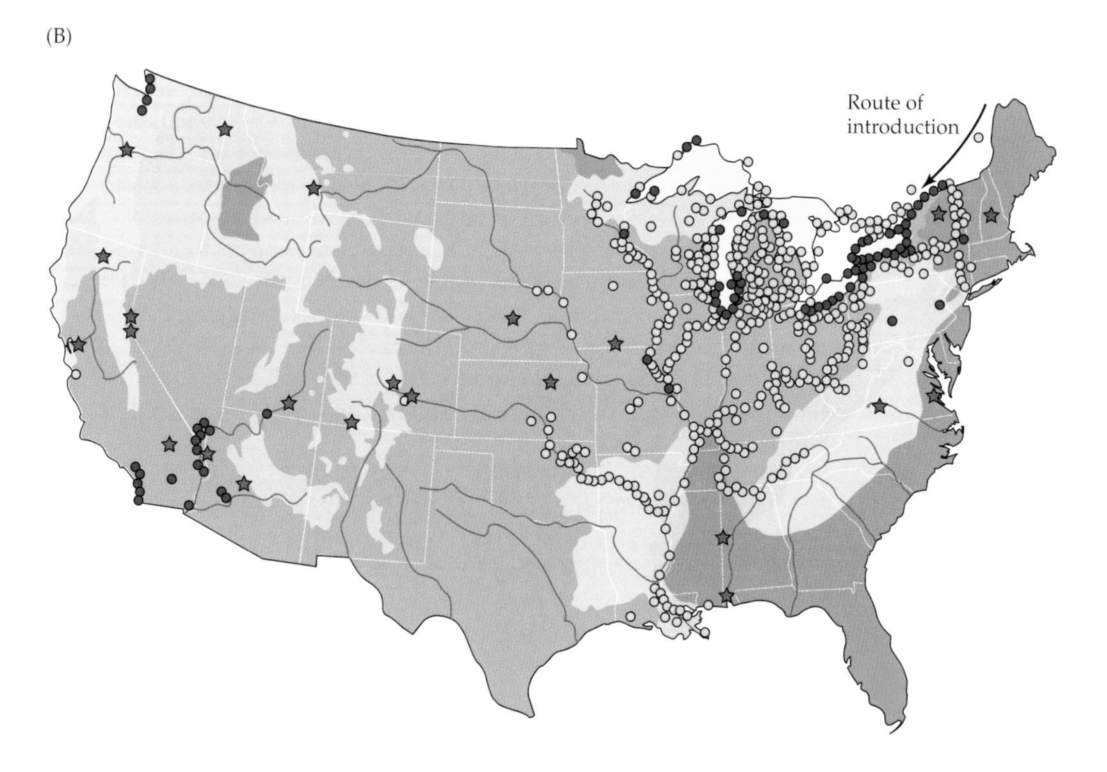
Route of introduction

River drainage, as well as in all other directions via the Great Lakes and their tributaries (**Figure 4.19B**). As it spreads (often via the boots or boat bottoms of unwitting local fishermen), this exotic species causes enormous economic damage to fisheries, dams, power plants, and water treatment facilities, as well as devastating the aquatic communities it encounters.

Many of the worst invasive species in aquatic environments are aquarium and ornamental species that are traded worldwide; the harmful impact of these species needs to be considered as part of the total cost of this trade. Invasions in freshwater environments are often more readily noticed than invasions in marine and estuarine ecosystems. A recent survey found that there are 329 invasive marine species, with 84% of marine areas worldwide affected by at least one (Molnar et al. 2008).

The ability of species to become invasive

Why are certain exotic species able to invade and dominate new habitats and displace native species so easily? One reason is the absence of their natural predators and parasites, which in their own habitat would control their population growth (Davis 2009). For example, rabbits introduced in Australia spread uncontrollably, grazing native plants to the point of extinction, because there were no effective checks on their numbers. Australian control efforts have focused in part on introducing diseases that help control the rabbit populations.

Exotic species also may be better suited to taking advantage of disturbed conditions than native species (Dukes et al. 2011). Human activity causes disturbances that may create unusual environmental conditions, such as higher mineral nutrient levels, increased incidence of fire, or enhanced light availability, to which exotic species sometimes are better adapted than are native species. In fact, the highest concentrations of invasive species are often found in the habitats that have been most altered by human activity. When habitats are altered by global climate change, they will become even more vulnerable to invasion (Willis et al. 2010; Huang et al. 2011; Bradley et al. 2012). In one of the key generalizations of this field, we can say that of the enormous number of introduced species, the relatively few species most likely to become invasive and exert strong impacts in a new location are those species that have already been shown to do so someplace else (Ricciardi 2003).

An *invasive species* is generally defined as an alien species that has proliferated outside its native range, but some native species dramatically flourish within their home ranges because they are suited to the ways in which humans have altered the environment and are therefore almost as large a source of concern as exotic invasive species. Within North America, fragmentation of forests, suburban development, and easy access to garbage have allowed the numbers of coyotes, red foxes, and certain gull species to increase. Native jellyfish have

> Invasive species and undesirable native species often thrive where human activities have changed the environment.

become far more abundant in the Gulf of Mexico because they use oil rigs and artificial reefs for spawning and feed on plankton blooms stimulated by nitrogen pollution. As these aggressive species increase, they do so at the expense of other local native species, such as the juvenile stages of commercially harvested fish. These unnaturally abundant native species represent a further challenge to the management of vulnerable native species and protected areas.

A special class of invasive species includes introduced species that have close relatives in the native biota. When invasive species hybridize with native species and varieties, unique genotypes may be eliminated from local populations, and species boundaries may become obscured (see Chapter 2; Laikre et al. 2010). This situation is sometimes called **genetic swamping**, and it appears to be the fate of native trout species when confronted by commercial species. In the American Southwest, the range of the Apache trout (*Oncorhynchus apache*) has been reduced by habitat destruction and by competition with introduced species. The species has also hybridized extensively with rainbow trout (*O. mykiss*), an introduced sport fish, blurring its identity as a distinct species.

Control of invasive species

Invasive species are considered the most serious threat facing the biota of the U.S. national park system. While the effects of habitat degradation, fragmentation, and pollution potentially can be corrected and reversed in a matter of years or decades as long as the original species are present, well-established exotic species may be impossible to remove from communities (Vince 2011). They may have built up such large numbers and become so widely dispersed and so thoroughly integrated into the community that eliminating them would be extraordinarily difficult and expensive (Rinella 2009).

The threats posed by invasive species are so severe that reducing the rate of their introduction needs to become a greater priority for conservation efforts (**Figure 4.20**) (Keller et al. 2008). Governments must pass and enforce laws and customs restrictions that prohibit the transport and introduction of exotic species. In some cases this may require restrictions and inspections related to the movement of soil, wood, plants, animals, and other items across international borders and even through checkpoints within countries. Better ecological information is required prior to deliberate introductions of species thought to be beneficial or potentially desirable (Gordon and Gantz 2008). Currently, vast sums are spent in controlling widespread outbreaks of exotics, but inexpensive, prompt control and eradication efforts at the time of first sighting can stop a species from getting established in the first place. Training citizens and protected-areas staff to monitor vulnerable habitats for the appearance of known invasive species, and promptly implementing intensive control efforts, can be an effective way to stop the establishment and

Countries need to prevent the introduction of new invasive species, to monitor the arrival and spread of invasives, and to eradicate new populations of invasives.

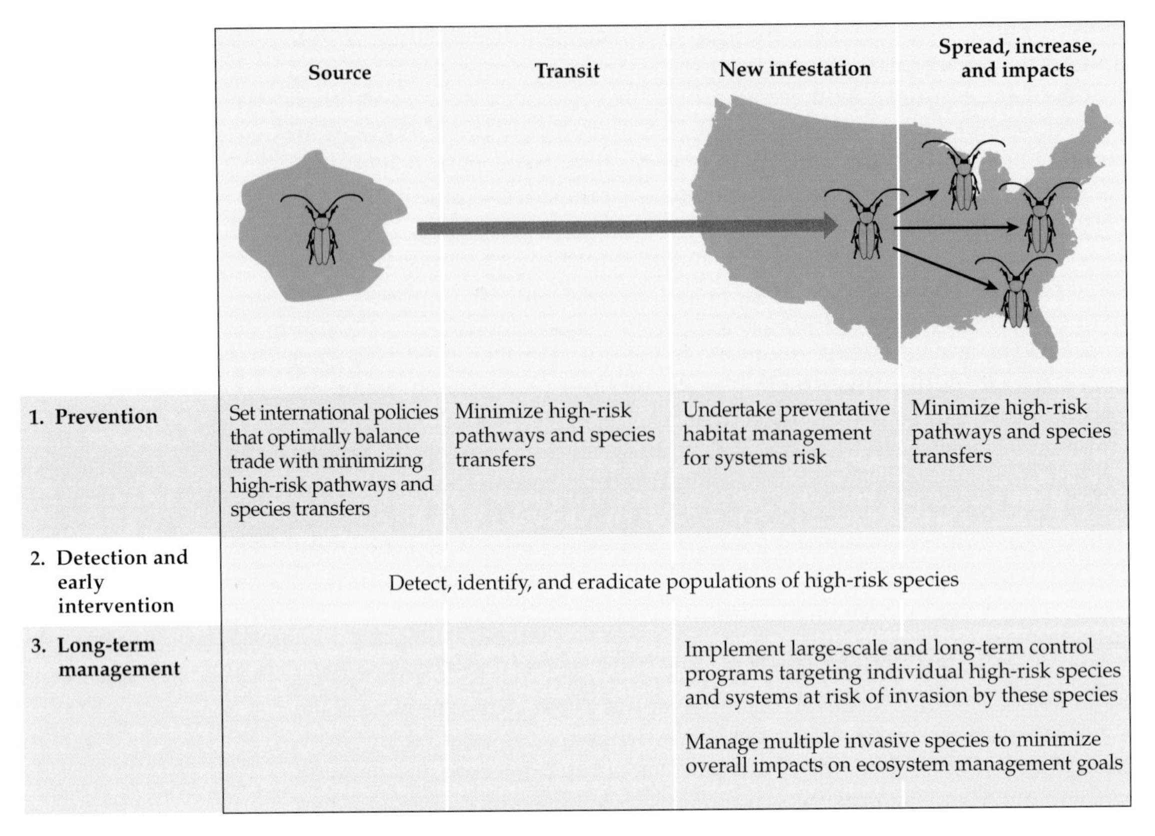

FIGURE 4.20 A strategy for reducing the impact of harmful invasive species involves a combination of prevention, detection and early intervention, and long-term management. This strategy is illustrated by the management of the Asian long-horned beetle (*Anoplophora glabripennis*) problem. This species arrived in North America in the late 1980s in wooden crates and packing material from its native Asia. The beetle infests and kills a wide range of trees, especially maples. The only effective treatment is to cut down infested trees and destroy the wood. (After Chornesky et al. 2005.)

early spread of a new exotic species. This may require a cooperative effort on the part of multiple levels of government and private landowners.

A variety of strategies for controlling and eliminating invasive species exists. In some cases, land-use practices will need to be changed in ways that favor the restoration of native species. In other cases, invasive species may be controlled through physical removal, trapping, and poisoning. When introduced rats were removed from the islands of New Zealand, many populations of seabirds showed dramatic improvement (Jones 2010). An extensive public education program is often necessary so that people are aware of why invasive species need to be removed or killed, especially when they are mammals such as goats, horses, and rabbits (Oppel et al. 2011).

A program of biological control, using species from the exotic's original range, may be necessary to control certain invasive species. Such programs require careful testing to determine the host specificity and likely ecological interactions of the biological control species. They also require careful monitoring after introduction of the control species, to determine its effectiveness in controlling the invasive species, as well as any nontarget effects on native species and communities.

Even though the impacts of invasive species are generally considered negative, they may provide some benefits as well. Invasive plant species can sometimes stabilize eroding lands, provide nectar for native insects, and supply nesting sites for birds and mammals (Chiba 2010). The trade-offs in such situations need to be evaluated to determine whether the potential benefits will outweigh the overall costs.

Genetically modified organisms

A special topic of concern for conservation biologists is the increasing use of **genetically modified organisms (GMOs)** in agriculture, forestry, aquaculture, and toxic waste cleanups (Snow et al. 2005). In such organisms, scientists have added genes from a different ("source") species into the genetic code of the GMO, using the techniques of recombinant DNA technology. Such gene transfer is done not only across species but across taxonomic domains, as when a bacterial gene that produces an insect toxin is transferred into a crop species such as corn. Enormous amounts of cropland—especially in the United States, Argentina, China, and Canada—have already been planted with GMOs, mainly soybeans, corn (maize), cotton, and oilseed rape (canola). Genetically modified animals are under development, with salmon and pigs showing commercial potential.

Humans have been genetically modifying domesticated crop and animal species since the dawn of civilization—by selective breeding, hybridization, and other forms of artificial selection. However, many species being investigated as sources for potential gene transfer, including viruses, bacteria, insects, fungi, and shellfish, have not previously been used in breeding programs. Fear of such "crossovers" between unrelated species has resulted in some governments implementing special controls on this type of research and its commercial applications. There is concern among some people, especially in Europe, that genetically modified crop species will hybridize with related species, leading to invasion by new, aggressive weeds and virulent diseases (Bagla 2010). Additionally, the use of GMOs could potentially harm noncrop species, such as insects, birds, and soil organisms that live in or near agricultural fields. Further, some people want assurances that eating food from GMO crops will not harm their health or cause unusual allergic reactions.

It is clear that GMO crop species have the potential to increase crop production to feed a growing human population; to produce new and cheaper medicines; and to reduce the use of chemicals on agricultural fields and the runoff associated with such use. (On the other hand, one hugely popular GMO—the Roundup-Ready soybean—has been genetically engineered by the herbicide manufacturer so that the crop can be treated with *more*—not less—glyphosate herbicide, commonly sold as the weed killer Roundup.)

In summary, the benefits of GMOs need to be examined and weighed against the potential risks. The best approach involves proceeding cautiously, investigating GMOs thoroughly before commercial releases are authorized, and monitoring environmental and health impacts after release.

Disease

The increased transmission of disease as a result of human activities and interaction with humans is a major threat to many endangered species and ecosystems. Disease-causing organisms such as bacteria, viruses, fungi, and protists can have major impacts on vulnerable species and even the structure of an entire ecosystem. Human activities may increase the incidence of disease-carrying vectors such as certain insects. For example, nutrient enrichment through fertilizer runoff, untreated sewage, and atmospheric nitrogen deposition increases populations of many disease-causing species, leading to outbreaks of animal and human disease. In addition, interaction with humans and their domestic animals exposes wild animals to diseases never previously encountered that can reduce the size and density of wild populations (**Figure 4.21**) (Jones et al. 2008).

Increased incidence of infectious disease threatens wild and domestic species as well as humans. Transfer of disease between different species is a subject of special concern.

Disease may be the single greatest threat to some rare species. The decline of numerous frog populations from pristine montane habitats across the world is apparently due in part to the introduction of an exotic fungal disease. The last population of black-footed ferrets (*Mustela nigripes*) known to exist in the wild was destroyed by the canine distemper virus in 1987, though a few healthy individuals were caught for a captive breeding program. One of the main challenges of managing the captive breeding program has been protecting the captives from canine distemper, human viruses, and other diseases; this is being done through rigorous quarantine measures and subdivision of the captive colony into geographically separate groups (see Figure 6.14).

White nose syndrome is another such disease that is currently killing millions of bats across the eastern United States (Foley et al. 2011). In some caves, 90% of the bats have died. The disease is recognized by the powdery white fuzz caused by the fungus on an infected bat's snout (**Figure 4.22**). Bats are apparently killed when the fungus causes skin irritation and the bat wakes up in midwinter instead of in spring, depleting its energy reserves and subsequently starving to death. Discovered in one cave in New York in 2006, this fungal disease has spread rapidly across the region, probably during bat migration. It is possible that cave explorers or bat researchers may have accidentally introduced the fungus to the United

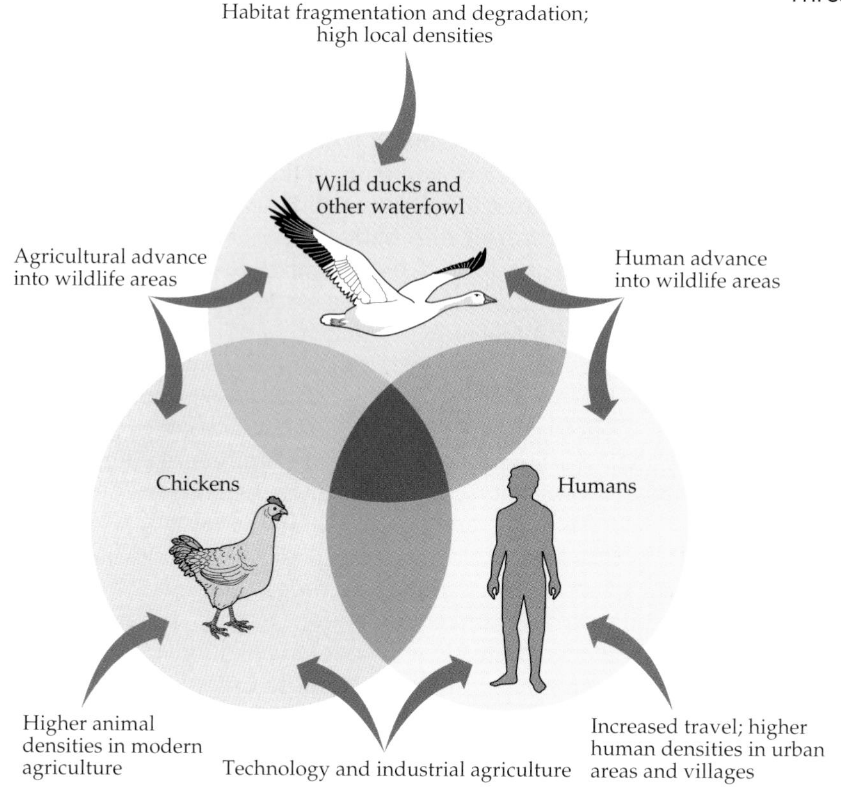

FIGURE 4.21 Infectious diseases—such as rabies, Lyme disease, influenza, bird flu, hantavirus, and canine distemper—spread among wildlife populations, domestic animals, and humans as a result of increasing population densities and the advance of agriculture and human settlements into wildlife areas. The figure illustrates the infection and transmission routes of bird flu—wild waterfowl, chickens, and humans are all susceptible to the virus. The shaded areas of overlap indicate that diseases can be shared among the three groups. Green arrows indicate factors contributing to higher rates of infection; blue arrows indicate factors contributing to the spread of disease among the three groups. (After Daszak et al. 2000.)

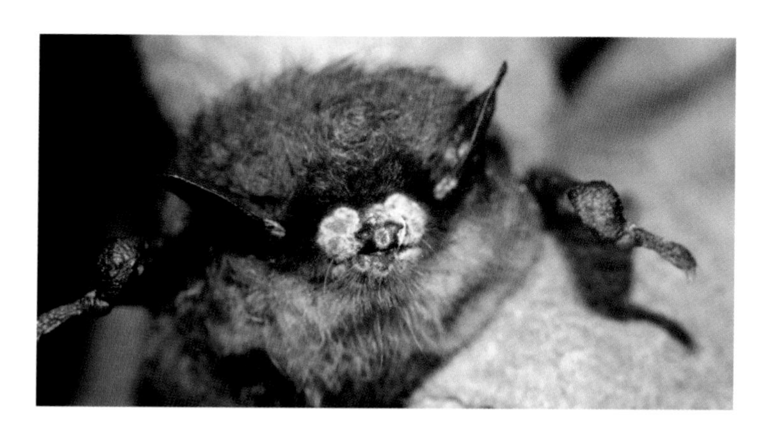

FIGURE 4.22 A little brown bat (*Myotis lucifugus*) infected with white-nose syndrome is being examined by a researcher. This fungal disease, which first appeared in the United States in 2006, affects a variety of bat species and has killed up to 90% of the bats in many caves. (Photograph courtesy of Ryan Von Linden/New York Department of Environmental Conservation.)

States as a contaminant on their clothes, boots, or equipment following a visit to a European bat cave. At this point, the only effective way to protect bats is to close bat caves to all human visitors except for scientists who sterilize their clothes and equipment before entering.

Three principles of epidemiology have obvious practical implications for the management and captive breeding of endangered species. First, both captive and wild animals in dense populations may face increased direct pressure from parasites and diseases. In fragmented conservation areas, populations of animals may temporarily build up to unnaturally high densities that promote high rates of disease transmission. In natural situations, the level of infection is typically reduced when animals migrate away from their droppings, saliva, old skin, and other sources of infection. In unnaturally confined situations, however, the animals remain in contact with the potential sources of infection, including other infected individuals, and disease transmission increases.

Second, the indirect effects of habitat destruction can increase an organism's susceptibility to disease. When a host population is crowded into a smaller area because of habitat destruction, habitat quality and food availability often deteriorate, leading to lowered nutritional status, weaker animals, and a greater susceptibility to infection. Crowding can also lead to social stress within a population, which lowers the animals' resistance to disease. Pollution may make individuals more susceptible to infection by pathogens, particularly in aquatic environments (Harvell et al. 2004). A recent insight is that biodiversity regulates disease—including diseases that can be transmitted to humans—by diluting the number of suitable host species or by constraining the size of host populations through predation and competition (Ostfeld 2009). For example, the increased incidence of Lyme disease and other tick-borne pathogens has been linked to the local abundance of certain host rodent species and the overall loss of local species diversity (**Figure 4.23**).

Third, in many conservation areas, zoos, national parks, and new agricultural areas, species come into contact with other species that they would rarely or never encounter in the wild—including humans and domestic animals—so infections such as rabies, Lyme disease, influenza, distemper, hantavirus, and bird flu can spread from one species to another. Infectious diseases can spread among wildlife populations, domestic animals, and humans as a result of increasing population densities and the advance of agriculture and human settlements into wildlife areas. The human immunodeficiency virus (HIV) and the deadly Ebola virus both appear to have spread from wildlife populations to humans and to domestic animals. Such examples are likely to become more common as a result of human-induced changes to the environment, the increased level of international travel, and globalization of the economy.

Steps must be taken to prevent the spread of disease in captive animals and to ensure that new diseases are not accidentally introduced into wild populations.

In zoos, colonies of animals are often caged together in small areas, and similar species are often housed close by.

(A)

(B)

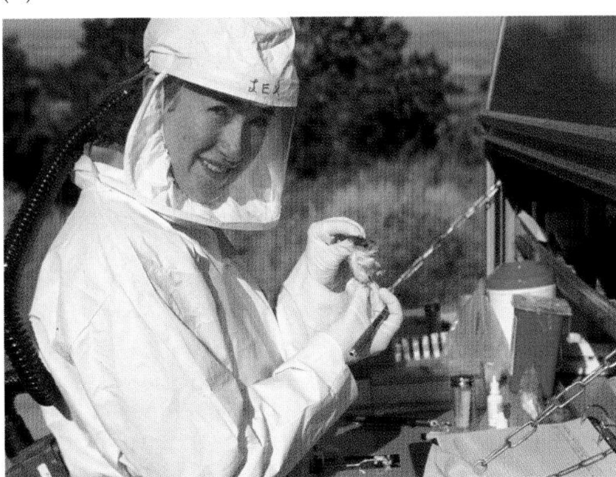

(C)

FIGURE 4.23 (A) A white-footed mouse (*Peromyscus leucopus*), one of the main hosts for Lyme disease, increases in abundance in habitat fragments created by suburban development. (B) Field biologists are sampling mice for the presence of infectious diseases, such as plague. (C) A black-legged tick, which can be up to 3 mm (0.12 inch) in length, can transfer Lyme disease to a human after acquiring the disease from an infected animal. (A, photograph © Rolf Nussbaumer/Naturepl.com; B, photograph from Crowl et al. 2008; C, photograph courtesy of Michael L. Levin/CDC.)

Consequently, if one animal becomes infected, the parasite or microbe can spread rapidly to other animals and to related species. Once they are infected with an exotic disease, captive animals cannot be returned to the wild without threatening the entire wild population. Also, a species that is both common and fairly resistant to a disease can act as a reservoir for the disease, which can then infect populations of susceptible species. For example, during the early 1990s in Tanzania's Serengeti National Park, about 25% of the lions were killed by canine distemper, apparently contracted from the 30,000 domestic dogs living near the park (Kissui and Packer 2004).

Conservation biologists have an obligation to help prevent the spread of potentially invasive and dangerous species that threaten both people and biological diversity. Conservation biologists also need to keep the public engaged in conservation-related activities, in part to counter media reports that exaggerate the dangers of the outdoors. People who have firsthand and frequent experience outdoors will be more likely to help protect the world's biodiversity from the threats described in this chapter.

A Concluding Remark

This chapter described seven major categories of threats faced by species and ecosystems. A study of 181 threatened and endangered species (Lawler et al. 2002) found that more than 85% of them faced at least four types of threats, the most common of which were related to habitat destruction and degradation, interactions with exotic species, and overexploitation. When the threats to biodiversity are best understood, protection and recovery efforts have the best chance for success.

Summary

- The biodiversity crisis is caused by an increasing human population, widespread poverty, and unequal consumption of resources. The major direct threats to biodiversity are habitat loss, fragmentation, pollution, and climate change. Other threats are overexploitation, invasive species, and disease. Rain forests, wetlands, coral reefs, and other species-rich communities are threatened by these external forces.
- Habitat fragmentation is the process whereby a large continuous area of habitat is both reduced and divided into two or more fragments. Habitat fragmentation leads to the rapid loss of remaining species because it creates barriers to the normal processes of dispersal, colonization, and foraging. Particular fragments may contain altered environmental conditions that make them less suitable for the original species.
- Environmental pollution eliminates many species from ecosystems even when the structure of the community is not obviously disturbed. The range of environmental pollution includes pesticide buildup due to excessive use; contamination of water with industrial wastes, sewage, and fertilizers; and air pollution resulting in acid rain, excess nitrogen deposition, photochemical smog, and high ozone levels.
- Global climate change, including warmer weather and changing precipitation, is already occurring because of the large amounts of carbon dioxide and other greenhouse gases produced by the burning of fossil fuels and deforestation. Predicted temperature increases may be so rapid in coming decades that many species will be unable to adjust their ranges and will become extinct.
- Overexploitation is driving many species to the point of extinction and causing the destruction of entire ecosystems. Overexploitation is caused by increasingly efficient methods of harvesting and marketing, increasing demand for products, and increased access to remote areas.
- Humans have deliberately and accidentally moved thousands of species to new regions of the world. Some of these exotic species have become invasive, greatly increasing their numbers at the expense of native species.
- Levels of disease often increase when animals are confined to nature reserves, zoos, or habitat fragments and cannot disperse over wide areas. In zoos and botanical gardens, diseases sometimes spread between related species of animals and plants. Diseases may also spread between domesticated species and wild species, and even between humans and wild animals.
- Species may be threatened by a combination of factors, all of which must be addressed in a comprehensive conservation plan.

For Discussion

1. Human population growth is often blamed for the loss of biological diversity. Is this valid? What other factors are responsible, and how do we weigh their relative importance? Is it possible to find a balance between providing for increasing numbers of people and protecting biodiversity?
2. Consider the most damaged and the most pristine habitats near where you live or go to school. Why have some been preserved and others been allowed to fragment and degrade?
3. Learn about one endangered species in detail. What is the full range of immediate threats to this species? How do these immediate threats connect to larger social, economic, political, and legal issues?

Suggested Readings

Bradley, B.A. and 11 others. 2012. Global change, global trade, and the next wave of plant invasions. *Frontiers in Ecology and the Environment* 10: 20–28. In a warming world, exotic plants from drier areas of the world may become invasive species.

Cox, T. M., R. L. Lewiston, R. Zydelis, L. B. Crowder, C. Safina, and J. Reed. 2007. Comparing effectiveness of experimental and implemented bycatch reduction measures: The ideal and the real. *Conservation Biology* 21: 1155–1164. Cooperation and monitoring are needed to translate experimental results into actual bycatch reductions.

Doney, S. C. 2010. The growing human footprint on coastal and open-ocean biogeochemistry. *Science* 328: 1512–1516. Human impacts are evident even in the vast ocean.

Hole, D. G. and 7 others. 2011. Toward a management framework for networks of protected areas in the face of climate change. *Conservation Biology* 25: 305–315. In coming decades, many species will not be able to survive in their present locations because of climate change.

Dukes, J. S, N. R. Chiariello, S. R. Loarie, and C. B. Field. 2011. Strong response of an invasive plant species (*Centaurea solstitialis* L.) to global environmental changes. *Ecological Applications* 21: 1887–1894. Invasive species are successful because they do well in the environments created by human activity.

Foley, J., D. Clifford, K. Castle, P. Cryan, and R. S. Ostfeld. 2011. Investigating and managing the rapid emergence of white-nose syndrome, a novel, fatal, infectious disease of hibernating bats. *Conservation Biology* 25: 223–231. Scientists struggle to understand and deal with a sudden new disease.

Intergovernmental Panel on Climate Change (IPCC). 2007. *Climate Change 2007: The Physical Science Basis*. Contribution of Working Group I to the Fourth Assessment Report. Cambridge University Press, Cambridge. Comprehensive presentation of the evidence for global climate change, along with predictions for the coming decades.

Jones, H. P. 2010. Seabird islands take mere decades to recover following rat eradication. *Ecological Applications* 20: 2075–2080. Control of invasive species can sometime lead to recovery of endangered species.

Laurance, W. F. and 15 others. 2011. The fate of Amazonian forest fragments: A 32-year investigation. *Biological Conservation* 144: 56–67. Results of a large-scale experiment.

Lotze, H. K. and B. Worm. 2009. Historical baselines for large marine mammals. *Trends in Ecology and Evolution* 24: 254–262. Large marine mammals have declined by 89% because of exploitation; certain groups such as whales and seals have started to recover because of conservation measures.

Primack, R. B. and A. J. Miller-Rushing. 2012. Uncovering, collecting, and analyzing records to investigate the ecological impacts of climate change: A template from Thoreau's Concord. *BioScience* 62: 170–181. We can see the effects of climate change happening all round us if we find the right tools and baseline studies.

Vila, M. and 10 others. 2010. How well do we understand the impacts of alien species on ecosystem services? A pan-European cross-taxa assessment. *Frontiers in Ecology and the Environment* 8: 135–144. Large numbers of invasive species have a broad and often negative impact on European biodiversity and society.

Willis, C. G., B. Ruhfel, R. B. Primack, A. J. Miller-Rushing, and C. C. Davis. 2008. Phylogenetic patterns of species loss in Thoreau's woods are driven by climate change. *Proceedings of the National Academy of Sciences USA* 105: 17,029–17,033. Climate change is already affecting the distribution and abundance of plants in a temperate ecosystem.

Wooldridge, S. A. and T. J. Done. 2009. Improved water quality can ameliorate effects of climate change on corals. *Ecological Applications* 19: 1492–1499. Much of the present damage to coral species is still coming from pollution.

The South Island giant moa (*Dinornis robustus*), a giant (up to 3.6 m [over 11 feet] in height) flightless bird, went extinct around 700 years ago after people arrived in New Zealand.

Chapter 5
Extinction Is Forever

The number and diversity of species found on Earth has been increasing since the origin of life. These increases have not been steady but have been characterized by periods of high rates of speciation followed by long periods of minimal change. We can also recognize five past episodes of mass extinction, the most massive of which took place at the end of the Permian period, about 250 million years ago (**Figure 5.1**). At that time, the fossil record indicates, more than 50% of animal families and 95% of marine species went extinct. It is likely that some massive perturbation, such as widespread volcanic eruptions and/or multiple asteroid collisions, caused such a dramatic change in the Earth's climate and sea levels that many species were no longer able to survive. Another speculation is that there might have been a massive release of methane gas from beneath the ocean floor—a "big burp," if you will. Such an event not only would have released toxic plumes but almost certainly would have affected the climate, since methane is an even more potent greenhouse gas than carbon dioxide (see Chapter 4). It took more than 50 million years to regain, through the process of evolution, the biodiversity that was lost during the Permian mass extinction.

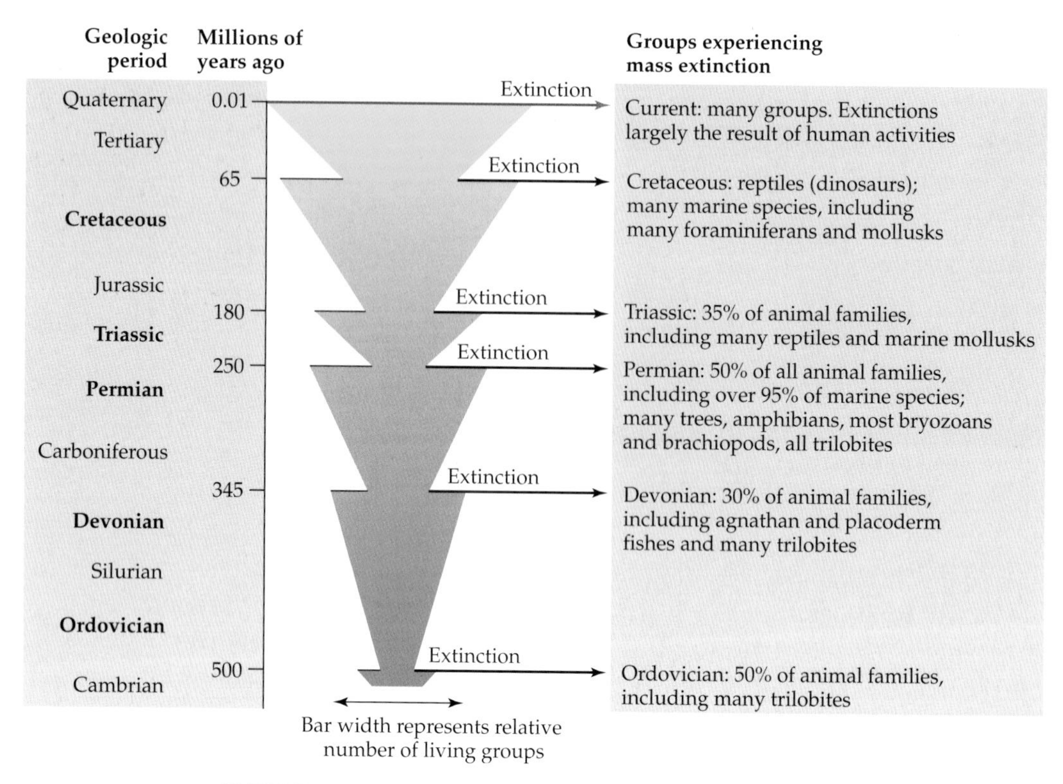

FIGURE 5.1 Although the total number of species groups on Earth has increased over the eons, during each of five episodes of natural mass extinction (named in bold at left), a large percentage of these groups disappeared. The most dramatic period of loss occurred about 250 million years ago, at the end of the Permian period. A sixth extinction episode (red arrow at top of figure) began during the present geological period and will continue for decades to come.

Species go extinct even in the absence of violent disturbance. One species may outcompete another for a vital resource, or predators may drive prey species to extinction. Extinction is as much a part of the natural life cycle as speciation is. If extinction is part of natural processes, why is the loss of species of such concern? The answer lies in the *relative rates* of extinction and speciation. Speciation is typically a slow process, occurring though the gradual accumulation of mutations and shifts in allele frequencies over thousands, if not millions, of years. As long as the rate of speciation equals or exceeds "background" extinction rates (i.e., the relatively constant rates observed over most of geological history), biodiversity will remain constant or increase. In past geological periods, the loss of existing species was eventually balanced and then exceeded by the evolution of new species. However, current extinction rates are believed to be 100 to 1000 times greater than background rates because of human influence, and more than 99% of modern species extinctions can be linked to human activity (Pimm and Jenkins 2005; Wake and Vrendenberg 2008). We are presently in the midst of a **sixth extinction episode**, this one caused by human activities rather than a natural disaster (Barnosky et al. 2011).

The current rate of species loss is unprecedented, unique, and irreversible. Ninety-nine percent of current extinctions are caused in some way by human activities.

Global biodiversity has undergone a progressive decline over the last 30,000 years as one species, *Homo sapiens*, has asserted its dominance. In our need to consume natural resources, humans have increasingly altered terrestrial and aquatic environments at the expense of other species. The first noticeable effect of human activity was the elimination of large mammals from Australia and North and South America. Shortly after humans arrived on each of these continents, between 74% and 86% of the megafauna (mammals weighing more than 44 kg, or about 100 pounds) became extinct. These extinctions were probably caused directly, by hunting, and indirectly, by burning and clearing forests and grasslands (Johnson 2009). On every continent except Antarctica and on many oceanic islands, paleontologists and archaeologists have found extensive records of prehistoric human alteration and habitat destruction that coincides with high rates of species extinctions. Human activities that result in species extinctions have only increased, and extinction rates have accelerated over the past two centuries.

The Meaning of "Extinct"

The term "extinct" has several nuances in conservation biology, and its meaning can vary somewhat depending on the context:

- A species is **extinct** when no member of the species remains alive anywhere in the world. Bachman's warbler, for example, is extinct (**Figure 5.2**).
- If individuals of a species remain alive only in captivity or in other human-controlled situations, the species is said to be **extinct in the wild**. The Franklin tree shown in Figure 5.2 is extinct in the wild but grows well under cultivation.

FIGURE 5.2 One of the first migrant songbirds to become extinct as a result of Neotropical deforestation was Bachman's warbler (*Vermivora bachmanii*), last sighted in the 1960s. The Cuban forests in which this species overwintered were almost entirely cleared for surgarcane plantations. This nineteenth-century Audubon print shows a pair of warblers in the flowering Franklin tree (*Franklinia alatamaha*), which is now extinct in the wild, although it can still be found in arboretums and other cultivated gardens. (Print after John James Audubon.)

- A species that no longer lives anywhere in the world is considered to be **globally extinct**.
- A species is **locally extinct** or **extirpated** when it is no longer found in an area it once inhabited but is still found elsewhere in the wild. The gray wolf (*Canis lupus*) once roamed throughout North America; it is now locally extinct in Massachusetts.
- A species is sometimes considered as being **ecologically extinct** if it persists at such reduced numbers that its effect on other species within its ecological community is negligible. Tigers are ecologically extinct in most of their range because so few remain in the wild that their impact on prey populations is insignificant.

In addition to the various nuances of the term "extinct," conservation biologists work with a variety of categories, including *endangered*, *vulnerable*, and *threatened*, that more specifically describe the status of species. These terms are discussed in detail in Chapter 6.

Rates of Extinction

Extinction rates are best known for birds and mammals because these species are conspicuous; that is, they are relatively large and well studied, and it is relatively easy to note when these species are no longer found in the wild. Extinction rates for the other 99% of the world's species are at present just rough guesses, and even the rates for some of the better known species, such as birds and mammals, reptiles, and amphibians are uncertain because some species that were considered extinct have been rediscovered. For example a species of giant tortoise (*Chelonoidis elephantopus*) known from only one island in the Galapagos was thought to have gone extinct shortly after Charles Darwin visited in 1835, but has recently been re-discovered on another island using DNA methods. It is also true that species presumed to be **extant** (still living) may actually be extinct; the Yangtze river dolphin, for example, has not officially been declared extinct, but no individuals could be found during an intensive survey in 2006 (Fisher and Blomberg 2012).

Sometimes it is difficult to determine whether a species truly is extinct. In 2004, for example, ornithologists in North America announced sightings of an ivory-billed woodpecker (*Campephilus principalis*) in an Arkansas swamp forest—decades after this bird was believed to have gone extinct. Since then, however, intensive efforts to find and conclusively identify existing individuals of the species have been unsuccessful (Stokstad 2007).

Estimates based on the best available evidence indicate that about 79 species of mammals and 136 species of birds have become extinct since the year 1600, representing 1.6% of known mammal species and 1.3% of known bird species (www.iucnredlist.org). While these numbers may not seem alarming, the trend of extinction rates is on the rise: the majority of extinctions occurred in the last 200 years (**Figure 5.3**). The extinction rate for birds and mammals was about one species every decade during the period 1600 to 1700 but rose

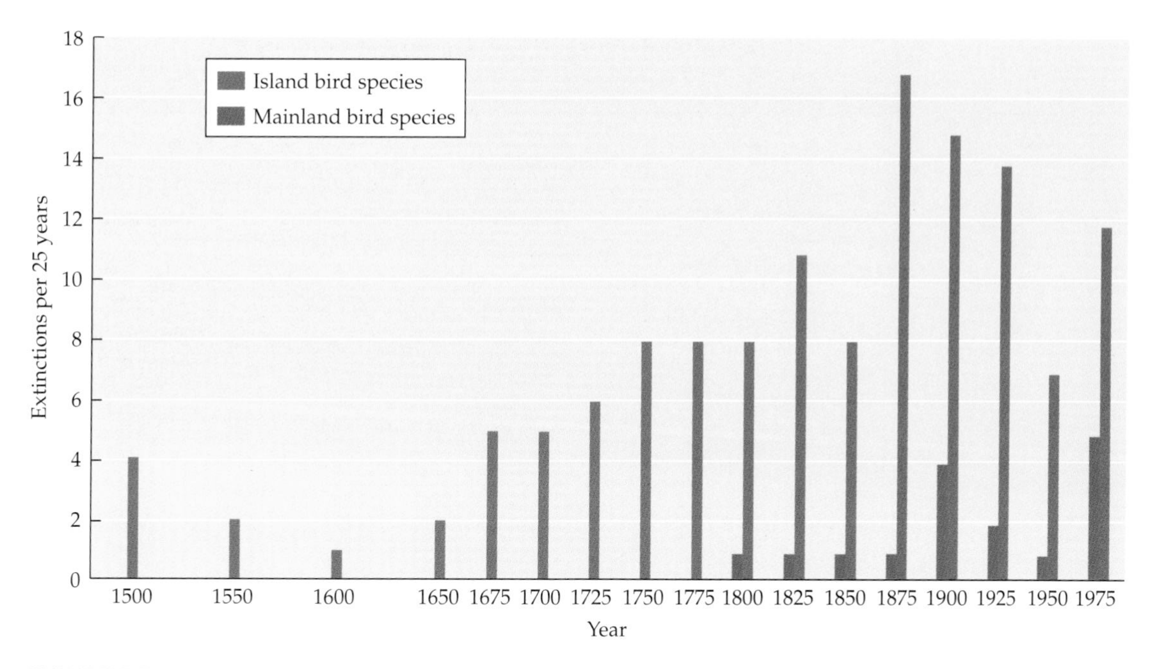

FIGURE 5.3 Rates of extinctions of bird species during 25-year intervals since 1500. Initially, extinctions were almost exclusively of island species, but extinctions of mainland species have increased since 1800. (After Baillie et al. 2004.)

to one species each year during the period 1850 to 1950. This increase in the extinction rate represents a serious threat to biological diversity.

The apparent decline in extinction rates since 1950 is due to the current practice of not declaring a species extinct until decades after it can no longer be found. In the coming years, numerous species will be declared to have gone extinct during the period 1950 to 2000. In the last decade, a number of species not found despite intensive searches were declared extinct, including the Monteverde golden toad of Costa Rica (*Bufo periglenes*) and the Hawaiian crow (*Corvus hawaiiensis*). Many species not yet listed as extinct—and some species that have not yet been documented at all—have been decimated by human activities and persist only in very low numbers. Our inability to locate any extant populations of many rare species is evidence that extinction rates are accelerating. In Britain, the country best known biologically, extinction rates for major groups of species have been in the range of 1% to 5% per century but are predicted to increase in the present century (Hambler et al. 2011).

Many species today are represented only by scattered populations with reduced numbers of individuals. Although these isolated populations could persist for years or decades, the ultimate fate of such species is extinction.

Many species can persist for years, decades, or centuries as a few individuals in scattered small populations (among woody plants in particular, isolated individuals can persist for hundreds of years). However, their ultimate fate is extinction

TABLE 5.1 Numbers of Species Threatened with Extinction in Major Groups of Animals and Plants

Group	Approximate number of species	Number of species threatened with extinction	Percentage of species threatened with extinction
Vertebrate animals			
Fishes	28,000	2084	7[a]
Amphibians	6338	2303	36
Reptiles	9400	1018	11[a]
Crocodiles	23	10	43
Turtles	228	170	75
Birds	10,052	2097	21
Penguins	18	11	61
Mammals	5499	1462	27
Primates	418	227	54
Manatees, dugongs	5	4	80
Horses, tapirs, rhinos	16	14	88
Plants			
Gymnosperms	18,000	521	3[a]
Angiosperms (flowering plants)	260,000	9614	4[a]
Palms	361	275	82
Fungi	100,000	3	0[a]

Source: Data from IUCN 2011 (www.iucnredlist.org). Data include the categories critically endangered, endangered, vulnerable, and near threatened.

[a]Low percentages reflect inadequate data due to the small number of species evaluated. For example, 11% of reptiles are listed as endangered, but only about one-third of species have been evaluated. For reptile species that have been evaluated, 31% are considered endangered.

(Janzen 2001). Such species, doomed to extinction by human activities, have been called "the living dead" or "committed to extinction." There are certainly many species in this category in the remaining fragments of species-rich tropical forests. The presumed eventual loss of species following habitat destruction and fragmentation is called the **extinction debt** (Cousins and Vanhoenacker 2011).

Extinction rates will remain high throughout the present century because the number of threatened species is so large. About 21% of the world's remaining bird species are threatened with extinction. There is even greater danger for other large groups with 27% of mammal species and 36% of amphibian species threatened (www.iucnredlist.org). There is lower level of threat for certain groups of species such as fish (7%) and reptiles (11%). These rates are likely to be much higher once poorly known species are assessed for their level of threat (Hoffman et al. 2010; McClenachan et al. 2012). **Table 5.1** shows certain animal groups for which the danger is even more severe, including turtles, manatees, rhinos, and penguins. Some plant species are also at risk with palms being especially vulnerable. Most species of plants, fungi, fishes, and insects are less well-known and for these the extinction risk has still not been determined.

Extinction rates in aquatic environments

In contrast to the large amount of information we have on extinct terrestrial species, only about 14 marine species—four mammal, five bird, one fish, and four mollusk species—are known to have gone extinct in the world's vast oceans during historic times (Régnier et al. 2009). This number is almost certainly an underestimate because marine species are not nearly as well-known as terrestrial species, but it may also reflect a greater resilience on the part of marine species in response to disturbance. The significance of these losses could be greater than the numbers suggest. Many marine mammals are top predators, and their loss could have a major impact on marine communities. It is also possible that what we currently recognize as individual widespread species could actually be composed of numerous similar-looking species, many of which are rare and threatened with extinction. In the past, the oceans were considered so enormous that it seemed unlikely that marine species could go extinct, and many people still share this viewpoint. However, as marine coastal waters become more polluted and species are harvested more intensely, even the vast oceans will not provide safety from extinction (Jackson 2008). Many species of whales and large fish have declined by 90% or more due to overharvesting and other human activities.

Also in contrast to terrestrial extinction, the majority of freshwater fish extinctions have occurred in mainland areas rather than on islands because of the vastly greater number of species in mainland waters. In North America, over one-third of freshwater fish species are in danger of extinction (Moyle et al. 2011). The fishes of California are particularly vulnerable because of the state's scarcity of water coupled with its intense development: 6% of California's 129 native fish species are already extinct, and 51% are in danger of extinction. Large numbers of fishes and aquatic invertebrates, such as mollusks, are in danger of extinction because of dams, pollution, irrigation projects, invasion of alien species, and general habitat damage (Hambler et al. 2011).

Extinction rates on islands

It should not come as a surprise that the highest species extinction rates during historic times have occurred on islands (see Figure 5.3). Island species often have limited areas, small population sizes, and small numbers of populations (Régnier et al. 2009; Clavero et al. 2009). The high extinction rates on islands include the extinctions of many species of endemic birds, mammals, reptiles, and plants during the last 350 years of European colonization. (Endemic species are species found in one place and nowhere else, so they are very vulnerable to extinction).

Island species have had higher rates of extinction than mainland species. Freshwater species are more vulnerable to extinction than marine species.

Of the terrestrial animal and plant species known to have gone extinct from 1600 to the present, almost half were island species even though islands represent only a tiny fraction of the Earth's land surface. Island species usually have evolved and undergone speciation with a limited number of competitors, predators, and diseases. When predatory species and new diseases from the mainland are introduced onto islands, they frequently decimate the endemic island species, which have not evolved any defenses against them. Animal species extinction rates peak soon after humans occupy an island and then decline after the most vulnerable species are eliminated. In general, the longer an island has been occupied by people, the greater the percentage of extinct biota. Island plant species are also threatened, mainly through habitat destruction, overgrazing by introduced animals, and competition from invasive plants. In Madagascar, 72% of the 9000 plant species are endemic, and 280 species are threatened with extinction. All lemur species are endemic to Madagascar, and most of these unique primates are threatened (see Figure 4.6).

Despite the documented danger to island species, however, in the coming decades a higher proportion of extinctions will occur in continental lowlands, especially tropical forest areas, where many species occur and where human alteration of the landscape is rapid and extensive. Hopefully, the lessons learned from island species may be applied to protect mainland species.

Island biogeography and extinction rate predictions

Biologists have observed that there is a relationship between the area of an island and the number of species it contains, synthesized as the **island biogeography model** (MacArthur and Wilson 1967). This model predicts that large islands will have more species than small islands (**Figure 5.4**). Species–area relationships have also been used to predict the numbers and percentages of species that would become extinct if habitats were destroyed (Koh and Ghazoul 2010). The calculation assumes that reducing the area of natural habitat on an island would result in the island's being able to support only the number of species corresponding to that on a smaller island.

The island biogeography model can be used to predict how many species will go extinct due to habitat loss. The model can also be used to predict how many species will remain in protected areas of different sizes.

The island biogeography model is extremely useful because it can be extended to national parks and nature reserves surrounded by damaged habitat (Chittaro et al. 2010). The reserves can be viewed as "habitat islands" in an inhospitable "sea" of unsuitable habitat. The model predicts that when 50% of an island (or a habitat island) is destroyed, approximately 10% of the species occurring on the island will be eliminated (**Figure 5.5**). If these species are endemic to an area, that is, found there and nowhere else, they will become extinct. When 90% of the habitat is destroyed, 50% of the species will be

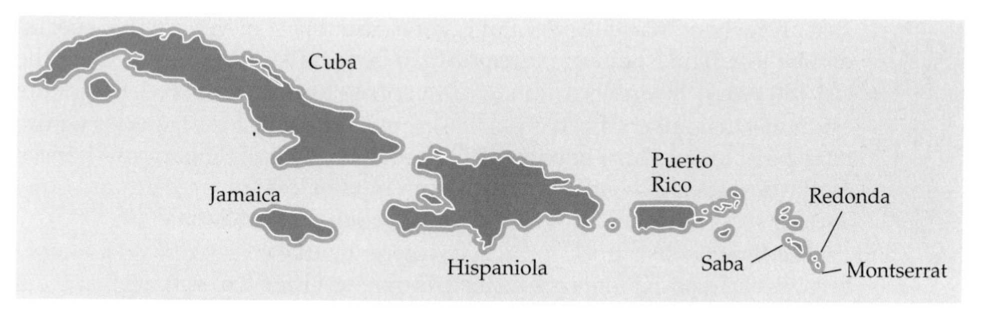

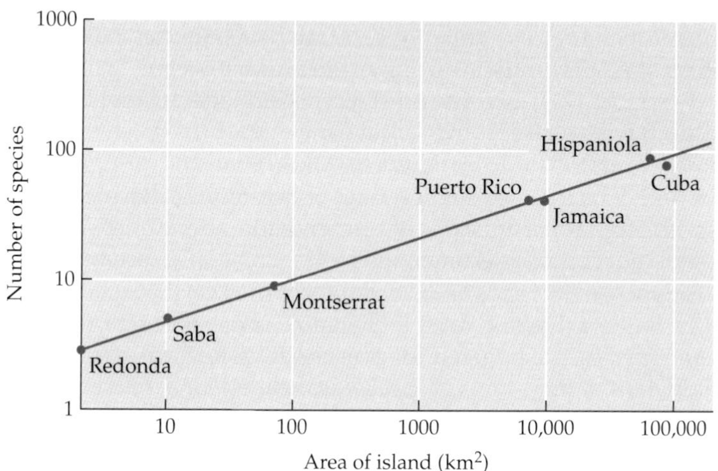

FIGURE 5.4 The number of species on an island can be predicted from the island's area. In this figure, the number of species of reptiles and amphibians is shown for seven islands in the West Indies. The number of species on large islands such as Cuba and Hispaniola far exceeds that on the tiny islands of Saba and Redonda. (After Wilson 1989.)

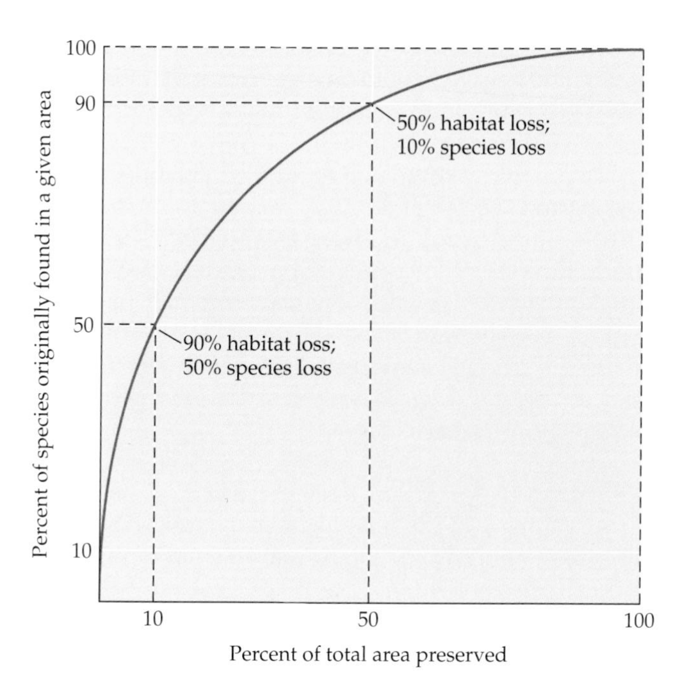

FIGURE 5.5 According to the island biogeography model, the number of species present in an area increases asymptotically—that is, it rises sharply and then levels off, as seen by the red curve in this example. The shape of the curve differs from region to region and for different species groups, but this model gives a general indication of the interrelationship of habitat loss and species loss. Here, if the area of habitat is reduced by 50%, then 10% of the species in the group will be expected to disappear; if the habitat is reduced by 90%, half the species will be lost. Stating this in another way, a system of protected areas covering 10% of a country could be expected to include 50% of the country's species.

lost; and when 99% of the habitat is gone, about 75% of the original species will be lost. The island of Singapore can be used as an example. Over the last 180 years, 95% of its original forest cover has been removed; the model estimates that, given this habitat loss, about 70% of its forest species would have been lost. In fact, between 1923 and 1998, 90% of Singapore's known native bird species were lost, with higher rates of loss for large ground birds and for insectivorous birds of the forest canopy (Castelletta et al. 2000).

Predictions of extinction rates based on habitat loss vary considerably because each species–area relationship is unique. Using the estimate that 1% of the world's rain forests is being destroyed each year, Wilson (1989) calculated that 0.2% to 0.3% of all species—roughly 10,000 to 15,000 species, based on a total of 5 million species worldwide (most of which are insects)—will be lost each year, or 34 species each day. The most recent estimates are that by 2050, species extinctions will reach 35% in tropical Africa, 20% in tropical Asia, 15% in tropical America, and 8% to 10% elsewhere (MEA 2005). The extinction rate may in fact be higher because the highest rates of deforestation are occurring in countries with large concentrations of rare, endemic species known to be threatened with extinction, such as Brazil and Indonesia, as large forest areas are increasingly fragmented by roads and development projects (Laurance 2007). There might be lower extinction rates if "hotspot" areas, those particularly rich in endemic species, are targeted for conservation. Regardless of which estimate is used, tens of thousands of species are headed for extinction in the next 50 years. Such an extinction event is without precedent since the mass extinction of the Cretaceous period, 65 million years ago.

The time required for a given species to go extinct following a reduction in area or fragmentation of its range is a vital question in conservation biology, and the island biogeography model makes no prediction about how long it will take. Small populations of some species may persist for decades or even centuries, even though their eventual fate is extinction. Amazon forest fragments of 100 ha lost half their bird species in 15 years, but even isolated forest fragments as large as 10,000 ha will lose numerous species over a 100-year period (Ferraz et al. 2003). In situations in which there is widespread habitat destruction followed by recovery, such as in New England and Puerto Rico over the last several centuries, species may be able to survive in small numbers in isolated fragments and then reoccupy adjacent recovering habitat. Even though 98% of the forests of eastern North America were cut down, the clearing took place in a patchwork fashion over hundreds of years, so forest always covered half the area, providing refuges for mobile animal species such as birds.

Local extinctions

In addition to the global extinctions that are a primary focus of conservation biology, many species are experiencing a series of local extinctions or extirpations across their ranges (Leidner and Neel 2011). Formerly widespread species are sometimes restricted to a few small pockets of their

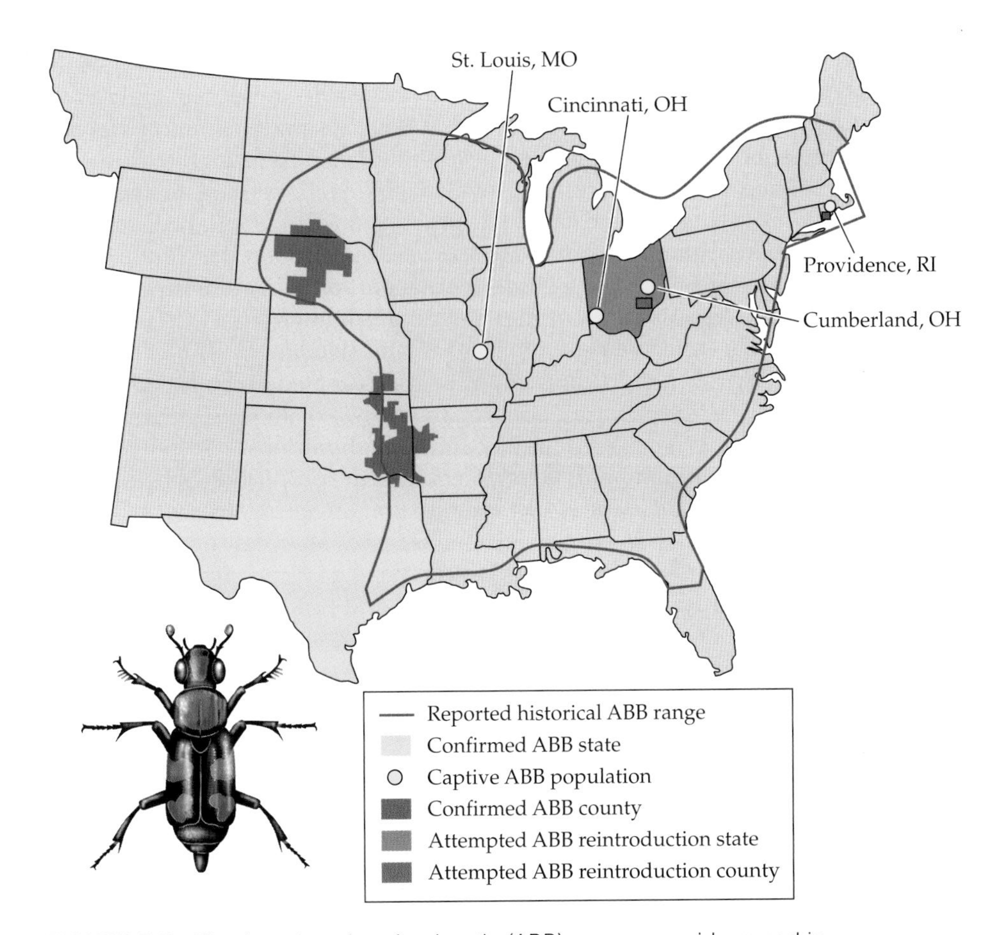

FIGURE 5.6 The American burying beetle (ABB) was once widespread in the eastern and central United States (outlined in green), but its range is now greatly reduced and it is found in the wild in only two separate areas of central United States and on Block Island in Rhode Island. Intensive efforts have been initiated to determine the cause of this decline and develop a recovery plan. The species is also being bred in captivity. (After O'Meilla 2004, with updates from the U.S. Fish and Wildlife Service.)

former habitats. For example, the American burying beetle (*Nicrophorus americanus*), once found all across central and eastern North America, is now found in the wild in only three isolated populations (**Figure 5.6**). Biological communities are impoverished by such local extinctions.

Concord, Massachusetts, an intensively surveyed town, is an example of a place with extensive local extinction. Concord was first assessed for wildflower species in the 1850s by the famous naturalist and philosopher Henry David Thoreau. Twenty-seven percent of the native species could not be found when Concord was surveyed 150 years later, even though these species still exist elsewhere in Massachusetts (Willis et al. 2008; Primack et al. 2009). A further 36% of Concord's species now persist in only one or two populations and are vulnerable to extinction. In some cases, only a few individual plants remain of species that were formerly common. Certain groups, such as orchids and lilies, have shown particularly severe losses. A combination of forest succession, invasions by exotic species, air and water pollution, grazing by deer, habitat destruction and fragmentation, and now climate change have contributed to species losses in Concord. Similar local extinctions were shown by a survey of one part of the Indonesian island of Sumatra; of 12 populations of Asian elephants known from the 1980s, only 3 populations were still present 20 years later (Hedges et al. 2005).

The world's 5 million known species are estimated to consist of 1 billion distinct populations, or an average of about 200 populations per species (Hughes and Roughgarden 2000). The loss of populations is roughly equal to the proportion of habitat that is lost, so the world's populations are being lost at a far higher rate than species are (see Figure 5.5). Because 98% of North America's tallgrass prairie has been destroyed, 98% of the populations of plant, animal, and fungus species there have also been lost. The situation in marine systems is comparable; approximately 85% of the world's oyster reefs have been lost (Beck et al. 2011). Tropical rain forests contain at least half of the world's species, and they are being lost at the rate of about 1% per year. This represents a loss of 5 million populations per year (1% of 500 million tropical forest populations), or about 13,500 populations per day.

> Species-rich tropical rain forests are being lost at a rate of 1% a year, a rate believed to result in the destruction of more than 13,500 biological populations each day. Population losses eventually result in species extinctions.

These large numbers of local extinctions serve as important biological warning signs that something is wrong with the environment. Action is needed to prevent further local extinctions, as well as global extinctions. The loss of local populations not only represents a loss of biological diversity; it also diminishes the value of an area for nature enjoyment, scientific research, regional ecosystem services, and the provision of crucial materials to people living in rural areas.

Vulnerability to Extinction

When ecosystems are damaged by human activity, the ranges and population sizes of many species are reduced, and some species go extinct. To prevent such extinctions, rare species must be carefully monitored and managed in conservation efforts. Ecologists have observed that five particular categories of species are most vulnerable to extinction (Grouios and Manne 2009):

- *Species with a very narrow geographical range*. Some species occur at only one or a few sites in a restricted geographical range, and if that whole range is affected by human activity, the species may become extinct (Lawler et al. 2010). Bird species on oceanic islands and tropical mountains are good examples of species with restricted ranges that have

FIGURE 5.7 The Iberian lynx (*Lynx pardinus*) is a critically endangered species. Once found over the entire Iberian peninsula, there are now believed to be fewer than 100 of these cats in a few scattered areas in Spain. The fragmentation of their shrubland habitat and the decline of their prey populations have contributed to the decline of this species. (Photograph © Carlos Sanz/VWPICS/Visual & Written SL/Alamy.)

become extinct or are in danger of extinction; many fish species confined to a single lake or a single watershed have also disappeared. In coming decades, climate change will be an increasing threat to such species.

- *Species with only one or a few populations.* Any one population of a species may become extinct as a result of chance factors, such as earthquakes, fire, an outbreak of disease, or human activity. Species with many populations are less vulnerable to extinction than are species with only one or a few populations. This category is linked to the previous category because species with few populations will also tend to have a narrow geographical range (**Figure 5.7**).
- *Species in which population size is small.* Small populations are more likely than large populations to go locally extinct because of their greater vulnerability to demographic and environmental variation and loss of genetic variation (see the next section); species that characteristically have small population sizes, such as large predators, are more likely to become extinct than species that typically have large populations. At the extreme are species whose numbers have declined to just a few individuals. A special category is species with a widely fluctuating population size in which the population is sometimes small.
- *Species in which population size is declining.* Population trends tend to continue, so a population showing signs of decline is likely to go extinct unless the cause of decline is identified and addressed by conservation management (Peery et al. 2004).
- *Species that are hunted or harvested by people.* Overharvesting can rapidly reduce the population size of a species. If hunting and harvesting are not regulated, either by law or by local customs, the species can be driven to extinction. The utility of a species has often been the prelude to its extinction.

Species most vulnerable to extinction have the following characteristics: narrow geographic range, only one or a few populations, small populations, declining population size, and being hunted or harvested by people.

The following species characteristics have also been linked to extinction, though they are not considered as all-encompassing as the previous five categories:

- *Species that need a large home range.* Species in which individual animals or social groups need to forage over a wide area are prone to die off when part of their range is damaged or fragmented by human activity.
- *Animal species with large body size.* Large animals tend to have large individual ranges, require more food, have low reproductive rates, and are more often hunted by humans. Top carnivores, especially, are often killed by humans for sport, or because they compete with humans for wild game and sometimes damage livestock. Within groups of species, the largest are often the most prone to extinction—that is, the largest carnivore, the largest lemur, and the largest whale will go extinct first.
- *Species that are not effective dispersers.* Species unable to adapt physiologically, genetically, or behaviorally to changing environments must either migrate to more suitable habitat or face extinction. The rapid pace of human-induced changes often prevents adaptation, leaving migration as the only alternative. Species that are unable to cross roads, farmlands, and disturbed habitats are doomed to extinction as their original habitat becomes affected by fragmentation, pollution, exotic species, and global climate change. Dispersal is important in the aquatic environment as well, where dams, eutrophic zones, and other human impacts can limit movement. Limited ability to disperse may explain in part why freshwater fauna such as mussels and snails are more likely to be extinct or threatened with extinction than dragonfly species, which are strong fliers.
- *Seasonal migrants.* Species that migrate seasonally depend on two or more distinct habitat types. If any of these habitat types is damaged, the species may be unable to persist (see Figure 5.2). The billion songbirds from 120 species that migrate each year between the northern United States and the American tropics depend on suitable habitat in both locations (as well as at stopover points along the migration route) to survive and breed. Also, if barriers to dispersal are created by roads, fences, or dams between the needed habitats, a species may be unable to complete its life cycle (**Figure 5.8**). Salmon species that are blocked by dams from swimming up rivers and spawning are striking examples of this problem.
- *Species with little genetic variability.* Genetic variation within a population can allow a species to adapt to a changing environment (see pp. 153–156). Species with little or no genetic variability may have a greater tendency to become extinct when there is a new disease, a new predator, or some other change in the environment.
- *Species with specialized niche requirements.* Once habitat is altered, the environment may no longer be suitable for specialized species (Van Turnhout et al. 2010). For example, wetland plants and animals such as frogs and fish that require very specific water conditions may be

rapidly eliminated when human activity affects the hydrology and chemistry of an area. Species with highly specific dietary requirements are also at risk; for instance, there are species of mites that feed only on the feathers of a single bird species. If the bird species goes extinct, so does its associated feather mite species.

- *Species that are found in stable, pristine environments*. Many species are found in environments where disturbance has been minimal, such as in old stands of tropical rain forests and in the interiors of old-growth temperate deciduous forests. When these forests are logged, grazed, burned, or otherwise altered, many native species are unable to tolerate the changed microclimatic conditions (more light, less moisture, greater temperature variation; see Figure 4.10) and influx of exotic species.
- *Species that form permanent or temporary aggregations*. Species that group together in specific places are highly vulnerable to local extinction (see Figure 5.8). Herds of bison, flocks of birds roosting in trees at night, and schools of spawning fish all represent aggregations that can be completely harvested, whether for food, sport, or some other reason. When population size or density falls below a certain number, many species of social animals are unable to forage, mate, or defend themselves, and their populations cannot persist; this is termed the **Allee effect** (see p. 163).

FIGURE 5.8 Migrating herd of wildebeests (*Connochaetes taurinus*) crossing a river in east Africa. Such large aggregations are under threat by illegal hunting and blocking of their traditional migration routes by fences and roads. Once their population size and density is reduced below a certain point, the group may no longer be able to defend its members against predators. (Photograph © ImageState/Alamy.)

- *Species that have had no prior contact with people.* Species encountering people (along with their associated domestic animals and plants) for the first time have a lower chance of surviving than species that have already experienced human disturbance and survived.
- *Species closely related to species that recently went extinct or are threatened with extinction.* Often groups of related species, such as the nonhuman primates, cranes, sea turtles, and cycads, are particularly vulnerable to extinction because they share some of the characteristics mentioned above.

Most of the threatened species that have been identified so far are in the best-studied groups of organisms (especially the vertebrates), highlighting the point that only when we are knowledgeable about a species can we recognize the dangers it faces. A lack of knowledge should not be taken to mean that a species or group is not threatened with extinction; rather, a lack of knowledge should be an argument for urgent study of those organisms. The conservation status of amphibians, for example, was relatively unknown until 20 years ago, when intensive study revealed that a high proportion of species were in danger of extinction.

Problems of Small Populations

An ideal conservation plan for an endangered species would protect as many individuals as possible within the greatest possible area of high-quality, protected habitat (Wilhere 2008). In practical terms, planners, land managers, politicians, and wildlife biologists often must attempt to achieve realistic goals, guided by general principles. For example, they need to know how much longleaf pine habitat a red-cockaded woodpecker population requires to persist. Is it necessary to protect habitat containing 50, 500, 5000, 50,000, or more individuals to ensure the survival of the species? Furthermore, planners must reconcile conflicting demands on finite resources—somehow a compromise must be found that allows the economic development required by society while at the same time providing reasonable protection for biological diversity. This problem is vividly demonstrated by the current debate in the United States over the need to protect caribou and other wildlife in the vast Arctic National Wildlife Refuge and the equally compelling need to utilize the considerable oil resources of the area.

Plans for protecting a species must determine the number of individuals—the minimum viable population (MVP)—necessary for a high survival probability for the foreseeable future. Protected habitats of adequate size to maintain the MVP can then be established.

Minimum viable population (MVP)

In a groundbreaking paper, Shaffer (1981) defined the number of individuals necessary to ensure the long-term survival of a species as the **minimum viable population**, or **MVP**: "A minimum viable population for any given species in any given habitat is the smallest isolated population having a 99% chance of remaining extant for 1000 years despite the foreseeable effects of demographic, environmental, and genetic stochasticity, and natural catastrophes." In other words, the MVP is the smallest population size that can be predicted to have a very high chance of persisting for a future period of time. Shaffer emphasized the tentative nature of this definition, saying that the survival probabilities could be set at 50%, 95%, 99%, or any other percentage and that the time frame might similarly be adjusted, for example, to 20, 100, or 500 years. The key point is that the MVP size allows a quantitative estimate to be made of how large a population must be to ensure long-term survival.

Shaffer (1981) compares MVP protection efforts to flood control. It is not sufficient to use average annual rainfall as a guideline when planning flood control systems and developing regulations for building on wetlands; instead, we must plan for extreme situations of high rainfall and severe flooding, which may occur only once every 50 years. In protecting natural systems, we understand that certain catastrophic events, such as hurricanes, forest fires, epidemics, and die-offs of food items, may occur at even greater intervals. To plan for the long-term protection of endangered species, we must provide for their survival, not only in average years but also in exceptionally harsh years. An accurate estimate of the MVP size for a species often requires a detailed demographic study of the population and an analysis of its environment. This can be expensive and require months, or even years, of research. Analyses of over 200 species for which adequate data were available indicated that most MVP values for a 90% probability of survival for 100 years or more fall in the range of 3000 to 5000 individuals, with a median of 4000 (Traill et al. 2007, 2010; Flather et al. 2011). In general, protecting a larger population increases the chance of the population persisting for a longer period of time (**Figure 5.9**). For species with extremely variable population sizes, such as certain

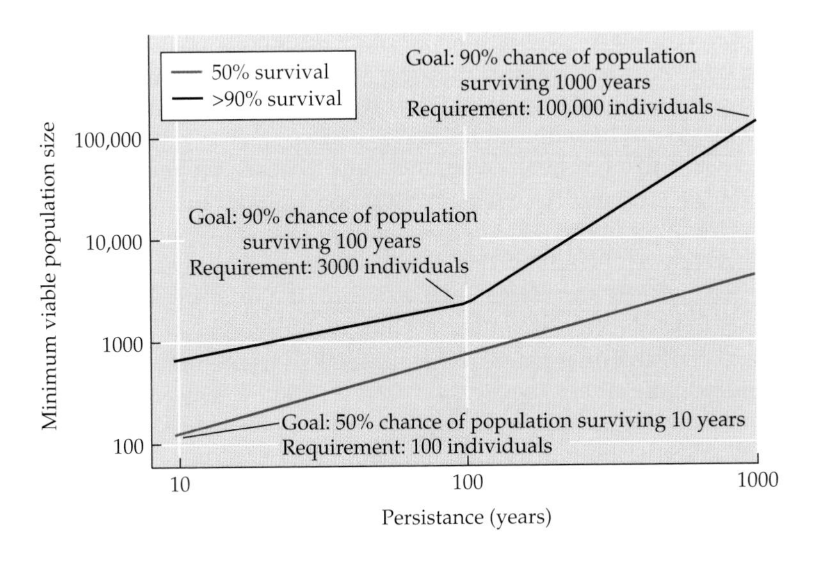

FIGURE 5.9 If the goal is persistence for a greater number of years, then a larger minimum viable population size (MVP) is needed. A greater MVP is needed to ensure a higher probability of persistence, as illustrated in this case by a 50% probability of survival and >90% probability of survival. Both axes are on log scales. The values were derived from changes in population size and persistence of 1198 species. (From Traill et al. 2010.)

invertebrates and annual plants, protecting a population of about 10,000 individuals might be an effective strategy.

Unfortunately, many species, particularly endangered species, have population sizes much smaller than these recommended minimums. For instance, a survey was done of two rare burrowing frog species in the genus *Geocrina*, which occur in swamps in southwestern Australia (Driscoll 1999). In one species, 4 of its 6 populations had fewer than 250 individuals, and in the other species, 48 of 51 populations had fewer than 50 individuals. Half of 23 isolated elephant populations remaining in west Africa have fewer than 200 individuals, a number considered to be inadequate for long-term survival of the population (Bouché et al. 2011).

Small populations are more likely to go extinct than large populations.

Field studies confirm that small populations are most likely to decline and go extinct (Grouios and Manne 2009). One of the best-documented studies of MVP size tracked the persistence of 120 bighorn sheep (*Ovis canadensis*) populations (some of which have been followed for 70 years) in the deserts of the southwestern United States (Berger 1999). The striking observation is that 100% of the unmanaged populations with fewer than 50 individuals went extinct within 50 years, while virtually all of the populations with more than 100 individuals persisted within the same time period (**Figure 5.10**). Despite the factors hindering the survival of small populations, habitat management by government agencies, as well as the release of additional animals, have allowed some other small populations of bighorn sheep to persist that might have gone extinct.

Long-term studies of birds on the Channel Islands off the California coast support the fact that large populations are needed to ensure population persistence; only bird populations with more than 100 breeding pairs had a greater than 90% chance of surviving for 80 years (Jones and Diamond 1976). However, in spite of significant odds against them, small populations sometimes prevail; many populations of birds apparently have survived for 80 years with 10 or fewer breeding pairs. Of course, birds are especially mobile and can readily recolonize areas following local extinction. Less mobile species do not have this ability.

Once an MVP size has been calculated for a species, the **minimum dynamic area (MDA)**—the area of suitable habitat necessary for maintaining the minimum viable population—can be estimated by studying the distribution of endangered species (Thiollay 1989). It has been estimated that reserves in Africa of 100 to 1000 km^2 are needed to maintain many small mammal populations. To preserve populations of large carnivores, such as lions, reserves of 10,000 km^2 are needed; this is equivalent to a square that is 100 km (or 60 miles) on a side.

Exceptions notwithstanding, large populations are needed to protect most species, and species with small populations are in real danger of going extinct. Small populations are subject to rapid decline in numbers and local extinction for three main reasons:

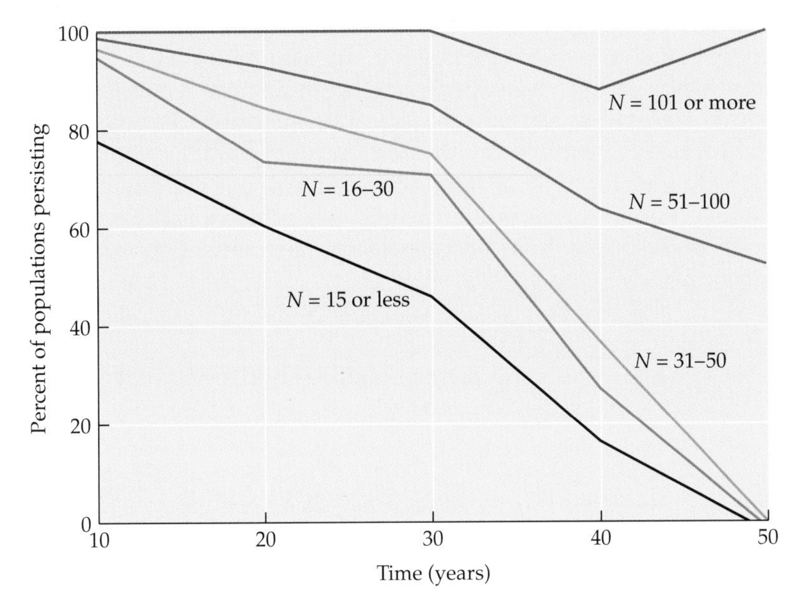

FIGURE 5.10 The relationship between initial population size (*N*) of bighorn sheep and the percentage of populations that persist over time. Almost all populations with more than 100 sheep persisted beyond 50 years, while populations with fewer than 50 individuals died out within 50 years. Not included are small populations that were actively managed and augmented by the release of additional animals. (Data from Berger 1990; photograph by Jim Peaco, courtesy of the National Park Service.)

1. Loss of genetic variation and inbreeding depression
2. Demographic fluctuations
3. Environmental fluctuations and natural catastrophes

We'll now examine in detail each of these causes for decline in small populations.

Loss of genetic variation

Genetic variation is important in allowing populations to adapt to a changing environment (see Chapter 2). Individuals with certain alleles or combinations of alleles may have just the characteristics needed to survive and reproduce under new conditions (Frankham 2005; Allendorf and Luikart 2007). Within a population, particular alleles may vary in frequency from common to very rare. In all populations, allele frequencies may change from one generation to the next simply by chance, depending on which individuals mate and produce offspring. This process is known as **genetic drift**, and it is a separate process from natural selection, in which the population changes in response to specific factors in the environment. When an allele occurs at a low frequency in a small population, it has a significant probability of being lost in each generation, due simply to chance.

Considering the general case of an isolated population in which there are two alleles of each gene in the gene pool (and no mutation), Wright (1931) proposed a formula to express the proportion of original heterozygosity (H), that is, individuals with two different allele forms of a gene, remaining after each generation. The formula includes the **effective population size** (**N_e**)—the size of the population as estimated by the number of its breeding individuals:*

$$H = 1 - 1/[2\ N_e]$$

According to this equation, a population of 50 breeding individuals would retain 99% of its original heterozygosity after one generation:

$$H = 1 - 1/100 = 1.00 - 0.01 = 0.99$$

The proportion of heterozygosity remaining after t generations (H_t) decreases over time:

$$H_t = H^t$$

For our population of 50 animals, then, the remaining heterozygosity would be 98% after two generations (0.99 × 0.99), 97% after three generations, and 90% after ten generations. A population of 10 individuals would retain 95% of its original heterozygosity after one generation, 90% after two generations, 86% after three generations, and 60% after ten generations (**Figure 5.11**).

This formula demonstrates that significant losses of genetic variation can occur in isolated small populations, particularly those on islands and fragmented landscapes. However, the migration of

*Factors affecting N_e, the effective population size, are discussed in detail beginning on page 158.

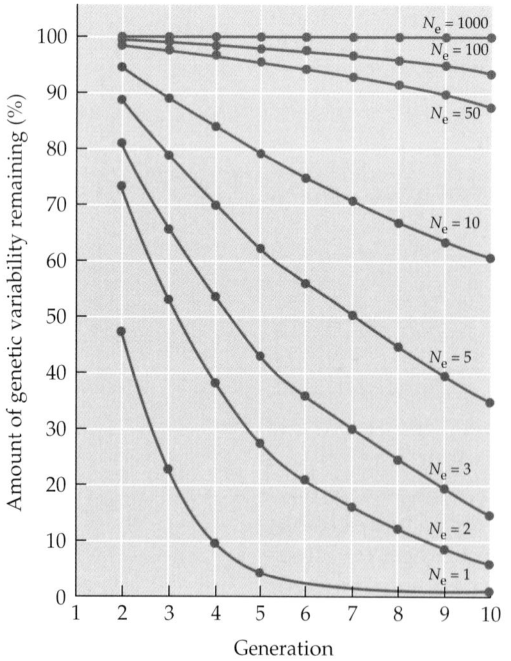

FIGURE 5.11 Genetic variability is lost randomly over time through genetic drift. This graph shows the average percentage of genetic variability remaining after 10 generations in theoretical populations of various effective population sizes (N_e). After 10 generations, there is a loss of genetic variability of approximately 40% with a population size of 10, 65% with a population size of 5, and 95% with a population size of 2. Blue lines indicate minimal loss of genetic variability in large populations; red lines indicate rapid loss of genetic variability in small populations. (After Meffe and Carroll 1997.)

individuals between populations and the regular mutation of genes tend to increase the amount of genetic variation within the population and balance the effects of genetic drift. The mutation rates found in nature—somewhere between 1 in 10,000 and 1 in 1,000,000 per gene per generation—may make up for the random loss of alleles in large populations, but they are insignificant in countering genetic drift in populations of 100 individuals or less. Fortunately, even a few individuals moving between populations minimizes the loss of genetic variation associated with small population size (Bell et al. 2010). In an isolated population of about 100 individuals, if even one or two immigrants arrive each generation, the impact of genetic drift will be greatly reduced.

Because of genetic drift, small populations lose genetic variation more rapidly than large populations. Some small populations may lack any genetic variation.

Field data also show that lower effective population size leads to a more rapid loss of alleles from the population (Evans and Sheldon 2008). For example, a broad survey of 89 bird species showed that there was a strong tendency for abundant birds to have more heterozygosity than species with small populations. Unfortunately, rare and endangered species often have small, isolated populations, leading to a rapid loss of genetic variation. Small populations sometimes lack genetic variation altogether. In the evolutionarily isolated Wollemi pine (*Wollemia nobilis*) in Australia, only 40 plants occur in two nearby populations. As might be predicted, an extensive investigation failed to find any genetic variation in this species (Peakall et al. 2003).

Small populations have greater susceptibility to a number of deleterious genetic effects such as inbreeding depression, outbreeding depression, and loss of evolutionary flexibility. These factors may contribute to a decline in population size, leading to an even greater loss of genetic variability, a loss of fitness, and a greater probability of extinction (Frankham et al. 2009).

INBREEDING DEPRESSION A variety of mechanisms prevents **inbreeding**, mating among close relatives, in most natural populations. In large populations of most animal species, individuals do not normally mate with close relatives; the mating of unrelated individuals of the same species is termed **outbreeding**. Individuals often disperse from their place of birth or are restrained from mating with relatives by behavioral inhibitions, unique individual odors, or other sensory cues. In many plants, numerous morphological and physiological mechanisms encourage cross-pollination and prevent self-pollination. In some cases, particularly when population size is small and no other mates are available, these mechanisms fail to prevent inbreeding.

Mating among parents and their offspring, siblings, and first cousins, and self-fertilization in hermaphroditic species, may result in **inbreeding depression**, a condition that occurs when an individual receives two identical copies of a deleterious allele, one from each of its parents. Inbreeding depression is characterized by higher mortality of offspring, fewer offspring, or offspring that are weak, sterile, or have low mating success (**Figure 5.12**).

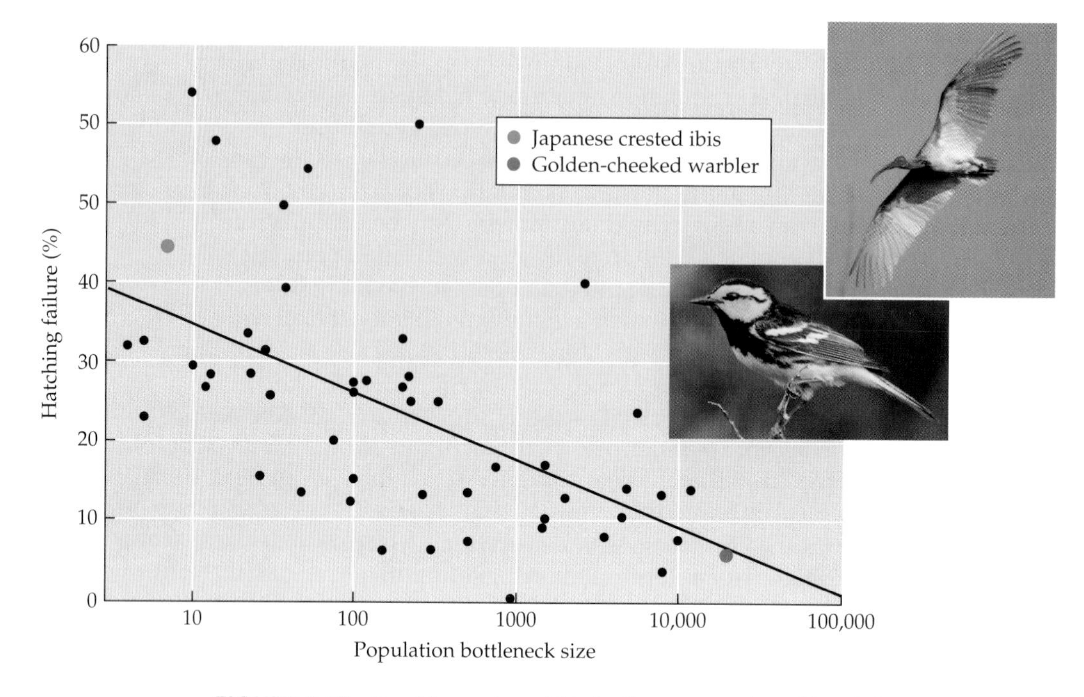

FIGURE 5.12 For 51 species of birds, the percent hatching failure is highest with the smallest population sizes (as expressed by the bottleneck size—the lowest size recorded for the population).The *x*-axis is on a log scale. The Japanese crested ibis (*Nipponia nippon*) (top photograph) represents one extreme, with fewer than 10 individuals in one year and hatching failure of about 45%; at the other extreme is the golden-cheeked warbler (*Dendroica chrysoparia*) (bottom photograph), with a population size always over 10,000 individuals and hatching failure of about 5%. (After Heber and Briskie 2010; Japanese crested ibis photograph © ChinaFotoPress/Zuma Press; golden-cheeked warbler photograph © Rolf Nussbaumer Photography/Alamy.)

These factors result in even fewer individuals in the next generation of small populations, leading to more pronounced inbreeding depression, smaller subsequent population size, and eventually extinction.

In one study, Bouzat et al. (2008) examined isolated small populations of greater prairie chickens (*Tympanuchus cupido pinnatus*) in Illinois. These populations were showing the effects of declining genetic variation and inbreeding depression, including lowered fertility and lowered rates of egg hatching. However, when individuals from large, genetically diverse populations were released into the small populations, egg viability was restored and the population sizes began to increase. This result demonstrates the importance of maintaining genetic variation in existing populations and of restoring genetic variation in genetically impoverished populations as a conservation strategy.

OUTBREEDING DEPRESSION Individuals of different species rarely mate in the wild; there are strong ecological, behavioral, physiological, and morphological isolating mechanisms that ensure mating occurs only between individuals of the same species. However, when a species is rare or its habitat is damaged, outbreeding—mating between individuals of different populations or species—may occur (**Figure 5.13**). Individuals unable to find mates within their own species may mate with individuals of related species. The resulting offspring sometimes exhibit **outbreeding depression**, a condition that results in weakness, sterility, or lack of adaptability to the environment. Outbreeding depression may be caused by incompatibility of the chromosomes and enzyme systems that are inherited from the different parents.

However, many other studies of animals have failed to demonstrate outbreeding depression, and other studies have shown that some hybrids are *more* vigorous than their parent species (McClelland and Naish 2007), a condition known as **hybrid vigor**. Thus, outbreeding depression may be considered of less concern for animals than inbreeding depression, the

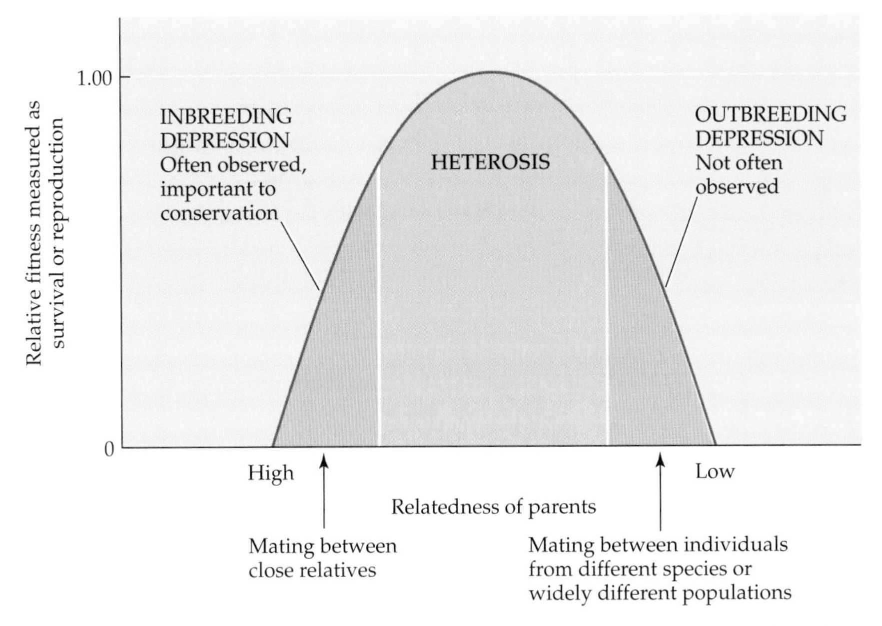

FIGURE 5.13 Mating between unrelated individuals of the same species often results in offspring with a high fitness (or heterosis) as measured by survival or high reproduction (number of offspring produced). Mating among close relatives (sibling–sibling or parent–offspring matings) or self-fertilization in hermaphroditic species leads to low fitness or inbreeding depression. Mating between individuals from widely different populations or even different species sometimes, but not always, results in lowered fitness or outbreeding depression. (After Groom et al. 2006.)

effects of which are well documented. It is also true that matings between species obscure species differences, resulting in problems of identification and legal protection.

LOSS OF EVOLUTIONARY FLEXIBILITY The existence of genetic variation within populations is extremely important to the long-term survival of a species. Rare alleles and unusual combinations of alleles that are harmless but confer no immediate advantage on the few individuals that carry them may turn out to be uniquely suited for a future set of environmental conditions. If such alleles and combinations do become advantageous, their incidence in the population will increase rapidly through natural selection, since the individuals that carry them will be those most likely to survive and reproduce successfully, passing on the formerly rare alleles to their offspring.

Loss of genetic variation in a small population may limit its ability to respond to changes and new conditions in the environment, such as polluting chemicals, new disease organisms, and global climate change (Willi et al. 2007). A small population is less likely than a large population to possess the genetic variation necessary to adapt to long-term environmental changes and thus is more likely to go extinct.

The effective population size, N_e, will be much smaller than the total population size, N, when there is a low percentage of individuals reproducing, great variation in reproductive output, an unequal sex ratio, or wide fluctuations in population size.

Effective population size

The degree to which genetic variation is maintained over time in a population is determined by the effective population size (N_e), which is an estimate of the number of breeding individuals in the population. The effective population size is almost always lower than the total population size because many individuals do not reproduce, due to factors such as inability to find a mate, being too old or too young to mate, poor health, sterility, malnutrition, small body size, and social structures (such as polygyny) that restrict which individuals can mate. Many of the factors are initiated or aggravated by habitat degradation and fragmentation. Furthermore, many plant, fungus, bacterium, and protist species have seeds, spores, or other structures in the soil that remain dormant unless stable conditions for germination appear. These individuals could be counted as members of the population, though they are obviously not part of the breeding population.

Because of these factors, the effective population size (N_e) is often substantially smaller than the total population size (N). Because the rate of loss of genetic variability is based on the effective population size, the loss of genetic variability can be quite severe, even in a large population. For example, consider a population of 1000 alligators with 990 immature animals and only 10 mature breeding animals: 5 males and 5 females. In this case, the effective population size is 10, not 1000.

In addition, the effective population is often lower than the actual number of breeding individuals because of unequal sex ratio, variation in reproductive output, and large annual changes in population size (Jamieson 2011). This reduced effective population size can lead to further population decline and extinction.

UNEQUAL SEX RATIO In certain animal species, social systems may result in unequal numbers of males and females mating, which can lower the effective population size. Among elephant seals, for example, a single dominant male usually mates with a large group of females and prevents other males from mating with them (**Figure 5.14**), whereas among African wild dogs, the dominant female in the pack often bears all of the pups.

The effect of unequal numbers of breeding males and females on N_e can be described by this formula:

$$N_e = [4(N_f \times N_m)] \;/\; (N_f + N_m)$$

where N_m and N_f are the numbers of adult breeding males and breeding females, respectively, in the population. In general, as the sex ratio of breeding individuals becomes increasingly unequal, the ratio of the effective population size to the number of breeding individuals goes down. This occurs because only a few individuals of one sex are making a disproportionately

FIGURE 5.14 Two large male southern elephant seals (*Mirounga leonina*) compete for the right to mate with the many smaller females surrounding them on the beach. The effective population size is reduced because only one male is providing genetic input to many females. (Photograph © National Geographic Image Collection/Alamy.)

large contribution to the genetic makeup of the next generation, rather than the equal contribution found in monogamous mating systems. In the case large of Asian elephants, for example, males are hunted by poachers for their tusks at the Periyar Tiger Reserve in India (Ramakrishnan et al. 1998). In 1997, there were 1166 elephants, of which 709 were adults. Of these adults, 704 were female and 5 were male. If all of these elephants were breeding, this would result in an effective population size of only 20 from a genetic perspective, using the equation shown above—dramatically less than the number of breeding individuals.

VARIATIONS IN REPRODUCTIVE OUTPUT In many species the number of offspring varies substantially among individuals. This phenomenon is particularly true of highly fecund species, such as plants and fish (see Hedrick 2005), where many or even most individuals produce a few offspring while others produce huge numbers. Unequal production of offspring leads to a substantial reduction in N_e because a few individuals in the present generation will be disproportionately represented in the gene pool of the next generation. In general, the greater the variation in reproductive output, the more the effective population size is lowered, with increased likelihood of population extinction.

POPULATION FLUCTUATIONS AND BOTTLENECKS In some species, population size varies dramatically from generation to generation. Good examples of this are butterflies, annual plants, and amphibians. In a population with extreme size fluctuations, the effective population size is much nearer the lowest than the highest number of individuals, and it tends to be determined by the years in which the population has the smallest numbers.

Even a single year of drastically reduced population numbers will substantially lower the value of N_e. This principle applies to a phenomenon known as a **population bottleneck**, which occurs when a population is greatly reduced in size and loses rare alleles if no individuals possessing those alleles survive and reproduce (Jamieson 2011).

A special category of bottleneck, known as the **founder effect**, occurs when a few individuals leave one population and establish another new population. The new population often has less genetic variability than the larger, original population. Bottlenecks can also occur when captive populations are established using relatively few individuals. For example, the captive population of the Speke's gazelle (*Gazella spekei*) in the United States was established from one male and three females.

The lions (*Panthera leo*) of Ngorongoro Crater in Tanzania provide a well-studied example of a population bottleneck (Munson et al. 2008). The lion population in the crater consisted of about 90 individuals until an outbreak of biting flies in 1962 reduced the lion population to 9 females and 1 male (**Figure 5.15**). Two years later, 7 additional males immigrated to the crater; there has been no further immigration since that time. Even though the population increased to 125 animals in 1983, the population later dropped

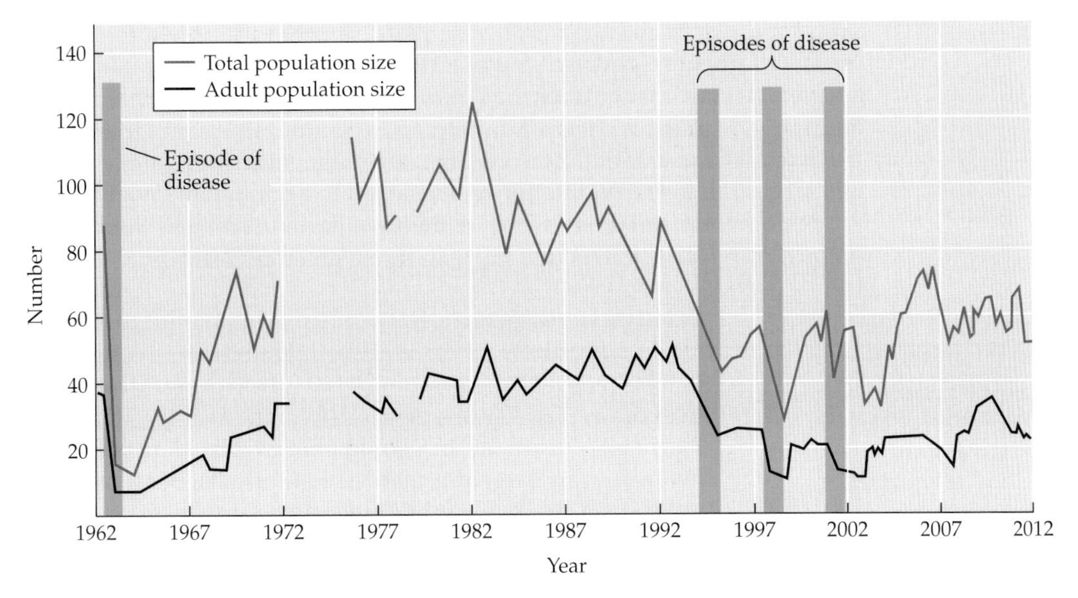

FIGURE 5.15 The Ngorongoro Crater lion population consisted of about 90 individuals in 1961 before crashing in 1962. Since that time, the population reached a peak of 125 individuals in 1983 before collapsing to 34 individuals (fewer than 20 of which were adults). Small population size, an isolated location, lack of immigration since 1964, and the impact of disease have contributed to the loss of genetic variation caused by a population bottleneck. A lack of census data for certain years is the cause of gaps in the lines. The four green bars represent episodes of disease outbreak. (After Munson et al. 2008, with updates from C. Packer.)

to 34 animals following an outbreak of canine distemper virus in the 1990s that spread from domestic dogs kept by people living just outside the crater area. The population has fluctuated between 50 and 75 animals since then.

These examples demonstrate that effective population size is often substantially less than the total number of individuals in a population. Particularly where there is a combination of factors, such as bad weather during a generation's reproductive season, numerous nonreproductive individuals, and an unequal sex ratio, the effective population size may be far lower than the number of individuals alive in a good year. As a consequence, such small populations will lose the genetic variation needed to adapt to a changing environment.

Demographic and environmental stochasticity

Random variation, or **stochasticity**, in the environment can cause variation in the population size of a species. For example, the population of an endangered butterfly species might be affected by fluctuations in the abundance of its food plants and the number of its predators. Variation in the physical environment might also strongly influence the butterfly population; in an average year, the weather may be warm enough for caterpillars to feed and grow, whereas a cold year might cause many caterpillars to become inactive and consequently starve. Such **environmental stochasticity** affects all individuals in the population and is linked to **demographic stochasticity** (or **demographic variation**), which is the variation in birth and death rates among individuals and across years within a given population.

Random fluctuations in birth and death rates, disruption of social behavior with lowered population density, and environmental stochasticity all contribute to further decrease in the size of populations, often leading to local extinction.

DEMOGRAPHIC VARIATION In any real population, individuals do not usually produce offspring in numbers equivalent to the average birth rate: some leave no offspring, some fewer than the average, and some more than the average. Similarly, the average death rate in a population can be determined only by examining large numbers of individuals, because some individuals die young while others live a relatively long time. As long as the population size is large, average birth or death rates provide an accurate and relatively stable description of the population. Once population size drops below about 50 individuals, however, demographic stochasticity becomes more significant, resulting in random fluctuations up or down (Schleuning and Matthies 2009). If population size fluctuates downward in any one year because environmental factors trigger a higher than average number of deaths or a lower than average number of births, the resulting smaller population will be even more susceptible to further fluctuations in subsequent years.

Random size fluctuations upward are eventually bounded by the carrying capacity of the environment (that is, the amount of resources available to sustain the population), and the population may fluctuate downward again. Once a population decreases because habitat destruction and fragmentation reduce their resources or increase potential pathogens, the population has an even higher probability of declining further, or even going extinct, as a result of chance alone (in a year with low reproduc-

FIGURE 5.16 In a declining population of Spanish imperial eagles (*Aquila adalberti*) the percent of mature birds breeding in the population decreased beginning in 1990 as mature male birds left the area to find food elsewhere. As a result, the percent of immature birds breeding at the site increased; these were mostly immature males mating with mature females (75% breeders in adult plumage means that probably half of the breeding males were immature). Matings involving immature males produce mainly male offspring, contributing to further population decline and increasing the probability of local extinction. Starting in 2004, supplemental food was provided at the site, more of the mature males remained at the site to breed, and the sex ratio of the offspring returned to approximately equal numbers of males and females. (After Ferrer et al. 2009, with updates by the author; photograph © imagebroker/Alamy.) ▶

tion and high mortality or sex ratios that deviate from equal numbers of males and females). Species with birth and death rates that are naturally highly variable, such as annual plants and short-lived insects, may be particularly susceptible to population extinction due to demographic stochasticity. The chance of extinction is also greater in species that have low birth rates, such as elephants, because these species take longer to recover from chance reductions in population size.

POPULATION DENSITY AND THE ALLEE EFFECT Many small populations are demographically unstable because social interactions (especially those affecting mating) can be disrupted once population density falls below a certain level (**Figure 5.16**) (Gascoigne et al. 2009). This interaction between population size, population density, and population growth rate is sometimes referred to as the Allee effect. Herds of grazing mammals and flocks of birds may be unable to find food and defend themselves against attack from predators when numbers fall below a certain level. Animals that

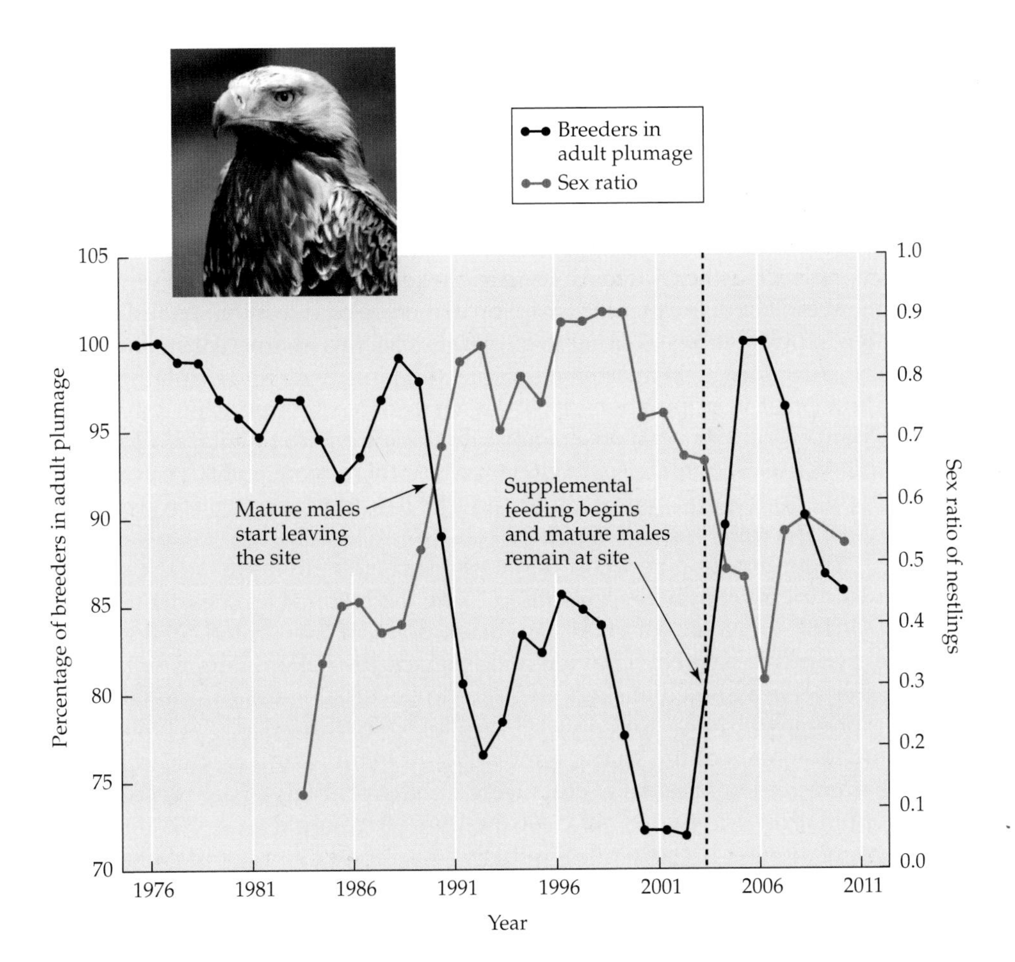

hunt in packs, such as wild dogs and lions, may need a certain number of individuals to hunt effectively. The population may not be able to remain viable when it is lower than this number.

Perhaps the most significant aspect of the Allee effect for small populations involves reproductive behavior: many species that live in widely dispersed populations, such as bears, spiders, and tigers, have difficulty finding mates once the population density drops below a certain point. Among plant species, as population size and density decrease, the distance between individual plants increases; pollinating animals may not visit isolated, scattered plants, resulting in insufficient transfer of compatible pollen and a subsequent decline in seed production. In such cases, the birth rate will decline, population density will become lower yet, problems such as unequal sex ratio will worsen, and birth rates will drop even more. Once the birth rate falls to zero, extinction is guaranteed; for example, the last five surviving individuals of the now extinct dusky seaside sparrow (*Ammodramus maritimus nigrescens*) were all males, so there was no opportunity to establish a captive breeding program to save the species.

ENVIRONMENTAL STOCHASTICITY AND CATASTROPHES Natural catastrophes that occur at unpredictable intervals, such as droughts, storms, earthquakes, and fires, along with cyclical die-offs of the surrounding biological community, can cause dramatic fluctuations in population levels. Natural catastrophes can kill part of a population or even eliminate an entire population from an area. Random environmental variation—including extended periods of unseasonable weather, excessive or insufficient rainfall, or catastrophic events such as hurricanes and earthquakes—is generally more important than random demographic variation in increasing the probability of extinction in populations of small to moderate size. Environmental stochasticity can substantially increase the risk of extinction, even for populations that show positive population growth when a stable environment is assumed (Mangel and Tier 1994). In general, introducing environmental stochasticity into population models—in effect making them more realistic—results in populations with lower average growth rates, lower population sizes, and higher probabilities of extinction.

The interaction between population size and environmental variation was experimentally demonstrated using the biennial herb garlic mustard (*Alliaria petiolata*), an exotic, invasive plant in the United States, as an experimental subject (Drayton and Primack 1999). Populations of various sizes were assigned at random either to be left alone as controls or to be experimentally eradicated by removal of every flowering plant in each of the four years of the study; removal of all plants could be considered an extreme environmental event. Overall, the probability of an experimental population's going extinct over the four-year period was 43% for small populations (<10 individuals initially), 9% for medium populations (10 to 50 individuals), and 7% for large populations (>50 individuals). For control populations, the probability of going extinct for small, medium, and large populations was 11%, 0%, and 0%. Large numbers of dormant seeds in the soil apparently allowed most experimental populations to persist even when every flowering plant was removed in four successive years. However, small populations were far more susceptible to extinction than large populations.

The extinction vortex

Intensive management is often required to prevent small populations from declining further in size and going extinct.

The smaller a population becomes, the more vulnerable it is to the combined effects of demographic variation, Allee effects, environmental variation, loss of genetic variability, and inbreeding depression that tend to lower reproduction, increase mortality rates, and so reduce population size even more, driving the population to extinction (**Figure 5.17**). This tendency of small populations to decline toward extinction has been likened to a vortex, a whirling mass of gas or liquid spiraling inward—the closer an object gets to the center, the faster it moves. At the center of an **extinction vortex** is oblivion: the local extinction of the species.

Small populations often require a careful program of population and habitat management, as described in later chapters, to increase population growth rate and allow the population to escape from the harmful effects of small population size.

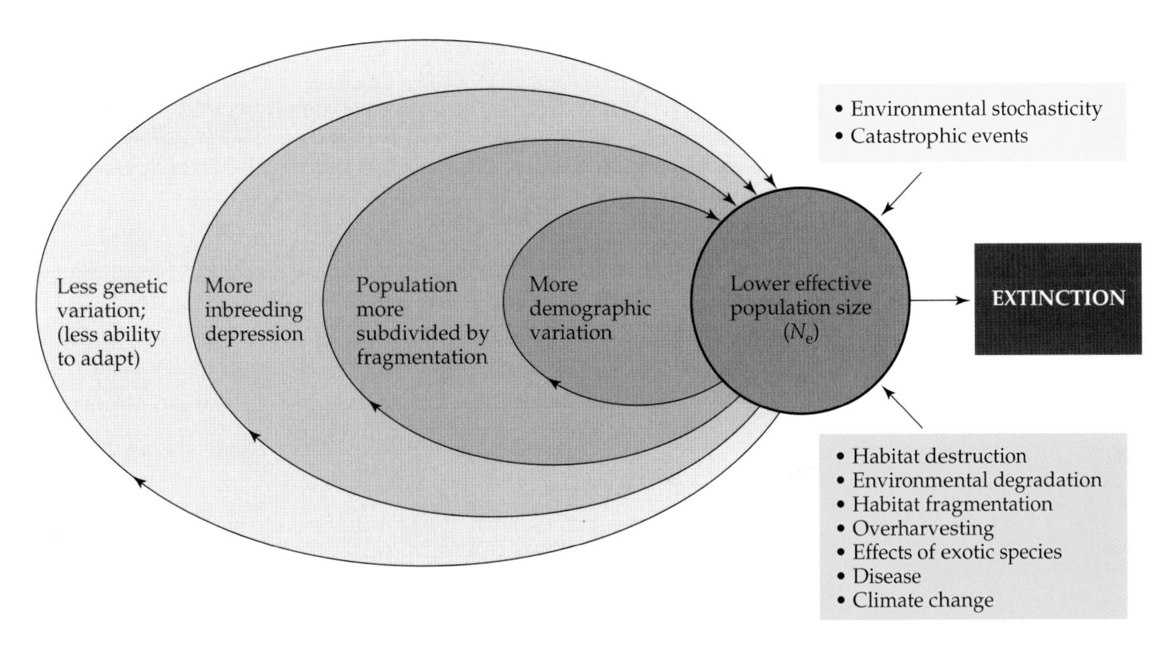

FIGURE 5.17 Once a population drops below a certain size, it enters an extinction vortex in which the factors that affect small populations tend to drive its size progressively lower. This downward spiral often leads to the local extinction of species. (After Gilpin and Soulé 1986 and Guerrant 1992.)

Summary

- Current rates of extinction for species are between 100 and 1000 times greater than background levels. More than 99% of modern species extinctions are attributable to human activity.
- Since 1600, about 1.6% of the world's mammal species and 1.3% of its bird species have gone extinct. Many extant species are on the brink of extinction, with 21% of bird species, 27% of mammal species, and 36% of amphibian species believed to be threatened.
- Island species have a higher rate of extinction than mainland species. Among aquatic species, freshwater species apparently have a higher extinction rate than marine species.
- The island biogeography model is used to predict the numbers of species that will persist in new national parks and other protected areas and that will go extinct elsewhere due to habitat destruction and other human activities.
- Species most vulnerable to extinction have particular characteristics, including a very narrow range, one or only a few populations, small population size, declining population size, and an economic value to humans, which leads to overexploitation.
- Small populations are vulnerable to further declines in size and eventual extinction due to genetic, demographic, and environmental factors. Intensive management of small populations may be required to prevent their extinction.

For Discussion

1. Why should conservation biologists, or anyone else, care if a species goes locally extinct if it is still found somewhere else?
2. Develop an imaginary animal, recently discovered, that is extraordinarily vulnerable to extinction. Give your species a whole range of characteristics that make it vulnerable; then, consider what could be done to protect it.
3. Find out about a species that is currently endangered in the wild. How might this species be affected by the problems of small populations? Address genetic, physiological, behavioral, and ecological aspects, as appropriate.

Suggested Readings

Fisher, D. O. and S. P. Blomberg. 2012. Inferring extinction of mammals from sighting records, threats, and biological traits. *Conservation Biology* 26: 57–67. Using past observations, a model can estimate the probability that an extinct species is still alive and that an endangered species is actually extinct.

Frankham, R., J. D. Ballou, and D. A. Briscoe. 2009. *Introduction to Conservation Genetics*, 2nd ed. Cambridge University Press, Cambridge, UK. Excellent introduction to the importance of genetics to conservation.

Hambler, C., P. A. Henderson, and M. R. Speight. 2011. Extinction rates, extinction-prone habitats, and indicator groups in Britain and at larger scales, *Biological Conservation* 144: 713–721. Extinction rates are estimated from the most well-studied region of the world.

Heber, S. and J. V. Briskie. 2010. Population bottlenecks and increased hatching failure in endangered birds. *Conservation Biology* 24: 1674–1678. Species with small populations have reduced reproductive success.

Hoffmann, M. and 165 others. 2010. The impact of conservation on the status of the world's vertebrates. *Science* 330: 1503–1509. Conservation actions can improve the status of species and entire regions.

Jamieson, I. G. 2011. Founder effects, inbreeding, and loss of genetic diversity in four avian reintroduction programs. *Conservation Biology* 25: 115–123. New birds populations are often established with surprisingly few individuals.

Laurance, W. F. 2007. Have we overstated the tropical biodiversity crisis? *Trends in Ecology and Evolution* 22: 65–70. The increasing rate of rain forest destruction threatens enormous numbers of species.

McClenachan, L., A. B. Cooper, K. E. Carpenter, and N. K. Dulvy. 2012. Extinction risk and bottlenecks in the conservation of charismatic marine species. *Conservation Letters* 5: 73–80. Most marine species have not been evaluated for their risk of extinction.

Primack, R. B., A. J. Miller-Rushing, and K. Dharaneeswaran. 2009. Changes in the flora of Thoreau's Concord. *Biological Conservation* 142: 500–508. Human impacts can cause extensive local extinction even in a well-protected suburban landscape.

Traill, L. W., B. W. Brook, R. R. Frankham, and C. J. A. Bradshaw. 2010. Pragmatic population viability targets in a rapidly changing world. *Biological Conservation* 143: 28–34. Thousands of individuals need to be protected to achieve a high probability of population survival.

A wildlife veterinarian and a conservation biologist examine a Magellanic penguin chick on the Argentinian coast.

Chapter 6
Conserving Populations and Species

Even without human disturbance, a population of any species can be stable, increasing, decreasing, or fluctuating in number. In general, widespread human disturbance destabilizes populations of many native species, often sending them into sharp decline. But how can this disturbance be measured, and what actions should be taken to prevent or reverse it? This chapter discusses management approaches for monitoring, managing, and protecting species and their populations.

In protecting and managing a rare or endangered species, it is vital to have a firm grasp of the ecology of the species, its distinctive characteristics (sometimes called its **natural history**), and the status of its populations, particularly the dynamic processes that affect population size and distribution (its **population biology**). With more information about the natural history and population biology of a rare species, land managers are able to more effectively maintain it, identify factors that place it at risk of extinction, and develop alternative management options.

Applied Population Biology

Scientists who study natural history and population biology have historically pursued knowledge for its own sake, but today these fields have gained very real and important conservation applications. To help implement effective population-level conservation efforts for rare and endangered species, conservation biologists need to know certain key facts. For many species, this information is not available and some management decisions have to be made before the information is available or while it is being gathered. Several types of natural history and population biology information are important:

- *Environment*. What are the habitat types where the species is found, and how much area is there of each type? How variable is the environment in time and space? How have human activities affected the environment? How might climate change affect the environment?
- *Distribution*. Where is the species found in its habitat? Are individuals clustered together, distributed at random, or spaced out regularly? Do individuals of the species move and migrate among habitats or to different geographical areas over the course of a day or over a year? How have human activities affected the distribution of the species (**Figure 6.1**)?
- *Biotic interactions*. What types of food and other resources does the species need and how does it obtain them? What other species compete with it for these resources? What predators, parasites, or diseases affect its population size? What mutualists (pollinators, dispersers, etc.) does it interact with? How have human activities altered the relationships among species in the community?
- *Morphology*. What does the species look like? What are the shape, size, color, surface texture, and function of its physical features? How do the characteristics of its body parts relate to their functions and help the species to survive in its environment? What are the characteristics that allow this species to be distinguished from species that are similar in appearance? Are there differences in appearance between the sexes and between juveniles and adults?
- *Physiology*. How much food, water, minerals, and other necessities does an individual need to survive, grow, and reproduce? How vulnerable is the species to extremes of climate, such as heat,

FIGURE 6.1 Researchers and staff from the Malaysian Department of Wildlife and National Parks attach a GPS-satellite collar to a wild elephant living in a small forest patch surrounded by rubber tree and oil palm plantations. The data obtained from collared elephants is helping to understand and reduce human–elephant conflicts. (Photograph courtesy of Ahimsa Campos-Arceiz.)

cold, wind, and rain? When does the species reproduce, and what are its special requirements during reproduction?

- *Demography*. What is the current population size, and what was it in the past? Are the numbers of individuals stable, increasing, or decreasing? Does the population have a mixture of adults and juveniles, indicating that recruitment of new individuals is occurring?
- *Behavior*. How do the actions of an individual allow it to survive in its environment? How do individuals in a population mate and produce offspring? In what ways do individuals of a species interact, cooperatively or competitively? How do individuals respond to human disturbance?

Knowledge of the natural history and population biology of a species is crucial to its protection, but urgent management decisions often must be made before all of this information is available, or while it is still being gathered.

- *Genetics*. How much genetic variation occurs in morphological, physiological, and behavioral characteristics? How is the variation spread across the species range? What percentage of the genes is variable? How many alleles does the population have for each variable gene? Are there genetic adaptations to local sites?
- *Interactions with humans*. How do human activities affect the species? What human activities are harmful or beneficial to the species? Do people harvest or use this species in any way? What do local people know about this species?

Methods for studying populations

Methods for the study of populations have developed largely from the study of land plants and animals. Small organisms such as protists, bacteria, and fungi have not been investigated in comparable detail. Species that inhabit soil, freshwater, and marine habitats are much less known with respect to population characteristics. In this section we will examine how conservation biologists undertake their studies of populations, recognizing that methods need to be modified for each species.

PUBLISHED LITERATURE Other people may have already investigated an ecosystem or studied the same species (or a related species). Library indexes such as BIOSIS and Biological Abstracts are often accessible by computer and provide easy access to a variety of books, articles, and reports relating to a particular topic. This literature may contain records of previous population sizes and distributions that can be compared with the current status of the species. Some sections of the library will have related material shelved together, so finding one book often leads to others. The Internet provides ever-increasing access to databases, websites, electronic bulletin boards, journals, news articles, specialized discussion groups, and subscription databases such as ScienceDirect and the Thomson Reuters (formerly ISI) Web of Science. Google Scholar is one of the best places to start searching on topics relating to conservation biology. Keep in mind that information on the Internet needs to be examined carefully to determine the accuracy and source of the data, because there is no control over what is posted. Asking biologists and naturalists for ideas on references is another way to locate published materials. Searching indexes of newspapers, magazines, and popular journals is also an effective strategy because results of important scientific research are often covered in the popular news media.

UNPUBLISHED LITERATURE An enormous amount of information on conservation biology is contained in unpublished reports by individual scientists, enthusiastic citizens, government agencies, and conservation organizations. This so-called gray literature is sometimes cited in published literature or mentioned by leading authorities in conversations, lectures, or articles. Often a report known through word of mouth can be obtained through direct contact with the author or from the Internet. Surveys of experts can also be used, to gather their collective knowledge about endangered species and ecosystems (Donlan et al. 2010).

FIELDWORK The natural history of a species usually must be learned through careful observations in the field (e.g., Leidner and Haddad 2011). Fieldwork is necessary because only a tiny percentage of the world's species have been adequately studied, and the ecology of a species often changes from one place to another. Only in the field can the conservation status of a species be determined, as well as its relationships to the biological and physical environment. Many of the technical methods for investigating populations are very specialized and are best learned by studying under the supervision of an expert or by reading manuals.

Monitoring populations

To learn the status of a species of special concern, scientists must survey its populations in the field and monitor them over time (**Figure 6.2**). Survey methods range from making a complete count of every individual, often called a census, to estimating population size using sampling methods or indexes. By repeatedly taking a survey of a population on a regular basis, one can determine changes in population size and distribution (McDonald-Madden et al. 2011; Holland et al. 2012). Long-term survey records can help to distinguish long-term population trends of increase or decrease (possibly caused by human disturbance) from short-term fluctuations caused by variations in weather or unpredictable natural events. Survey records can also determine whether an endangered species is showing a positive response to conservation management or is responding negatively to present levels of harvest or the arrival of invasive species.

Observing a long-term decline in a species under study often motivates biologists to take vigorous action to conserve it. Monitoring efforts can be targeted at particularly sensitive species, such as butterflies, using them

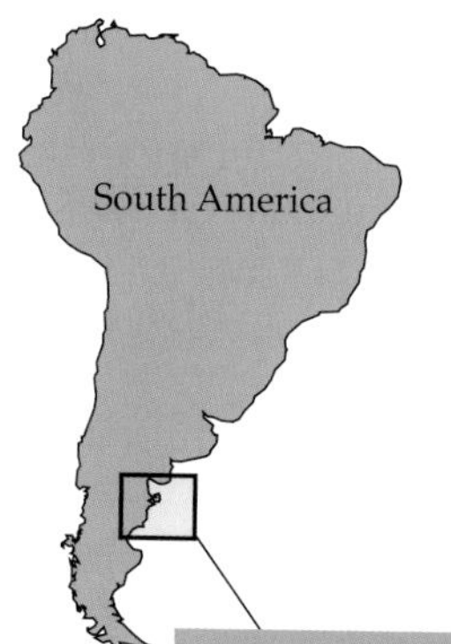

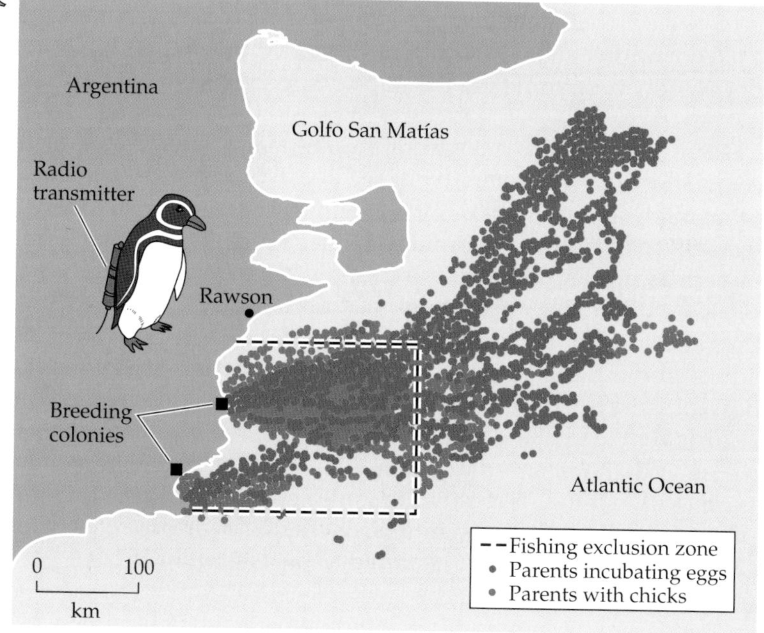

FIGURE 6.2 Satellite tracking of Magellanic penguins (*Spheniscus magellanicus*) fitted with radio transmitters shows that penguins incubating eggs forage up to 600 km from their breeding colonies. When penguins are feeding chicks, foraging takes place mainly within a seasonal fishing exclusion zone that was established to protect spawning fish. Fieldwork provided this vital information about the penguins' foraging habits, which led to the fishing zone's remaining closed until the chicks left their nests. (After Boersma et al. 2006.)

as indicator species of the long-term stability of ecological communities (Wikström et al. 2008) (see Chapter 7).

The number of monitoring studies has been increasing dramatically as governmental and conservation agencies have become more concerned with protecting rare and endangered species. Some of these studies are mandated by law as part of management efforts. The geographical range and intensity of monitoring has often been greatly extended through the use of volunteers, often called "citizen scientists" (Mueller et al. 2010). Training and educating volunteers not only expands the data available to scientists but often transforms these people into advocates for conservation. Local people, in particular hunters and fishermen, sometimes have detailed knowledge of the distribution of species around where they live and of changes in abundance and occurrence over time; this type of information can be valuable in conservation planning (Azzurro et al. 2011) (see Chapter 9).

Population monitoring often needs to be combined with monitoring of other parameters of the environment to understand the reasons behind population changes. The long-term monitoring of ecosystem processes (e.g., temperature, rainfall, humidity, soil acidity, water quality, discharge rates of streams, and soil erosion) and community characteristics (e.g., species present, percentage of land with plant cover, and amount of biomass present at each trophic level) allows scientists to determine the health of the ecosystem and the status of species of special concern. Monitoring these parameters allows managers to determine whether the goals of their projects are being achieved or whether adjustments must be made in the management plans (called adaptive management), as discussed in Chapter 7. Three types of population monitoring are the census, survey, and demographic study.

CENSUS A **census** is a count of the number of individuals present in a population. It is a comparatively inexpensive and straightforward method of monitoring populations. By repeating a census over successive time intervals, biologists can determine whether a population is stable, increasing, or decreasing in number. In one example of a monitoring study, population censuses of the Hawaiian monk seal on the beaches of several islands in the Kure Atoll of the South Pacific documented a decline, from almost 100 adults in the 1950s to fewer than 14 in the late 1960s (**Figure 6.3**). On the basis of these trends, the Hawaiian monk seal was declared endangered in 1976 under the U.S. Endangered Species Act (discussed later in this chapter) (Baker and Thompson 2007). Conservation efforts were implemented that reversed the trend, but only for some seal populations. The Tern Island population showed a substantial recovery following the closing of a Coast Guard station in 1979, but it started to decline again in the 1990s because of high juvenile mortality.

Censuses of a community can be conducted to determine what species are currently present in a locality; a comparison of current occurrences with past censuses can highlight species that have been lost. Censuses conducted over a wide area can help to determine the range of a species and its areas of local abundance and to highlight changes in the range of a species.

SURVEY A **survey** of a population involves using a repeatable sampling method to estimate the number of individuals or the density of a species in an ecosystem. An area can be divided into sampling segments and the number of individuals in certain segments counted. These counts can then be used to estimate the actual population size.

For example, a rabbit population can be surveyed by walking transects and recording the number of animals observed (Petrovan et al. 2011). Survey methods are used when a population is very large or its range is extensive. Such methods are particularly valuable when the species being studied has stages in its life cycle that are inconspicuous, tiny, or hidden, such as the seedling stages of many plants or the larval stages of aquatic invertebrates.

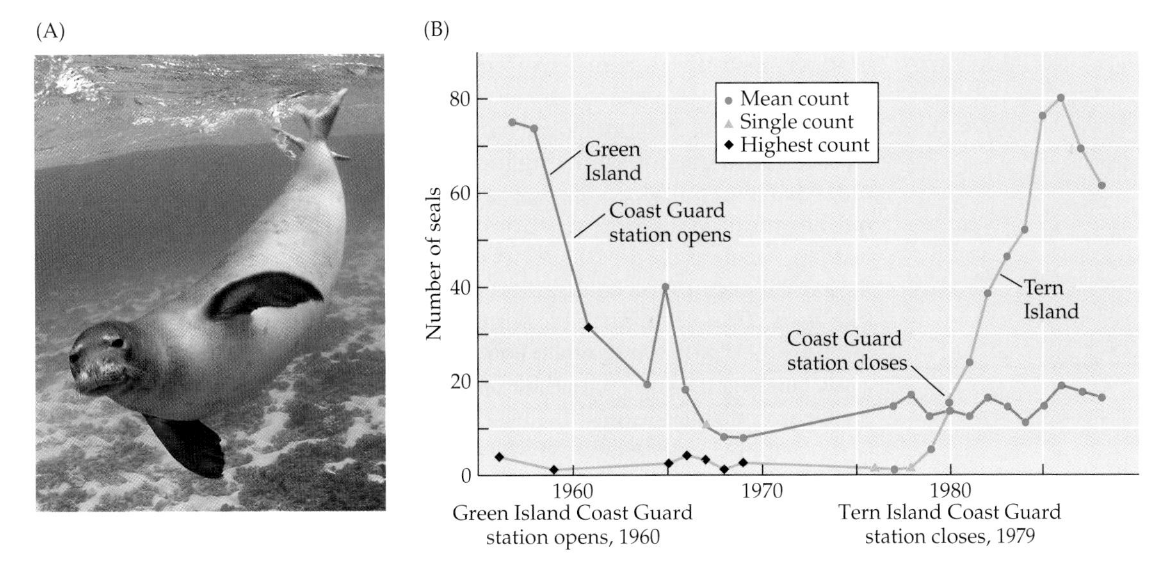

FIGURE 6.3 (A) A Hawaiian monk seal (*Monachus schauinslandi*). (B) Censusing the seal populations on Green Island of Kure Atoll (blue trace) and on Tern Island of French Frigate Shoals (green trace) revealed that this species was in danger of extinction. Population counts were plotted from a single count, the mean of several counts, or the maximum of several counts. Seal populations declined when a Coast Guard station was opened on Green Island in 1960, because of disturbance by people and dogs; populations increased on Tern Island after the closing of a Coast Guard station in 1979, when there was less disturbance to seals. (A, photograph by James D. Watt, courtesy of U.S. Department of the Interior; B, after Gerrodette and Gilmartin 1990.)

Surveys have expanded in recent years to include analysis of scat and hair samples (Vynne et al. 2011). Noninvasively collecting samples such as hair and feces enables researchers to gather valuable information about health, endocrinology, and genetics from free-ranging species without having to capture, tranquilize, or disturb individuals. In some cases, specially trained dogs are used to locate scat samples of rare animal species. DNA studies using dung have revealed that population size is often larger than previously estimated from traditional survey data, because some individuals have never been seen.

> Demographic studies provide data on the numbers, ages, sexes, conditions, and locations of individuals within a population. These data indicate whether a population is stable, increasing, or declining and are the basis for statistical models used to predict the future of a species.

DEMOGRAPHIC STUDY **Demographic studies** follow known individuals of different ages and sizes in a population to determine their rates of growth, reproduction, and survival (Saraux et al. 2011). Either the whole population or a subsample can be followed. In a complete population study, all individuals are counted, aged if possible, measured for size, sexed, and tagged or marked for future identification; their position on the site is mapped, and tissue samples sometimes are collected for genetic analysis. The techniques used to conduct a population study vary depending on the characteristics of the species and the purpose of the study. Each discipline has its own technique for following individuals over time; ornithologists band birds' legs, mammalogists often attach tags to animals' ears, and botanists nail aluminum tags to trees. Information from demographic studies can be used in standard mathematical formulas (life history formulas) to calculate the rate of population change and to identify critical stages in the life cycle (De Roos 2008). Demographic studies provide the most information of any monitoring method and, when analyzed thoroughly, suggest ways in which a site can be managed to ensure population persistence.

Demographic studies can also provide information on the age structure of a population. A stable population typically has an age distribution with a characteristic ratio of juveniles, young adults, and older adults. The absence or low representation of any age class, particularly of juveniles, may indicate that the population is declining. Conversely, a large number of juveniles and young adults may indicate that the population is stable or even expanding.

Population viability analysis

PVA uses mathematical and statistical methods to predict the probability that a population or species will go extinct within a certain time period. PVA is also useful in modeling the effects of habitat degradation and management efforts.

Predictions of whether a species has the ability to persist in an environment can be made using **population viability analysis (PVA)**, an extension of demographic studies PVA can be thought of as risk assessment—using mathematical and statistical methods to predict the probability that a population or a species will go extinct at some point in the future. By looking at the range of requirements a species has and the resources available in its environment, one can identify the vulnerable stages in the natural history of the species. PVA can be useful in considering the effects of habitat loss, habitat fragmentation, and habitat deterioration on a rare species, such as the European bison (Beissinger et al. 2009).

An important part of PVA is estimating how management efforts such as reducing (or increasing) hunting or increasing (or decreasing) the area of protected habitat will affect the probability of extinction (Sebastián-González et al. 2011; Molano-Flores and Bell 2012). PVA can model the effects of augmenting a population through the release of additional individuals caught in the wild elsewhere or raised in captivity.

Such statistical models must be used with caution and a large dose of common sense. The results of some models can change dramatically with different model assumptions and slight changes in parameters. Generally, about 10 years of data are needed to obtain a PVA with good predictive power (McCarthy et al. 2003). PVA also has value in demonstrating the possible effectiveness of alternative management strategies (Traill et al. 2010). For this reason, attempts to utilize PVA as part of practical conservation efforts are increasingly common in management planning, as the following examples demonstrate:

- The Hawaiian stilt (*Himantopus mexicanus knudseni*) is an endangered, endemic bird of the Hawaiian islands. Hunting and coastal development 70 years ago reduced the number of stilts to 200, but protection has allowed recovery to the present population size of about 1400 individuals (Reed et al. 2007). The goal of government protection efforts is to allow the population to increase to 2000 birds. A PVA was made of the ability of the species to have a 95% chance of persisting for the next 100 years. The model predicted that stilt numbers would increase until they occupied all available habitat but that they would show a rapid decline if nesting failure and mortality rates of first-year birds exceeded 70% or if the mortality rate of adults increased above 30% per year. To keep mortality rates below these values would require the control of exotic predators and the restoration of wetland habitat. And most important, additional wetland would need to be protected if the goal of protecting 2000 stilts was to be achieved.
- In another example, the marsh fritillary butterfly (*Euphydryas aurinia*) is declining in abundance in the United Kingdom, where it occupies lightly grazed grasslands. The average area occupied by the six extant populations studied is larger than the average area formerly occupied by six extinct populations. A PVA showed that an area of at least 100 ha is necessary to ensure a 95% probability that a population will persist for 100 years (**Figure 6.4**). Only two of the extant populations have access to areas of this magnitude. The other four populations face a high probability of extinction unless the habitat is enlarged in area and managed to encourage the growth of food plants (Bulman et al. 2007).

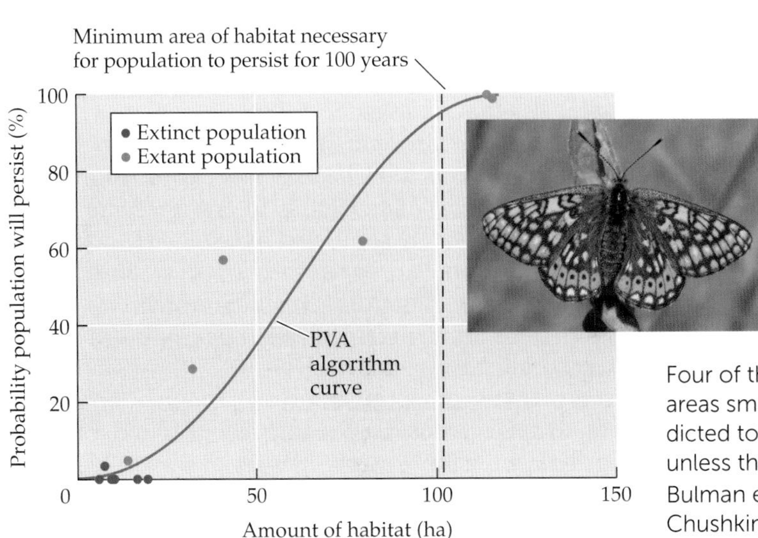

FIGURE 6.4 Population viability analyses predict that it takes 100 ha to ensure (at 95% likelihood) the persistence of a marsh fritillary butterfly population for 100 years. All of the extinct populations occupied areas much smaller than 100 ha. Four of the six extant populations occupy areas smaller than 100 ha and are predicted to go extinct in the coming decade unless their habitat is increased. (After Bulman et al. 2007; photograph © Sergey Chushkin/shutterstock.)

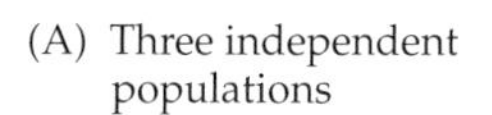

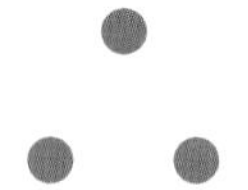

(B) Simple metapopulation of three interacting populations

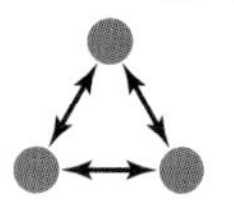

(C) Metapopulation with a large core population and three satellite populations

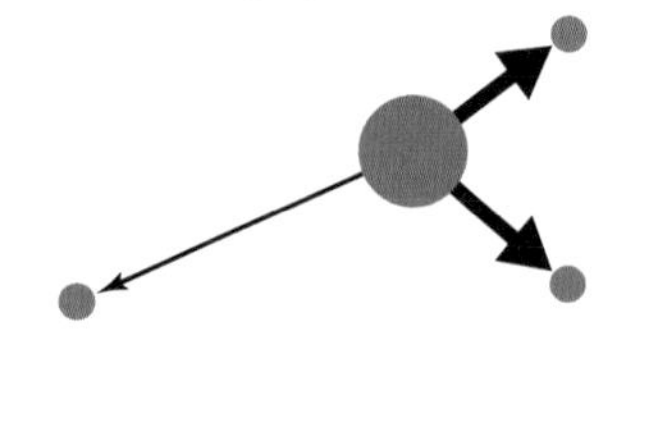

(D) Metapopulation with complex interactions

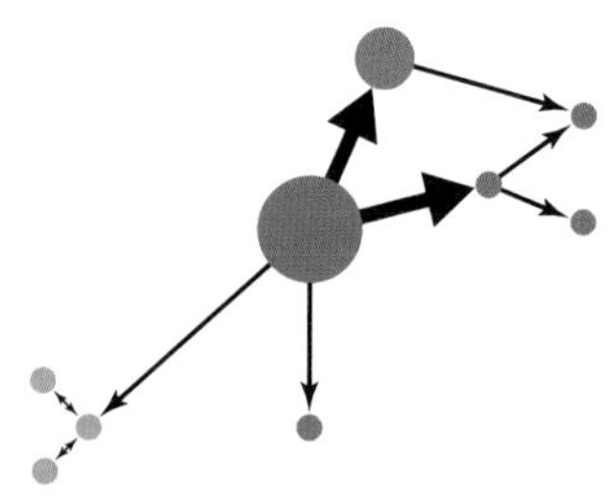

FIGURE 6.5 Possible metapopulation patterns, with the size of a population indicated by the size of the circle. The arrows indicate the direction and intensity of migration between populations. (After White 1996.)

Metapopulations

Over time, populations of a species may become extinct on a local scale, while new populations may form nearby on other suitable sites. Often a species found in an ephemeral habitat, such as a a small wildflower that occurs on riverbanks, is better characterized by a **metapopulation** (a "population of populations") that is made up of a shifting mosaic of populations linked by some degree of migration (Holt and Barfield 2010). In some species, every population in the metapopulation is short-lived, and the distribution of the species changes dramatically with each generation. In other species, the metapopulation may be characterized by one or more **source populations** (core populations) with fairly stable numbers and several **sink populations** (satellite populations) that fluctuate with arrivals of immigrants. Populations in the satellite areas may become extinct in unfavorable years, but the areas are recolonized, or rescued, by migrants from the more permanent core population when conditions become more favorable (**Figure 6.5**). Metapopulations might also involve relatively permanent populations that individuals occasionally move between.

Populations of a species are often connected by dispersal and can be considered a metapopulation. In such a system, the loss of one population can negatively affect other populations.

Metapopulations also lend themselves to modeling efforts, and various programs have been developed for simulating them (Donovan and Welden 2002). In one approach, metapopulation dynamics are simulated by using PVA combined with dispersal rates and spatial information on multiple populations. A typical population study targets one or several populations, but a metapopulation study may produce a more accurate portrayal of the species because it includes the dispersal between populations. Metapopulation models allow biologists to consider the impacts of founder effects, genetic drift, and gene flow on the species.

Bighorn sheep (*Ovis canadensis*) in the desert of southeastern California offer a well-studied example of metapopulation

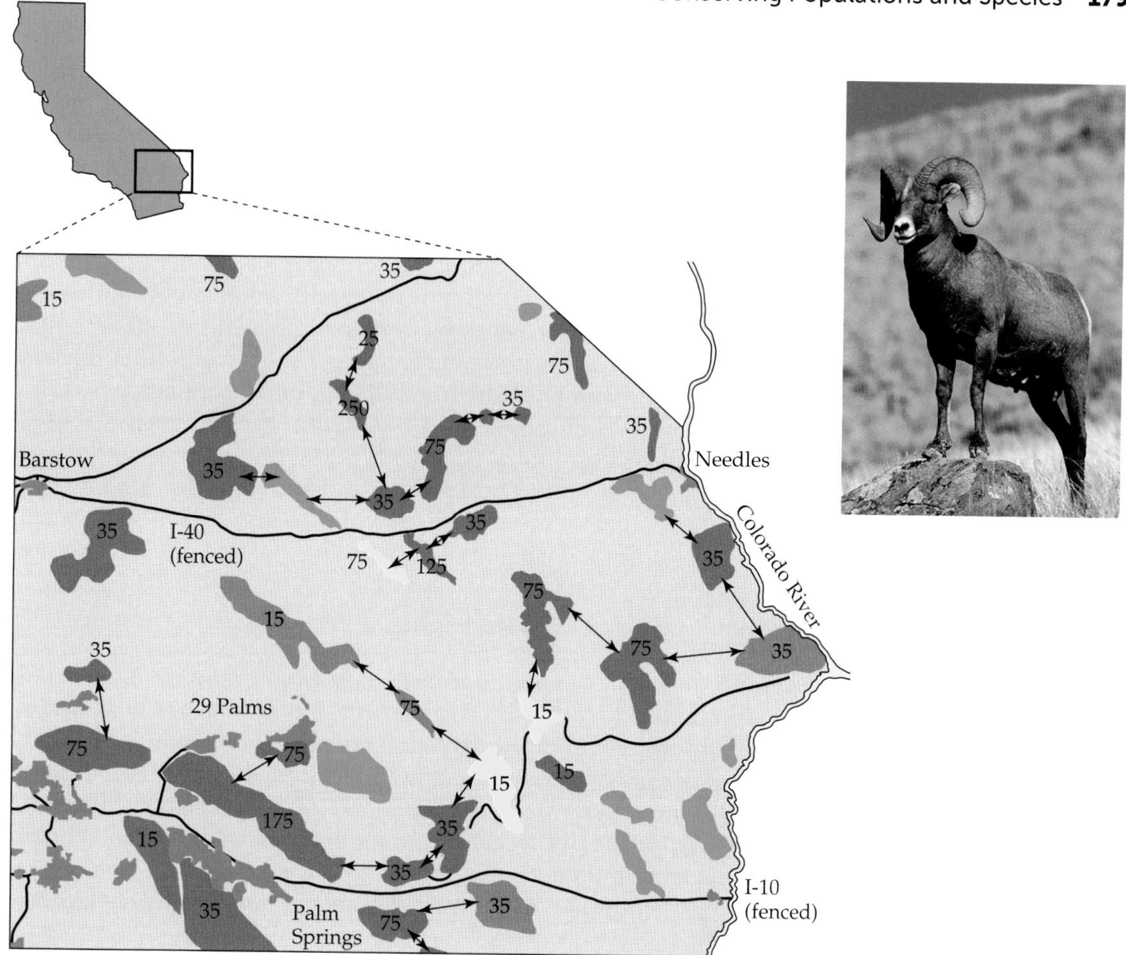

FIGURE 6.6 Bighorn sheep (*Ovis canadensis*) in the southeastern California desert are an example of a metapopulation. The species has permanently occupied the mountain ranges shown in red, with populations of the sizes indicated. Mountain ranges shown in orange do not currently have permanent mountain sheep populations, though they may have been occupied in the past. The species has been reintroduced into the mountain ranges shown in purple; yellow indicates areas where natural recolonization has occurred in the past 15 years. Arrows indicate observed sheep migrations. Human settlements, major highways, and canals—all of which are barriers to the animals' movement—are shown in green or as solid black lines. (After Epps et al. 2007; photograph © Paul Tessier/istock.)

dynamics, exhibiting the shifting mosaic of populations that defines a metapopulation. These sheep have been observed dispersing from occupied mountain ranges to nearby unoccupied mountain ranges, forming new populations in the process (**Figure 6.6**). Migration occurs primarily between populations less than 15 km apart and is greater when the intervening countryside is more hilly (Epps et al. 2005, 2007). Human-made barriers such as highways, irrigation canals, and urban areas almost completely eliminate movement between populations. Maintaining dispersal routes between existing population areas and potentially suitable sites is important in managing this species.

In a metapopulation, destruction of the habitat of one central, core population might result in the extinction of numerous smaller populations that depend on the core population for periodic colonization (Gutiérrez 2005). Habitat fragmentation resulting from roads, fences, and other human activities sometimes has the effect of changing a large, continuous population into a metapopulation in which small, temporary populations occupy habitat fragments. When population size within each fragment is small and the rate of migration among fragments is low, populations within each fragment gradually go extinct and recolonization does not occur. Effective management of a species often requires an understanding of these metapopulation dynamics and a restoration of lost habitat and dispersal routes.

Conservation Categories

Organizing the vast amounts of information from population-, species-, and ecosystem-level studies is an expensive, labor-intensive activity, but it is a crucial component of conservation efforts. It is imperative to know what species are in danger of extinction and where they occur, in order to protect them (**Figure 6.7**) (Mace and Baillie 2007). The **International Union for Conservation of Nature (IUCN)** has designated the status of rare and endan-

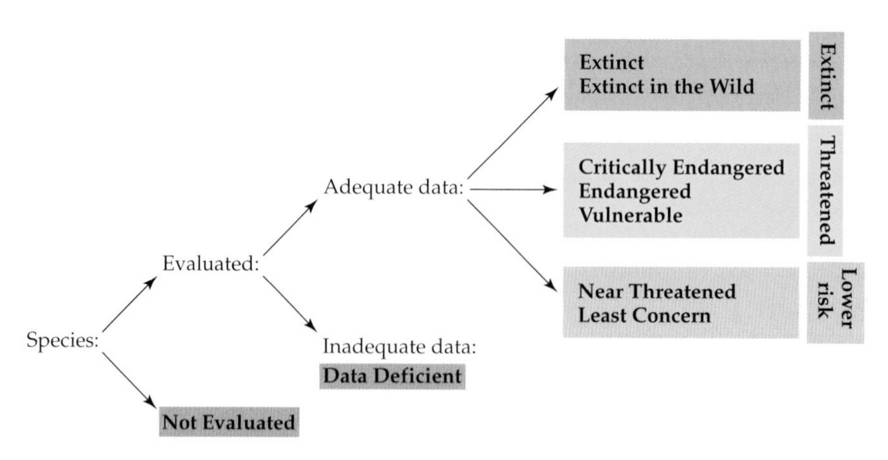

FIGURE 6.7 The IUCN categories of conservation status. This chart shows the distribution of the categories. Reading from left to right, they depend on (1) whether a species has been evaluated or not and (2) how much information is available for the species. If data are available, the species is then put into a category of lower risk, threatened, or extinct. (After IUCN 2001.)

gered species using an internationally accepted standard of conservation categories (www.iucn.org):

- *Extinct (EX)*. The species (or other taxon, such as subspecies or variety) is no longer known to exist. The IUCN currently lists 717 animal species and 87 plants species as extinct.
- *Extinct in the wild (EW)*. The species exists only in cultivation, in captivity, or as a naturalized population well outside its original range. The IUCN currently lists 37 animal species and 28 plant species as extinct in the wild.
- *Critically endangered (CR)*. These species have an extremely high risk of going extinct in the wild, according to any of the criteria A to E in Table 6.1.
- *Endangered (EN)*. These species have a very high risk of extinction in the wild, according to any of the criteria A to E.
- *Vulnerable (VU)*. These species have a high risk of extinction in the wild, according to any of the criteria A to E.
- *Near threatened (NT)*. The species is close to qualifying for a threatened category (see Figure 6.7) but is not currently considered threatened.
- *Least concern (LC)*. The species is not considered near threatened or threatened. (Widespread and abundant species are included in this category.)
- *Data deficient (DD)*. Inadequate information exists to determine the risk of extinction for the species.
- *Not evaluated (NE)*. The species has not yet been evaluated against the Red List criteria.

The IUCN uses quantitative information, including the area occupied by the species and the number of mature individuals presently alive, to assign species to conservation categories.

Species in the critically endangered, endangered, and vulnerable categories are considered **threatened** with extinction. For these three categories, the IUCN has developed quantitative measures of threat based on the probability of extinction. These **Red List criteria**, described in **Table 6.1**, are based on the developing methods of population viability analysis. These criteria focus on population trends and habitat conditions. The advantage of this system is that it provides a standard method of classification by which decisions can be reviewed and evaluated according to accepted quantitative criteria, using whatever information is available.

The IUCN system has been used to identify Red Lists of threatened species and to determine whether species are responding to conservation efforts.

Using habitat loss as a criterion in assigning categories is particularly useful for many species that are poorly known biologically, because species can be listed as threatened if their habitat is being destroyed even if scientists know little else about them. In practice, a species is most commonly assigned to an IUCN category based on the area it occupies, the number of mature individuals it has, or the rate of decline of the habitat or population; the probability of extinction is least commonly used (van Swaay et al. 2011).

TABLE 6.1 IUCN Red List Criteria for the Assignment of Conservation Categories

Red List criteria A–E	Quantification of criteria for Red List category "critically endangered"[a]
A. Observable reduction in numbers of individuals	The population has declined by 80% or more over the last 10 years or 3 generations (whichever is longer), either based on direct observation or inferred from factors such as levels of exploitation, threats from introduced species and disease, or habitat destruction and/or degradation
B. Total geographical area occupied by the species	The species has a restricted range (<100 km^2 at a single location) *and* there is observed or predicted habitat loss, fragmentation, ecological imbalance, or heavy commercial exploitation
C. Predicted decline in number of individuals	The total population size is less than 250 mature, breeding individuals and is expected to decline by 25% or more within 3 years or 1 generation
D. Number of mature individuals currently alive	The population size is less than 50 mature individuals
E. Probability the species will go extinct within a certain number of years or generations	Extinction probability is greater than 50% within 10 years or 3 generations

[a]A species that meets the described quantities for *any one* of criteria A–E may be classified as critically endangered. Similar quantification for the Red List categories "endangered" and "vulnerable" can be found at www.iucnredlist.org.

Using criteria in Table 6.1 and the categories in Figure 6.7, the IUCN has evaluated and described the threats to plant and animal species in its series of **Red Data Books** and **Red Lists** of threatened species; these detailed lists of endangered species by group and by country are available at www.iucn.org. Species listed as threatened include 1462 of 5499 described mammal species, 2097 of 10,052 bird species, and 2303 of 6338 amphibian species (see Table 5.1).

Although the IUCN evaluations have included numerous species of fish (2084), reptiles (1018), mollusks (2197), insects (1259), crustaceans (1735), and plants (12,041), they are still not extensive enough. Most bird, amphibian, and mammal species have been evaluated using the IUCN system, but the levels of evaluation are lower for reptiles, fish, and flowering plants. The evaluations of insects and other invertebrates, mosses, algae, fungi, and microorganisms are even less adequate despite their importance to ecosystem health and human well-being (Régnier et al. 2009). A recent survey of dragonflies suggests that about 10% to 15% can be considered threatened (Clausnitzer et al. 2009). Evaluating a greater number of species using the IUCN system is an urgent priority.

By tracking the conservation status of species over time, it is possible to determine whether species are responding to conservation efforts or are continuing to be threatened (Butchart and Bird 2010). For example, the Red List Index has been used to demonstrate that the conservation status of certain animal groups has continued to decline during the past two to three decades. However, the rates of extinction of critically endangered species have been lower than predicted, most likely because of the positive effects of conservation interventions. Another measure, the Living Planet Index, follows population sizes for 1686 vertebrate species from 1970 to 2007; this index showed sharp population declines in tropical regions and moderate increases in temperate regions (wwf.panda.org).

In Switzerland, a different approach is being used to identify those threatened (or Red List) species that are responding to conservation efforts (Gigon et al. 2000). The 317 species that have stable populations or are increasing in abundance as of 2007 are listed in a Blue List. The Blue List highlights successful conservation efforts and suggests further projects that might succeed (www.bluelists.ethz.ch). While the Blue List approach has not been widely adopted, it remains an important concept in showing the way forward.

A program similar to the efforts of the IUCN is the NatureServe network of Natural Heritage programs that covers all 50 U.S. states, 3 provinces in Canada, and 14 Latin American countries (www.natureserve.org/explorer). This network, strongly supported by The Nature Conservancy, gathers, organizes, and manages information on the occurrence of "elements of conservation interest"—more than 64,000 species, subspecies, and biological communities (in addition to half a million carefully mapped populations) (**Figure 6.8**) (De Grammont and Cuarón 2006). Elements are given status

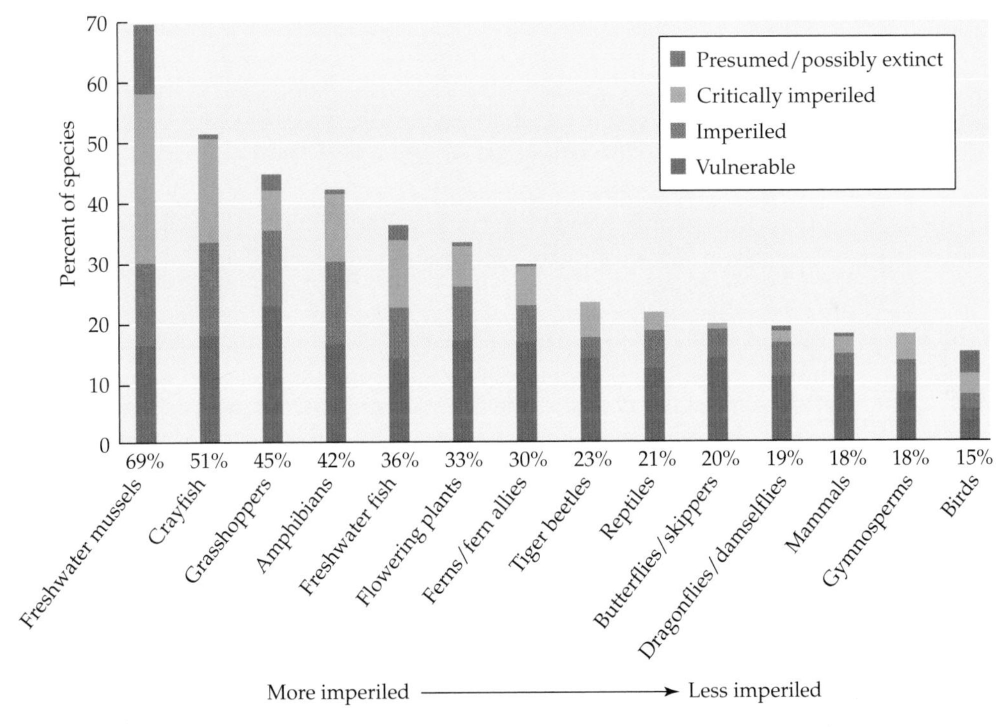

FIGURE 6.8 Some species groups from the United States ranked as presumed extinct, critically imperiled, imperiled, or vulnerable according to criteria endorsed by The Nature Conservancy and coordinated by NatureServe. The groups are arranged with those at greatest risk on the left. Freshwater species are at greater risk of extinction than terrestrial species. (After Wilcove and Master 2005.)

ranks based on a series of standard criteria: number of remaining populations or occurrences, number of individuals remaining (for species) or extent of area (for communities), number of protected sites, degree of threat, and innate vulnerability of the species or community. When methods of the Natural Heritage system and the IUCN categories are used to evaluate the same species, the rankings of threat are quite similar.

Legal Protection of Species

Once conservation biologists have identified a species as needing protection, laws can be passed and treaties can be signed to implement conservation efforts. National laws protect species within individual countries; international agreements provide a broader framework for conservation.

National laws

National governments protect designated endangered species within their borders, establish national parks, and enforce legislation on environmental protection.

People in many countries recognize that preserving a healthy environment and protecting species are linked to sustaining human health. National governments and national conservation organizations in such countries acknowledge this and play an important role in the protection of all levels of biological diversity. Laws are passed to establish national parks and other protected areas; to regulate activities such as fishing, logging, and grazing; and to limit air and water pollution. International treaties that restrict trade in endangered animals are implemented at the national level and enforced at borders. The true measure of a nation's commitment to protecting biodiversity is the effectiveness with which these laws are enforced.

Countries in the European Union carry out endangered species conservation through domestic enforcement of international agreements such as CITES (the Convention on International Trade in Endangered Species) and the Ramsar Convention on Wetlands. Endangered species on International Red Lists prepared by the IUCN are protected through legislation (Fontaine et al. 2007). In addition, the Fauna Europaea database provides information, critical to conservation efforts, on the distribution and abundance of 130,000 terrestrial and freshwater species; this information is used to establish national Red Data Books of endangered species that are protected by law and the target of conservation efforts. Some countries may have additional laws, such as the Wildlife and Countryside Act of 1981 in the United Kingdom, which protects habitat occupied by endangered species.

Many of the factors described so far contributed to the recovery of green turtle (*Chelonia mydas*) populations at Tortuguero Beach, on the Caribbean coast of Costa Rica (**Figure 6.9**). Following decades of overcollection of sea turtle eggs and adults, the Costa Rican government undertook a series of actions to protect this endangered species. First, the government banned the

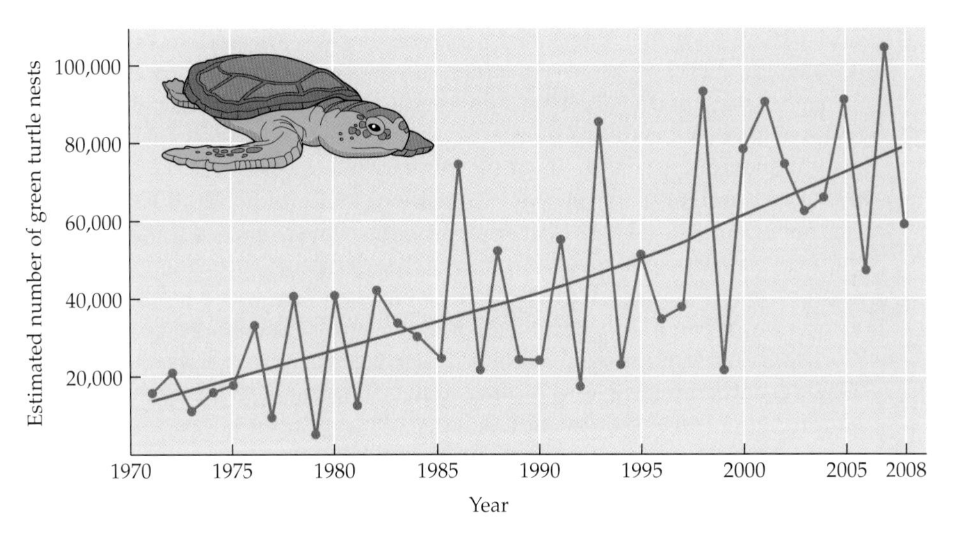

FIGURE 6.9 Greater numbers of green turtles have been nesting on the beach in Tortuguero in Costa Rica since a series of protective measures implemented by the government started in 1963. Nest counts (dots) are variable from year to year. The red curve tracks the general trend of increase in numbers. (After Troëng and Rankin 2005, with updates provided by Caribbean Conservation Corporation.)

collecting of eggs and adults at Tortuguero Beach in 1963. Then, it stopped exports of turtle products in 1969. In 1970, the government established Tortuguero National Park to protect the whole area. Protection has gradually been extended by a ban on turtle fishing and the recognition of how valuable nesting turtles are to the tourist industry. Nicaragua and neighboring countries have signed the CITES treaty and are also implementing protection measures. As a result, the nesting sea turtle population has more than tripled over the last 35 years (Troëng and Rankin 2005).

Despite the fact that many countries have enacted legislation to preserve biodiversity, it also is true that national governments are sometimes unresponsive to requests from conservation groups to protect the environment. In some cases national governments have acted to decentralize decision making, relinquishing control of natural resources and protected areas to local governments, village councils, and conservation organizations.

The U.S. Endangered Species Act

In the United States, the principal conservation law protecting species is the **Endangered Species Act (ESA)**, enacted by the U.S. Congress in 1973 to "provide a means whereby the ecosystems upon which endangered species and threatened species depend may be conserved [and] to provide a program for the conservation of such species." Species are protected under the ESA if they are on the official list of endangered and threatened species. In addition, a recovery plan is generally required for each listed species. As defined by the ESA, "endangered species" are those likely to become extinct as a result of human activities and/or natural causes in all or a significant portion of their range; "threatened species" are those likely to become endangered in the near future. Since 1973, more than 1300 U.S. species have been added to the list, including many well-known species such as the whooping crane (*Grus americana*) and the manatee (*Trichechus manatus*). In addition 576 endangered species from elsewhere in the world that face special restrictions when they are imported into the United States are also on the list.

The ESA prohibits all U.S. government agencies from activities, such as logging, cattle grazing, and mining, that will harm listed species and their habitats—a critical feature, since many of the threats to species occur on federal lands. By protecting habitats, the ESA in effect uses listed species as indicator species to protect entire biological communities and the thousands of species that they contain. The ESA also prevents private individuals, businesses, and local governments from harming or "taking" listed species—the definition of which can include damaging their habitat—and prohibits all trade in listed species. The restrictions on private land are important to species recovery because about 10% of endangered species are found exclusively on private land. Although the ESA provides legal recourse, obtaining the goodwill and cooperation of private landowners is important for recovery efforts (Wilcove and Lee 2004).

Many species are listed under the ESA only when they have fewer than 100 individuals remaining, making recovery difficult (see Chapter 5). An early listing of a declining species might allow it to recover and thus become a candidate for removal from the list sooner than if authorities wait for its status to worsen before adding it to the list. The great majority of U.S. species listed under the ESA are plants (797 species) and vertebrates (over 386 species), despite the fact that most of the world's species are insects and other invertebrates. Clearly, greater efforts must be made to study the lesser-known and underappreciated invertebrate groups, especially freshwater species, and to extend listing to those endangered species whenever necessary.

The ESA has sometimes become a source of contention between conservationists and some business interests in the United States. The protection afforded to species listed under the ESA is so strong that business interests and landowners often lobby strenuously against the listing of species in their area. At the extreme are landowners who destroy endangered species and their potential habitat on their property to evade the provisions of the ESA, a practice informally known as "shoot, shovel, and shut up." Such was the fate of a quarter of the sites that contained habitat suitable for the threatened Preble's meadow jumping mouse (*Zapus hudsonius preblei*) that lives in streamside habitats in Colorado and Wyoming (Brook et al. 2003). Clearly, landowners must be compensated in some

The U.S. Endangered Species Act mandates such strong protection for species that conservation and business groups often agree to compromises that allow some species protection along with limited development.

way for losses incurred because of the protection of a species or its habitat, and they must be encouraged publicly to support the provisions of the ESA.

Wildlife biologists and government officials have also confronted the difficulty of implementing the recovery of listed species—rehabilitating species or reducing the threats to species to the point where they can be removed from listing under the ESA, or "delisted." So far, only about 20 of more than 1300 listed U.S. species have been delisted, though another 20 species have shown enough recovery to be recategorized from endangered to threatened (Schwartz 2008). The most notable successes include the brown pelican (*Pelecanus occidentalis*), the American peregrine falcon (*Falco peregrinus*), and the American alligator (*Alligator mississippienis*). In 2007, the bald eagle (*Haliaeetus leucocephalus*) was removed from the federal list of threatened and endangered species because its numbers in the lower 48 states had increased from 400 breeding pairs in the 1960s to the current 7000 pairs. Overall, most listed species are still declining in range and abundance, and most surprising, for around 20% of listed species there are insufficient data to determine whether their populations are changing over time (Leidner and Neel 2011). Due to their low numbers and consequent vulnerability, there is now recognition that even species that are candidates for delisting will still require some degree of conservation management to maintain their populations (**Figure 6.10**) (Redford et al. 2011).

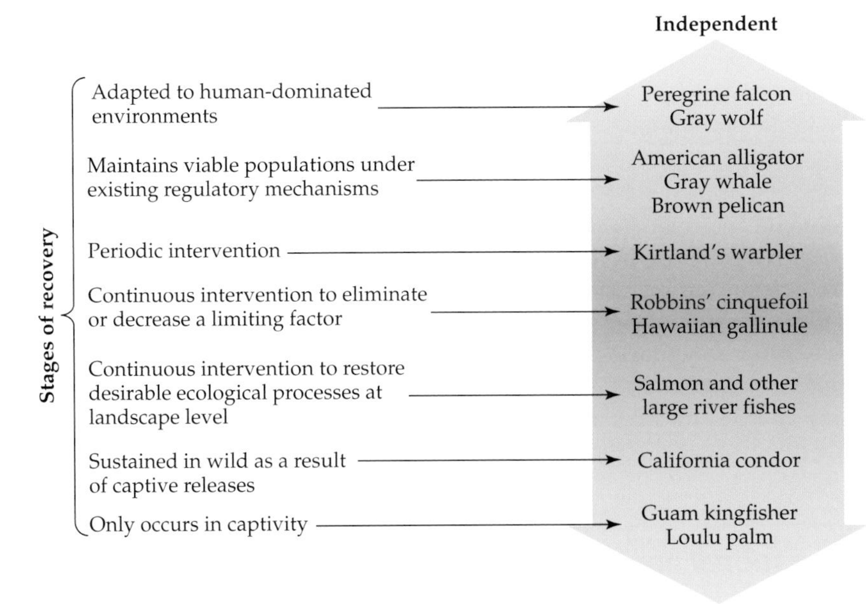

FIGURE 6.10 Endangered species often require active management and intervention as part of the recovery process. There is a continuum, with some species independent of humans and others dependent on human intervention. (After Scott et al. 2005.)

The difficulty of implementing recovery plans for so many species is often not primarily biological but, rather, political, administrative, and ultimately financial (Briggs 2009). For example, an endangered river clam species might need to be protected from pollution and the effects of an existing dam. Installing sewage treatment facilities and removing a dam are theoretically straightforward actions, but they are expensive and difficult to carry out in practice. The U.S. Fish and Wildlife Service (FWS), the principal agency responsible for preserving species and habitats under the ESA, would need to double its budget to create and implement a program that would effect recovery of all listed species (Taylor et al. 2005). The cost would be even higher if private landowners were given financial compensation for ESA-imposed restrictions on the use of their property.

Funding for the ESA has grown steadily over the past 20 years, but the number of species protected under the act has grown even faster. As a result, there is less money available per species. The importance of adequate funding for species recovery is shown by a study demonstrating that species receiving higher proportions of the funding requested for their recovery plans are more likely to reach stable or improved status than species that receive lower proportions of funding (Miller et al. 2002).

Concerns about the implications of ESA protection have frequently forced business organizations, conservation groups, and governments to develop compromises that reconcile both conservation and business interests. To provide a legal mechanism to achieve this goal, Congress amended the ESA in 1982 to allow the design of **habitat conservation plans** (**HCPs**). HCPs are regional plans that allow development in designated areas but also protect remnants of ecosystems that contain groups of actual or potentially endangered species. These plans are drawn up by the concerned parties—developers, conservation groups, and local governments—and given final approval by the FWS. About 650 HCPs covering more than 16 million ha and over 500 species have been approved so far. In one case, an innovative program in Riverside County, California, allows developers to build within the historical range of the endangered Stephen's kangaroo rat (*Dipodomys stephensi*) if they contribute to a fund that will be used to buy wildlife sanctuaries. While HCPs are not perfect, they are attempts to create the next generation of conservation planning. They seek to protect many species, entire ecosystems, or whole communities; extend over a wide geographical region; and include many projects, landowners, and jurisdictions.

International agreements

International agreements to protect species and other aspects of biodiversity and ecosystems are needed for several reasons: species migrate across borders, there is international trade in biological products, the benefits of biodiversity are of international importance, and the threats to biodiversity are often international in scope. International agreements have provided a framework for countries to cooperate in protecting species, ecosystems, and genetic variation. Treaties are negotiated at international conferences and come into force when they are ratified by a certain number of countries, often under the authority of international bodies, such as the United Nations Environment Programme (UNEP), the Food and Agriculture Organization of the United Nations (FAO), and the IUCN.

One of the most important treaties protecting species at an international level is the **Convention on International Trade in Endangered Species (CITES)**, established in 1973 in association with UNEP (Smith et al. 2011; www.cites.org). The treaty has currently been ratified by 175 countries. CITES establishes lists (known as Appendices) of species for which international trade is to be controlled or monitored. Member countries agree to restrict trade in and destructive exploitation of these species. Regulated plants include horticultural species such as orchids, cycads, cacti, carnivorous plants, and tree ferns; timber species and wild-collected seeds are increasingly being considered for regulation as well. For animals, closely regulated groups include parrots, large cats, whales, sea turtles, birds of prey, rhinos, bears, and primates. Species collected for the pet, zoo, and aquarium trades and species harvested for their fur or other commercial products are closely monitored also.

FIGURE 6.11 A customs official shows wildlife products seized at the U.S. border. For some products, such as the sea turtle, the type of animal involved can be easy to identify, but for other products, such as bags and shoes, the type of animal used to make them may be hard to determine. (Photograph by John and Karen Hollingsworth, courtesy of U.S. Fish and Wildlife Service.)

International treaties such as CITES are implemented when a country ratifies the treaty and passes laws to enforce it. Nongovernmental organizations such as the IUCN specialist groups, the World Wildlife Fund TRAFFIC network, and the UNEP World Conservation Monitoring Centre provide technical advice regarding legal and enforcement aspects of CITES to national governments. Countries may also protect species listed by national Red Data books. Once species protection laws are passed within a country, then police, customs inspectors, wildlife officers, and other governmental agents can arrest and prosecute individuals possessing or trading in protected species, and they can seize the products or organisms involved (**Figure 6.11**). One notable success of CITES was been a global ban on the ivory trade, implemented after poaching caused severe declines in African elephant

The Convention on International Trade in Endangered Species (CITES) has developed extensive lists of species for which trade is prohibited, controlled or monitored. Many countries have established their own Red Data list of species that they protect within their borders.

populations (Wasser et al. 2010). Since the initial ban, limited commercial trade in ivory has been permitted in a few African countries with healthy and expanding elephant populations.

Another key treaty is the Convention on the Conservation of Migratory Species of Wild Animals, often referred to as the Bonn Convention, signed by 116 countries, with a primary focus on bird species (www.cms.int). This convention complements CITES by encouraging international efforts to conserve bird species that migrate across international borders and by emphasizing regional approaches to research, management, and hunting regulations. The convention now includes protection of bats and their habitats and of cetaceans in the coastal zones and seas around Europe. Several other important international agreements protect species, including the following:

- Convention on the Conservation of Antarctic Marine Living Resources (www.ccamlr.org)
- International Convention for the Regulation of Whaling, which established the International Whaling Commission (www.iwcoffice.org)
- International Convention for the Protection of Birds, and the Benelux (Belgium/Netherlands/Luxembourg) Convention Concerning Hunting and the Protection of Birds
- Convention for the Conservation and Management of Highly Migratory Fish Stocks in the Western and Central Pacific Ocean (www.wcpfc.int)
- Additional agreements protecting specific groups of animals, such as prawns, lobsters, crabs, fur seals, Antarctic seals, salmon, and vicuñas

A number of international agreements with broader focuses are also increasingly seeking direct protection of endangered species. The Convention on Biological Diversity, described in Chapter 9, for example, now includes recommendations for the protection of IUCN Red Listed species (www.iucnredlist.org).

A weakness of all these international treaties is that they operate through consensus, so necessary strong measures often are not adopted if one or more countries are opposed to the measures. Also, any nation's participation is voluntary, and countries can ignore these conventions to pursue their own interests when they find the conditions of compliance too difficult (Carraro et al. 2006). This flaw was highlighted when several countries decided not to comply with the International Whaling Commission's 1986 ban on whale hunting, and the Japanese government announced its fleet would continue hunting whales under the dubious claim that further data were needed to evaluate the status of whale populations. There is frequently no monitoring mechanism in place to determine whether countries are even enforcing the treaties. Persuasion and public pressure are the principal means used to induce countries to enforce treaty provisions and prosecute violators.

Establishing New Populations

An important approach to assisting the recovery of endangered species involves establishing new wild and semiwild populations of rare and endangered species and increasing the size of existing populations (Armstrong and Seddon 2008). Such establishment programs offer the hope that species now living only in captivity or small wild populations can regain their ecological and evolutionary roles within their biological communities. Populations in the wild may have less chance of being destroyed by natural or human-caused catastrophes (such as epidemics or wars) than confined captive populations. Furthermore, simply increasing the number and size of its populations generally lowers the probability a species will go extinct.

Establishment programs are unlikely to be effective, however, unless the factors leading to the decline of the original wild populations are clearly understood and eliminated, or at least controlled (Houston et al. 2007). For example, the peregrine falcon declined throughout its range because of the harmful effects of DDT. Before new populations could be established, DDT first had to be banned. Then, starting with birds raised in captivity, peregrine falcons were released within their former range, and they are now recovering across North America, with notable increases in cities.

Three basic approaches, all involving relocation of existing captive-bred or wild-collected individuals, have been used to establish new populations of animals and plants and to enlarge existing populations. Many of these are coordinated by the IUCN's Re-introduction Specialist Group (www.iucnsscrsg.org):

- A **restocking program** involves releasing individuals into an existing population to increase its size and gene pool. These released individuals may have been raised in captivity or may be wild individuals collected elsewhere.
- A **reintroduction program*** involves releasing individuals into an ecologically suitable site within their historical range where the species no longer occurs.
- An **introduction program** involves moving individuals to areas outside their historical range but suitable for the species.

Establishing new populations of endangered species can benefit the species itself, other species, and the ecosystem. Such programs must identify and attempt to eliminate the factors that led to the original population's decline.

The main objectives of a reintroduction program are to create a new population in its original environment and to help restore a damaged ecosystem. For example, a program initiated in 1995 to reintroduce gray wolves into Yellowstone National Park aims to restore the equilibrium of

*Unfortunately, some confusion exists about the terms denoting the establishment of populations. Restocking programs are sometimes called "augmentations." Reintroduction programs sometimes are called "reestablishments" or "restorations." Another term, "translocation," usually refers to moving individuals from a location where they are about to be destroyed to another site that, hopefully, provides a greater degree of protection.

(A)

(B)

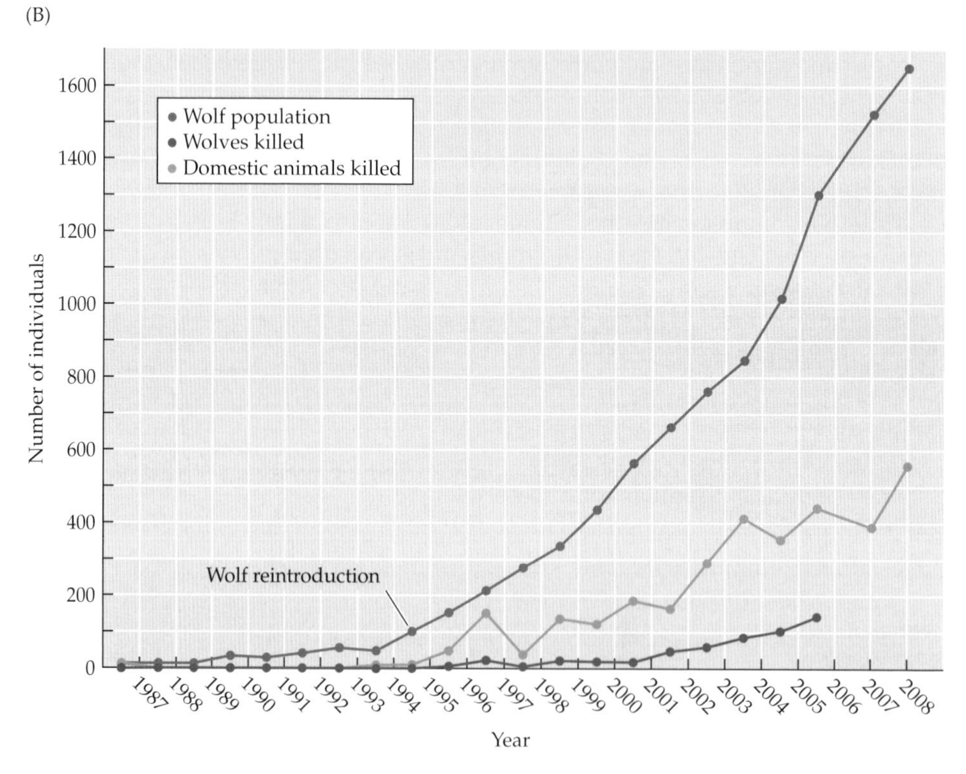

FIGURE 6.12 (A) A gray wolf (*Canis lupus*) in Yellowstone National Park wears a radio transmitter collar that allows researchers to follow its movements. (B) The number of wolves in Wyoming, Idaho, and Montana increased following the reintroduction of wolves to the Yellowstone area in 1995. There has also been an increase in the number of domestic animals, primarily sheep, killed by wolves and an increase in the number of problem wolves killed by government authorities. (A, photograph courtesy of William Campbell/U.S. Fish and Wildlife Service; B, after Musiani et al. 2003, with updates courtesy of M. Musiani and from Clark and Johnson 2009.)

predators, herbivores, and plants that existed prior to human intervention in the region (**Figure 6.12**) (Hamlin et al. 2008). Wild-collected individuals are also sometimes caught and then released elsewhere within the range of their species when a new protected area has been established, when an existing population is under a new threat in its present location, or when natural or artificial barriers to the normal dispersal tendencies of the species exist. If possible, individuals are released near the site where they or their ancestors were collected, to ensure genetic adaptation to their environment.

In contrast to reintroduction, *introduction* to an entirely new location outside of the existing range of a species may be appropriate when the environment within its known historical range has deteriorated to the point where the species can no longer survive there, or when reintroduction is impossible because the factors that caused the original decline are still present. In the near future, introductions of some species may be necessary if they can no longer survive within their current ranges because of a changing climate, especially warming temperatures. This is sometimes referred to as assisted colonization (Loss et al. 2011).

The reintroduction and introduction of a species to an entirely new site must be carefully thought out so that the released species does not damage its new ecosystem or harm populations of any local endangered species (Olden et al. 2011). Care must be taken that released individuals have not acquired any diseases that could spread to and decimate wild populations.

One special method used in establishing new populations and restocking is "head-starting," an approach in which animals are raised in captivity during their vulnerable young stages and then are released into the wild. The release of sea turtle hatchlings produced from eggs collected from the wild and raised in nearby hatcheries is an example of this approach (see Chapter 1).

Considerations for animal programs

Establishing new populations is often expensive and difficult because it requires a serious, long-term commitment. The programs to capture, raise, monitor, and release rare species such as California condors, peregrine falcons, and black-footed ferrets have cost millions of dollars and have required years of work (Grenier et al. 2007). Establishment programs can become highly emotional public issues and are often criticized on many fronts. They may be attacked as a waste of money ("Millions of dollars for a few ugly birds!"), unnecessary ("Why do we need wolves here when there are plenty of them somewhere else?"), intrusive ("We don't want the government telling us what to do!"), poorly run ("Most of the animals died after they were released"), or unethical ("Why can't the last animals just be allowed to live out their lives in peace?"). The answer to all of these criticisms is straightforward. Although not appropriate for every endangered species, a well-run, well-designed captive breeding and establishment program may be the best hope for preservation.

Because of the conflicts and high emotions, it is crucial that establishment programs include local people so that (ideally) the community has a stake in a program's success. (Indeed, this is true of any conservation project.) At a minimum, it is necessary to explain the need for the program and its goals, to convince local people to support it—or at least not to oppose it. In many cases, such programs have considerable educational value (Ausband and Foresman 2007). Programs are often more successful if they provide incentives to affected people rather than impose rigid restrictions and laws. For example, direct payments are made to Wyoming residents whose farm animals and dogs are injured or killed by reintroduced wolves, and the few wolves that repeatedly attack livestock are either killed or moved; these arrangements help the program to retain local support (Haney et al. 2007).

It is imperative that captive-bred mammals and birds learn predator avoidance and species-appropriate social behavior if they are to survive and reproduce after being released into the wild. They may also require some support after release.

To be successful, reestablishment programs need to consider the social organization and behavior of the animals that are being released (Buchholz 2007). When social animals (particularly mammals and some birds) grow up in the wild, they learn about their environment and how to interact socially from other members of their species. Animals raised in captivity may lack the skills needed to survive in their natural environment, as well as the social skills necessary to find food cooperatively, sense danger, find mating partners, raise young, and migrate. To overcome these socialization problems, captive-raised mammals and birds may require training in feeding and predator avoidance both before and after their release into the environment (Nicholson et al. 2007). In some cases, human trainers use puppets or wear costumes to mimic the appearance and behavior of wild individuals so that young animals learn to identify with their own species rather than with humans (**Figure 6.13**). In other cases, wild individuals are used as "instructors" for captive individuals of the same species. Wild golden lion tamarins (*Leontopithecus rosalia*) are caught and held with captive-bred tamarins so that the captive-bred tamarins will learn from the wild ones. After they form social groups, they are released together. Wild-caught, African wild dogs (*Lycaon pictus*) that gave birth and bonded together in holding areas

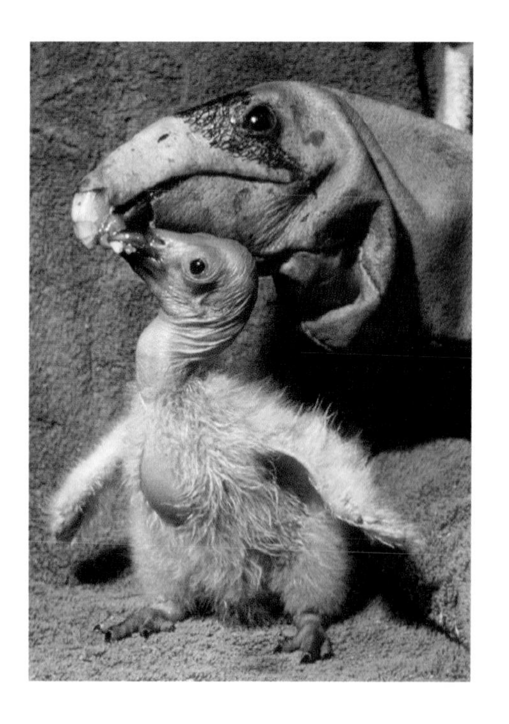

FIGURE 6.13 California condor chicks (*Gymnogyps californianus*) raised in captivity are fed by researchers using puppets that look like adult birds. Conservation biologists hope that minimizing human contact with the birds will improve their chances of survival when they are returned to the wild. (Photograph by Ron Garrison, courtesy of U.S. Fish and Wildlife Service.)

(A)

(B)

FIGURE 6.14 (A) Within a protected range, cages allow black-footed ferrets (*Mustela nigripes*) to experience the environment into which they will eventually be released. The ferrets' caretaker is wearing a mask to reduce the chance of exposing the animals to human diseases. (B) A black-footed ferret raised at the captive colony in Colorado. (A, photograph by M. R. Matchett, courtesy of U.S. Fish and Wildlife Service; B, photograph by Ryan Hagerty/U.S. Fish and Wildlife Service.)

prior to release had a higher success rate after reintroduction in multiple sites in South Africa (Gusset et al. 2008).

Some animal species are much more likely to survive if they receive special care and assistance upon release (Mitchell et al. 2011). This approach is known as **soft release**. Animals may have to be fed and sheltered at the release point until they are able to subsist on their own, or they may need to be caged temporarily at the release point and introduced gradually, as they become familiar with the area (**Figure 6.14**). Social groups abruptly released from captivity without assistance such as food supplementation (**hard release**) may disperse explosively from the protected area, resulting in a failed establishment effort. Intervention may be necessary if animals appear unable to survive, particularly during episodes of drought or low food abundance (Blanco et al. 2011). In every case, a decision has to be made as to whether it is better to give occasional temporary help to the species or to force the individuals to try to survive on their own. It is especially important to analyze the impact of human activities in the release area (including farming and hunting) and to minimize the effects of such activities.

Establishment programs for common bird and mammal game species have always been widespread and have contributed a great deal of knowledge to biologists trying to establish new populations (Fischer and Lindenmayer 2000). Analysis of many of these game release programs shows that success is generally greater for releases in excellent-quality habitat than in poor-quality habitat and when wild-caught rather than captive-reared animals are used. The probability of establishing a new population generally increases with the number of animals being released, up to about 100. The key measure of success of a reintroduction program is the establishment of a self-maintaining population. The success rate of reintroduction projects of endangered mammals, birds, reptiles, amphibians, and fish is generally low, suggesting that improved methods need to be developed (Griffiths and Pavajeav 2008; Germano and Bishop 2009). This low rate of success emphasizes that protecting existing populations of endangered species and improving their habitat are still the highest priority.

Clearly, monitoring such establishment programs is crucial in determining whether the efforts are achieving their stated goals (Armstrong and Seddon 2008; Gusset et al. 2008). Monitoring may need to be carried out over many years, even decades, because many reintroductions that initially appear successful eventually fail. The costs of reintroduction also need to be tracked and published so it can be determined whether reintroduction represents a cost-effective strategy. For example, in the case of wild dogs in South Africa, the initial reintroduction target of nine self-sustaining packs was achieved in half the allotted time of 10 years, but the reintroduction cost 20 times more than conserving existing packs in protected areas (Lindsey et al. 2005). Scientists also need to monitor ecosystem elements to determine the broader impacts of a reintroduction—for example, when a predator is introduced, it is crucial to determine its direct impact on prey species and competing species, as well as its indirect impact on vegetation.

New plant populations

Methods used to establish new populations of rare and endangered plant species are fundamentally different from those used to establish terrestrial vertebrate animal species. Plants cannot actively disperse to new locations the way animals do; instead, seeds are dispersed by agents such as wind, animals, and water. Once a seed lands on the ground, it is unable to move, even if a suitable microsite exists just a few meters away. The immediate microsite is thus crucial for plant survival. If the site is too sunny, too shady, too wet, or too dry, either the seed will not germinate or the resulting young plant will die. Plant ecologists are investigating the effectiveness of site treatments prior to and after planting—such as burning the leaf litter, removing competing vegetation, digging up the ground, and excluding grazing animals—as a means of enhancing population establishment (Donath et al. 2007). They are also investigating the most effective way to establish a species at a new site; seeds can be sown and adult individuals and seedlings can be transplanted into the enhanced site (**Figure 6.15A**). For example, during the introduction of Mead's milkweed (*Asclepias meadii*), a threatened perennial plant of tallgrass prairies in the midwestern

> New plant populations are established by sowing seeds or transplanting seedlings or adults. Site treatments such as burning off or physically removing competing plants are often necessary for success.

(A)

(B)

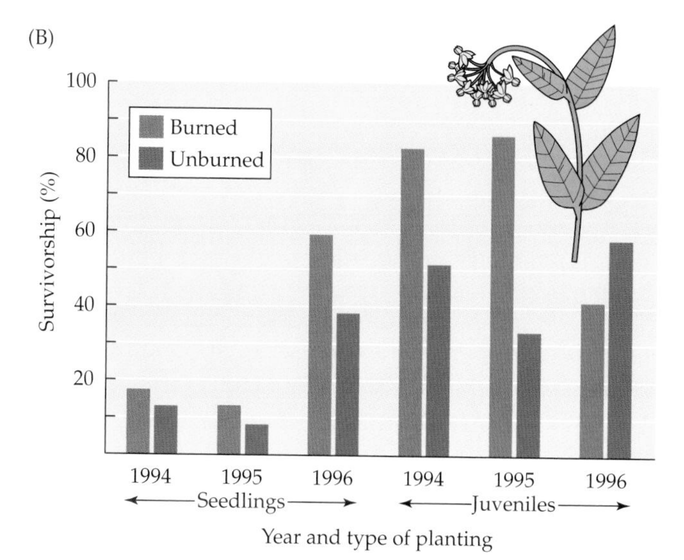

FIGURE 6.15 Several methods are being used to create new populations of rare wildflower species on U.S. Forest Service land in South Carolina. (A) Seeds are being planted in a pine forest from which the oak understory has been burned away. Wire cages will be placed over some plantings to determine whether excluding rabbits, deer, and other animals will help in plant establishment. (B) Introduced seedlings and juvenile plants of Mead's milkweed (*Asclepias meadii*) were evaluated in a reintroduction experiment. Survivorship was greater for juvenile plants than for seedlings and greater in burned habitat than in unburned habitat. Seedling survivorship was greatest in 1996, a year with high rainfall. (A, photograph by R. Primack; B, after Bowles et al. 1998.)

United States (Bell et al. 2003), it was found that survivorship was greater for juvenile plants than for seedlings, and survival was higher in burned habitat than in unburned habitat (**Figure 6.15B**). Seedling survival was also higher in years with greater than average rainfall.

A recent review argues that the success rates of plant reintroduction are lower than the published literature would suggest, as researchers do not publish failed experiments (Godefroid et al. 2011a). Also, many apparently successful reintroductions either failed after additional years or never established a second generation of plants, which is an important indicator of success. As research on this developing topic is published and synthesized, hopefully the chances for successful plant reintroduction will improve. Because of climate change, certain plant species may no longer be genetically suited to their present sites, and conservation biologists may have to look elsewhere in the range of a species for the genotypes to plant (Gray et al. 2011).

The status of new populations

The establishment of new populations raises some novel issues at the intersection of scientific research, conservation efforts, government regulation, and ethics. These issues need to be addressed because reintroduction, introduction, and restocking programs will increase in the coming years as the biodiversity crisis eliminates more species and populations from the wild. In addition, many species may need to be moved if their present ranges become too hot or dry because of global climate change (McLachlan et al. 2007). Many of the reintroduction programs for endangered species are mandated by official recovery plans set up by national governments. For these plans to be formulated and implemented, conservation biologists must be able to explain the benefits of reintroduction programs in a way that government officials and the general public can understand, and they must address the legitimate concerns of those groups (Seddon et al. 2007).

Sometimes stakeholders' concerns can be addressed by giving varying degrees of protection to new populations. For example, populations can be designated "experimental essential" or "experimental nonessential." Experimental essential populations are regarded by the U.S. Endangered Species Act as critical to the survival of the species, and they are as rigidly protected as naturally occurring populations. Experimental nonessential populations have less protection under the law; designating populations as nonessential often helps to overcome the fear of local landowners that having endangered species on their property will restrict how their land can be managed and developed.

Legislators and scientists alike must understand, however, that reintroduction programs are not substitutes for protecting existing wild populations and their habitat. The original populations are more likely to have the most complete gene pools and the most intact interactions with other members of the biological community. In many cases, developers propose creating new habitat or new populations to compensate for what has occurred or is about to occur—habitat damage or the eradication of endangered species populations during development projects. This is generally referred to as **mitigation**. Mitigation is often directed at legally protected species and habitats and includes (1) reduction in the extent of damage, (2) establishment of new populations and habitat as compensation for what is being destroyed, and (3) enhancement of what remains after development. Given the poor success of most attempts to create new populations of endangered

> The establishment of new populations through reintroduction programs in no way reduces the need to protect the original populations of endangered species.

species, protection of existing populations of rare and endangered species should be given the highest priority.

Ex Situ Conservation Strategies

The best strategy for the long-term protection of biodiversity is the preservation of existing species and ecosystems in the wild, known as **in situ**, or on-site, conservation. However, if the last remaining populations of a rare and endangered species are too small to maintain the species, if they are declining despite conservation efforts, or if the remaining individuals are found outside protected areas, then in situ preservation may not be effective. It is likely that the only way to prevent species in such circumstances from going extinct is to maintain individuals in artificial conditions under human supervision (Bowkett 2009). This strategy is known as **ex situ**, or off-site, preservation. Already a number of species that are extinct in the wild survive in captive colonies or reintroduced populations, including Père David's deer (*Elaphurus davidianus*) and Przewalski's horse (*Equus caballus przewalski*). The beautiful Franklin tree (see Figure 5.2) grows only in cultivation and is no longer found in the wild.

Ex situ and in situ conservation are complementary strategies (Zimmermann et al. 2007). The long-term goal of many ex situ conservation programs is the establishment of new populations in the wild, once sufficient numbers of individuals and a suitable habitat are available. Individuals from ex situ populations can periodically be released into the wild to augment in situ conservation efforts (**Figure 6.16**). In situ preservation of species, in turn, is vital to the survival of species that are difficult to maintain in captivity.

> Integrated with efforts to protect existing populations and to establish new populations, ex situ conservation involving zoos, aquariums, and botanical gardens is an important conservation strategy to protect endangered species and educate the public.

Research on captive populations can provide insight into the basic biology, physiology, and genetics of a species through studies that may not be possible to perform on wild animals. Results of these studies can suggest new conservation strategies for in situ populations. Long-term, self-sustaining ex situ populations can also reduce the need to collect individuals from the wild for display and research. Ex situ facilities for animal preservation include zoos, game farms, and aquariums, as well as the facilities of private breeders, while plants are maintained in botanical gardens, arboretums, and seed banks.

Ex situ conservation has several limitations. First, it is not cheap. The cost of maintaining zoos is enormous in comparison with many other conservation activities, such as the budgets of national parks in developing countries; in the United States alone, zoos cost about $1 billion per year to run. Also, ex situ programs protect only one species at a time. In contrast, when a species is preserved in the wild, an entire community of species—perhaps consisting of thousands of species—may be preserved, along with a range of ecosystem services.

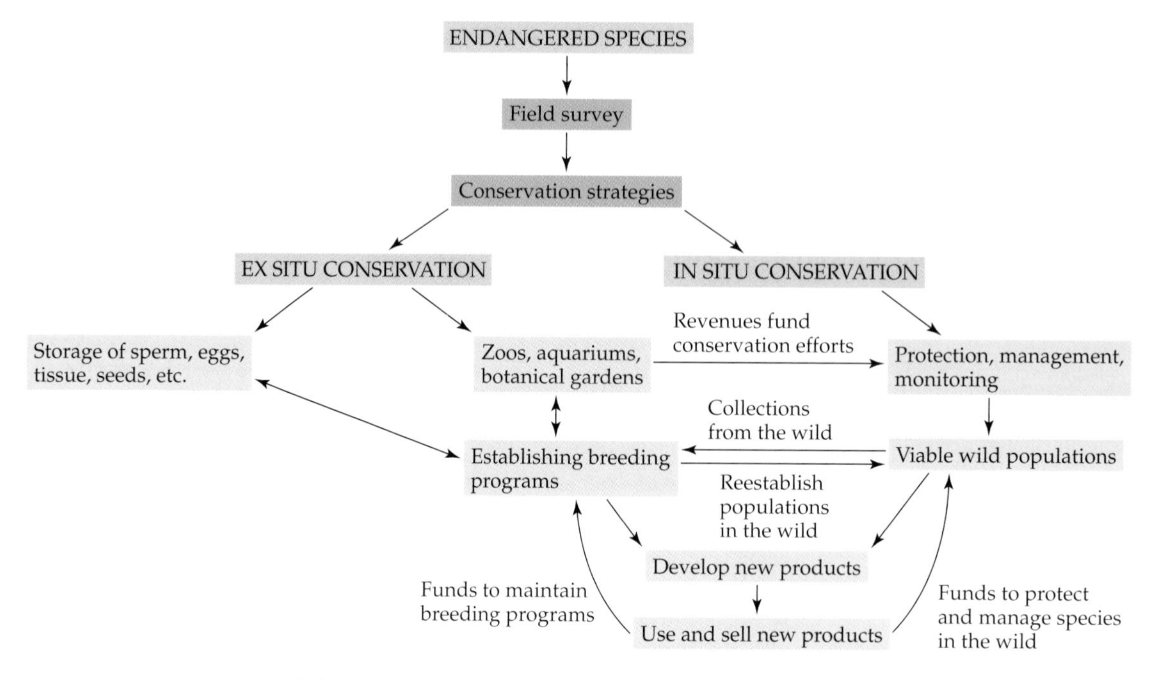

FIGURE 6.16 This model shows ways in which in situ (on-site) and ex situ (off-site) conservation efforts can benefit each other and provide alternative conservation strategies. While no species conforms exactly to this idealized model, the giant panda program has many of its elements. (After Maxted 2001.)

Zoos

Zoos, along with affiliated universities, government wildlife departments, and conservation organizations, presently maintain over 500,000 terrestrial vertebrate individuals, representing almost 8000 species and subspecies of mammals, birds, reptiles, and amphibians (**Table 6.2**). While this number of captive animals may seem impressive, it is trivial in comparison with the number of domestic cats, dogs, and fish kept by people as pets. Zoos could establish breeding colonies of even more species if they directed more of their efforts to smaller-bodied species such as insects, amphibians, and reptiles, which are less expensive to maintain in large numbers than large-bodied mammals. A better balance must be reached between displaying large animals that draw many visitors and displaying smaller, less-well-known animals that appeal less to the public but represent a greater proportion of the world's biodiversity. Zoos can effectively work together to conserve some of these smaller species. For instance, seven North American zoos are joining with universities and the Defenders of Wildlife to form breeding colonies of endangered amphibians as part of the Panama Amphibian Rescue and Conservation Project (www.aza.org).

TABLE 6.2 Number of Terrestrial Vertebrates Currently Maintained in Zoos, according to the International Species Information System (ISIS)

Location	Mammals	Birds	Reptiles	Amphibians	Total
Europe	93,482	109,903	26,778	13,661	243,824
North America	54,393	57,668	29,967	25,208	167,236
Central America	11,630	4175	1195	65	17,065
South America	2372	3927	1682	177	8158
Asia	8437	22,624	3637	529	35,137
Australasia	6266	8629	3188	1288	19,371
Africa	6235	15,018	1278	293	22,824
Totals					
All species	182,725	221,944	67,725	41,221	513,615
Number of taxa[a]	2238	3753	969	544	7486
Percent wild-born[c]	5%	9%	15%	5%	
Rare species[b]	59,030	37,748	22,474	3398	122,650
Number of taxa[a]	527	344	207	29	1107
Percent wild-born[c]	7%	9%	18%	7%	

Source: Data from ISIS as of February 2009, provided by Laurie Bingaman Lackey.

[a]The number of taxa is not exactly equivalent to species, because many species have more than one subspecies listed.

[b]Rare species are those covered by CITES (the Convention on International Trade in Endangered Species).

[c]The percentage of individuals born in the wild is approximate (particularly for reptiles and amphibians), since the origin of the animals is often not reported.

Ex situ conservation efforts have been increasingly directed at saving endangered species of invertebrates as well, including butterflies, beetles, dragonflies, spiders, and mollusks. A current goal of most major zoos is to establish viable, long-term captive breeding populations of rare and endangered animals (Zimmermann et al. 2008). Other important targets for ex situ conservation efforts are rare breeds of domestic animals on which human societies depend for animal protein, dairy products, leather, wool, agricultural labor, transport, and recreation (Ruane 2000). Secure populations of these breeds are a potential genetic resource for the improvement and long-term health of our supplies of pigs, cattle, chickens, sheep, and other domestic animals.

Zoos traditionally focus on maintaining large vertebrates—especially mammals and birds—because these species are of greatest interest to the general public, whose entrance fees fund zoo budgets. Ex situ animals serve as "ambassadors" for the plight of their free-ranging counterparts. In fact, over 90% of families enjoy seeing biodiversity at zoos and aquariums and believe these places teach children about protecting species and habitat (www.aza.org). Such charismatic species influence public opinion favorably toward conservation. As part of the World Zoo Conservation Strategy, which seeks to link zoo programs with conservation efforts in the wild, the

(A)

(B)

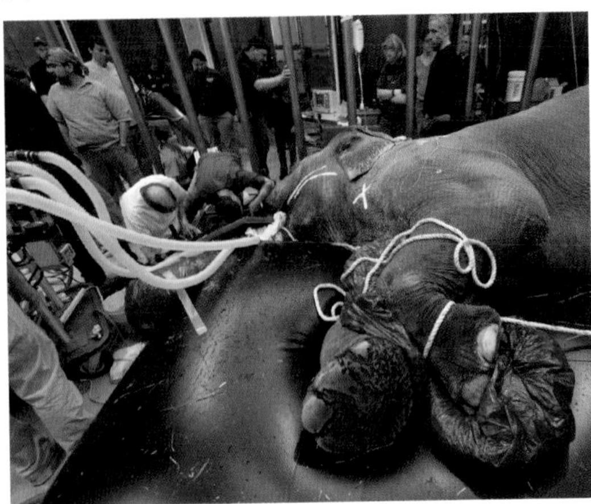

FIGURE 6.17 (A) Visitors to the Asahiyama Zoo on Japan's northern island of Hokkaido enjoy a parade of king penguins (*Aptenodytes patagonicus*). Zoos can serve to educate the public about the need to protect wildlife as well as providing research facilities for in situ species preservation. (B) Veterinarians carry out dental surgery on a captive Asian elephant. The knowledge gained can then potentially assist in helping the species in the wild. (A, Photograph © JTB Photo Communications, Inc./Alamy; B © Richard Clement/Zuma Press.)

world's 2000 zoos and aquariums are increasingly incorporating ecological themes and information about the threats to endangered species in their public displays and research programs. Zoos have an important role to play in educating the public about the need to protect endangered species in the wild. The number of people visiting zoos is enormous; in the United States alone, over 120 million people visit zoos every year; worldwide, 600 million people annually visit zoos (**Figure 6.17A**). The funds raised by zoos from visitor fees and other programs can also directly fund in situ conservation activities.

Zoos have the necessary knowledge and experience in animal care, veterinary medicine, animal behavior, reproductive biology, and genetics to establish captive animal populations of endangered species (**Figure 6.17B**) (Zimmerman et al. 2008). Zoos and related conservation organizations have been building facilities and developing the technology necessary for maintaining captive populations, and have also been developing the methods and programs to protect and reintroduce species in the wild (**Figure 6.18**). Some of these facilities are highly specialized, such as one run by the International Crane Foundation in Wisconsin, which is attempting to establish captive breeding colonies of all crane species. Currently, only around 10% of the terrestrial vertebrates kept in zoos were collected in the wild, and this number is declining as zoos gain more experience in captive breeding (see Table 6.2).

FIGURE 6.18 China's giant panda (*Ailuropoda melanoleuca*) is one of the world's most charismatic animals and has become emblematic of the fight to save endangered species. Wolong National Reserve and other facilities in China have been successful at breeding giant pandas using artificial insemination and hand rearing. Wolong Reserve has established a reintroduction program, but the loss of habitat and the fact that captive-raised individuals may lack the needed behavioral skills to survive in the wild combine to make reintroduction of pandas particularly problematic. (Photograph © LMR Group/Alamy.)

The success of captive breeding programs has been enhanced by efforts to collect and disseminate the knowledge and experience acquired by the world's zoos. Organizations such as the Species Survival Commission's Conservation Breeding Specialist Group, the Association of Zoos and Aquariums, and the European Association of Zoos and Aquaria provide zoos with information for proper care and handling of endangered species, as well as updates on the status and behavior of animals in the wild (www.aza.org). Information is gathered on nutritional requirements, anesthetic techniques to immobilize animals and reduce stress during transport and medical procedures, optimal housing conditions, vaccinations and antibiotics to prevent the spread of disease, and breeding records. This effort is being aided by a central database called ZIMS, the Zoological Information Management System, maintained by the International Species Information System (ISIS), which keeps track of all relevant information on 2 million animals belonging to 10,000 species of animals at 852 member institutions in 76 countries.

Most species reproduce with abandon in good captive conditions—so much so that the use of contraceptives and other management programs are required to control populations. However, some rare animal species do not adapt or reproduce well in captivity. Zoos conduct extensive research and are constantly identifying management conditions to overcome these problems and promote successful reproduction by genetically appropriate mates (Wildt et al. 2009). Some management methods come directly from human and veterinary medicine, while others are novel methods, developed for particular species at special research facilities such as the Smithsonian Conservation Biology Institute and the San Diego Zoo Institute for Conservation Research. These techniques include **cross-fostering**, in which common species raise the offspring of rare

Zoos often use the latest methods of veterinary medicine to establish healthy breeding colonies of endangered animals.

FIGURE 6.19 A bongo calf (*Tragelaphus euryceros*, an endangered species) produced by embryo transfer using an eland (*Taurotragus oryx*, a common species) as a surrogate mother at the Cincinnati Zoo's Center for Conservation and Research of Endangered Wildlife. Bongos also breed successfully on their own in captive herds. (Photograph © Cincinnati Zoo and Botanical Garden.)

species; **artificial incubation** of eggs in ideal hatching conditions; **artificial insemination** when adults do not show interest in mating or are living in different locations; and **embryo transfer**, which involves implanting fertilized eggs of rare species into surrogate mothers of common species (**Figure 6.19**). One of the most unusual techniques utilizes a **genome resource bank** (**GRB**) and involves freezing purified DNA, eggs, sperm, embryos, and other tissue of species so they can be used to contribute to breeding programs, maintain genetic diversity, and perhaps even to reestablish extinct species in the future. However, many of these techniques are expensive and species-specific and, in any case, are no substitute for in situ conservation programs that preserve ecological relationships and behaviors that are necessary for survival in the wild.

Raising animals in captivity is recognized as having limitations to the animals' conservation value. Animals living in captive breeding facilities may lose the behaviors they need to survive in the wild. Further, populations raised in captivity may undergo genetic, physiological, and morphological changes that make them less able to tolerate the natural environment if they are returned to the wild. Diseases acquired in captivity may render them unsuitable for release. Consequently, when researchers establish an ex situ program to preserve a species, they must address a series of ethical questions:

- How will establishing an ex situ population benefit the wild population?

- Is it better to let the last few individuals of a species live out their days in the wild or to breed a captive population that may be unable to adapt to wild conditions?
- Does a population of a rare species that has been raised in captivity and does not know how to survive in its natural environment really represent a victory for the species?
- Are rare individuals held in captivity primarily for their own benefit or the benefit of their entire species, for the economic benefit of zoos, or for the pleasure of zoo visitors?
- Are the animals in captivity receiving appropriate care based on their biological needs?
- Are sufficient efforts being made to educate the public about conservation issues?

Aquariums

Currently, approximately 600,000 individual fish are maintained in aquariums throughout the world (**Figure 6.20**). Most of these fish have been obtained from the wild; however, many fresh-water and some marine species are now being bred in captivity, reducing the need to collect in the wild. The hope is

FIGURE 6.20 Public aquariums participate in both in situ and ex situ conservation programs and provide a valuable function by educating people about marine conservation issues. Here a mother and son interact with a Cuban hogfish (*Bodianus pulchellus*). (Photograph by Julie Larsen Maher © Wildlife Conservation Society.)

that eventually aquarium-bred populations of endangered freshwater and marine species can be released into the wild to reestablish former populations. However, for this to succeed, the habitats of endangered species must first be restored, and this is rarely accomplished.

Fish breeding programs utilize indoor aquarium facilities, seminatural water bodies, and fish hatcheries and farms. Many of the techniques were originally developed by fisheries biologists for large-scale stocking operations involving trout, bass, salmon, and other commercial and game species. Other techniques were discovered in the aquarium pet trade, when dealers attempted to propagate tropical fish for sale. Currently, both public and private groups are making impressive efforts to unlock the secrets of breeding some of the more difficult species, particularly marine fishes, coral species, and other invertebrates.

Aquariums have a particularly important role to play in the conservation of whales, dolphins, and other marine mammals. Aquarium personnel often respond to public requests for assistance in handling whales stranded on beaches or disoriented in shallow waters. The lessons learned from working with common species may be used by the aquarium community to develop programs to aid endangered species.

The ex situ preservation of aquatic biodiversity takes on additional significance because of the dramatic increase in aquaculture, which currently represents 30% of fish and shellfish production worldwide. This aquaculture includes extensive salmon, carp, and catfish farms in the temperate zones, shrimp farms in the tropics, and the 12 million tons of aquatic products grown in China and Japan. As fish, frogs, mollusks, and crustaceans increasingly become domesticated and are raised to meet human needs, it becomes necessary to preserve the genetic stocks needed to continue improvements in these species—and to protect them against disease, climate change, and unforeseeable threats. A challenge for the future will involve balancing the need to increase human food production by aquaculture with the need to protect aquatic biodiversity from increasing human impacts.

Botanical gardens

The world's 1775 botanical gardens contain major collections of living plants and are a crucial resource for plant conservation, agriculture, forestry, and horticulture. An **arboretum** is a specialized botanical garden focusing on trees and other woody plants. Botanical gardens currently contain about 4 million living plants, representing 80,000 species—approximately 30% of the world's flora (Guerrant et al. 2004; www.bgci.org). When we add in the species grown in greenhouses, subsistence gardens, and hobby gardens, the numbers increase. The world's largest botanical garden—the Royal Botanic Gardens, Kew, in England—has an estimated 25,000 species of plants under cultivation, about 10% of the world's total; 2700 of these are listed as threatened under the IUCN categories. One of the most exciting new botanical gardens is the Eden Project in southwest England, which focuses

FIGURE 6.21 The Eden Project in Cornwall, England, cultivates more than 5000 plant species of economic importance in a series of greenhouses that are giant geodesic domes. The project has an appealing public image, as seen by these visitors to the project's Mediterranean Biome. (Photograph © Jack Sullivan/Alamy.)

on displaying and explaining over 5000 species of economically important plants in a series of giant greenhouse domes (**Figure 6.21**). Botanical gardens increasingly focus their efforts on cultivating rare and endangered plant species, and many specialize in particular types of plants. Many botanical gardens are involved in plant conservation, especially in the reintroduction of rare and endangered plants back to the wild and the restoration of degraded ecosystems (Hardwick et al. 2011).

Staff members are often recognized authorities on plant identification, distribution, and conservation status. In addition, botanical gardens are able to educate an estimated 200 million visitors per year about conservation issues. At an international level, Botanical Gardens Conservation International (BGCI) represents and coordinates the conservation efforts of over 700 botanical gardens. Priorities of this program involve creating a worldwide database to coordinate collecting activity and identify important species that are underrepresented or absent from living collections. One project involves creating an online PlantSearch database that currently lists over 575,000 species and varieties growing in botanical gardens, of which about 3000 are rare or threatened.

Botanical gardens have living collections and seed banks that provide ex situ protection and knowledge of endangered and economically important plants.

Also, because most existing botanical gardens are located in the temperate zone, establishing botanical gardens in the tropics is a primary goal of the international botanical community.

Seed banks

In addition to growing plants, botanical gardens and research institutes have developed collections of seeds, sometimes known as **seed banks**, collected from the wild and from cultivated plants, which provide a crucial backup to their living collections. This is especially important for rare and endangered species that may need to be reintroduced back into the wild. Seeds of approximately 10% of the world's species are stored in seed banks, with the figure approaching 70% for European plants (Godefroid et al. 2011b). When seeds are collected, efforts are made to include the range of genetic variation found in the species by collecting seeds from populations growing across the range of the species.

The seeds of most plant species can be stored in cold, dry conditions in these seed banks for long periods of time and then later germinated (**Figure 6.22**). The ability of seeds to remain dormant is extremely valuable for ex situ conservation efforts because it allows the seeds of large numbers of rare species to be frozen and stored in a small space, with minimal supervision and at a low cost. More than 60 major seed banks exist in the world, many of them in developing countries, with their activities coordinated by the Consultative Group on International Agricultural Research (CGIAR). However, if power supplies fail, equipment breaks down, or funding runs out at a facility, its entire frozen collection could be damaged or destroyed. To prevent such a loss, Norway has recently established the newest seed bank, the Svalbard Global Seed Vault, in which frozen material will be stored below permanently frozen ground.

Seed banks have been embraced by the international agricultural community as an effective way of preserving the genetic variability that exists in agricultural crops such as rice, wheat, and corn. This genetic variability is often crucial to the agricultural industry in its efforts to maintain and increase the high productivity of modern crops and to respond to changing environmental conditions, such as acid rain, drought, and soil salinity. Researchers are in a race against time to preserve this genetic variability of major food crops because traditional farmers throughout the world are abandoning local crop varieties in favor of standard, high-yielding varieties (Altieri 2004). Also, wild relatives of crop plants are not adequately represented in seed banks, even though these species are extremely useful in crop improvement programs.

A major controversy involving seed banks is ownership and control of the genetic resources of crop plants (Brush 2007). In the past, researchers

(A)

(B)

FIGURE 6.22 (A) At seed banks, seeds of many plant varieties are sorted, cataloged, and stored at freezing temperatures. (B) Seeds come in a wide variety of sizes and shapes. Each such seed represents a genetically unique, dormant individual. (Photographs courtesy of U.S. Department of Agriculture.)

from international seed banks, often from developed countries, freely collected seeds and plant tissues from developing countries and gave them to research stations and seed companies. But once seed companies had used the donated seeds to develop new, "elite" strains through sophisticated breeding programs and field trials, they sold the resulting seeds at high prices to developing countries to maximize profits. Developing countries are now questioning why they should share their genetic resources freely but then have to pay for advanced seeds based on those genetic resources. One solution to this dilemma involves negotiating agreements using the framework of the Convention on Biological Diversity (see Chapter 9), in which countries agree to share their genetic resources in exchange for receiving new products and a share of the profits. A few such contracts have been negotiated, such as the one between the Costa Rican government and Merck for the development of products based on species collected in the wild (see Chapter 3). These agreements will be followed carefully to determine whether they are mutually satisfactory and can serve as models for future contracts.

Summary

- Protecting and managing a rare or endangered species requires a firm grasp of its ecology and its distinctive characteristics, sometimes called its natural history. Long-term monitoring of a species in the field can determine if it is stable, increasing, or declining in abundance over time.
- Population viability analysis (PVA) uses demographic, genetic, and environmental data to estimate how various management actions will affect the probability that a population will persist until some future date.
- A species may be best described as a metapopulation made up of a shifting mosaic of populations that are linked by some degree of migration.
- The IUCN has developed quantitative criteria for populations and ecosystems to assign species to conservation categories: extinct, extinct in the wild, critically endangered, endangered, vulnerable, near threatened, least concern, data deficient, and not evaluated.
- National governments protect biodiversity by establishing national parks and refuges, controlling imports and exports at their borders, and creating regulations for air and water pollution. The most effective law in the United States for protecting species is the Endangered Species Act.
- At the international level, the Convention on International Trade in Endangered Species (CITES) allows governments to regulate, monitor, and sometimes prohibit trade in individuals and products from endangered species. Many countries, particularly those in the European Union, use CITES and similar international agreements and Red Data Book lists to protect species within their own borders and to cooperate in conservation activities.
- New populations of rare and endangered species can be established in the wild using either captive-raised or wild-caught individuals. Animals sometimes require behavioral training before release as well as maintenance after release.
- Some species that are in danger of going extinct in the wild can be maintained in zoos, aquariums, botanical gardens, and seed banks; this strategy is known as ex situ conservation. These captive colonies can sometimes be used later to establish new populations in the wild.

For Discussion

1. How do you judge whether a reintroduction project is successful? Develop simple and then increasingly detailed criteria to evaluate a project's success.
2. Would it be a good idea to create new wild populations of African rhinos, elephants, and lions in Australia, South America, the southwestern United States, or other areas outside their current range, as described by Donlan et al. (2006)? What would be some of the legal, economic, and ecological issues?
3. Would biodiversity be adequately protected if every species were raised in captivity? Is this possible? Is it practical? How would

freezing a tissue sample of every species help to protect biodiversity? Again, is this possible and/or practical?

4. A wide range of laws protect endangered species. Why don't species covered by such laws quickly recover?

Suggested Readings

Donlan, J. and 11 others. 2006. Re-wilding North America. *Nature* 436: 913–914. Some conservation biologists have suggested establishing large African game animals on the American plain.

Holland, G. J., J. S. A. Alexander, P. Johnson, A. H. Arnold, M. Halley, and A. F. Bennett. 2012. Conservation cornerstones: Capitalising on the endeavours of long-term monitoring projects. *Biological Conservation* 145: 95–101. Monitoring can help to inform management decisions.

Miller, B. and 9 others. 2004. Evaluating the conservation mission of zoos, aquariums, botanical gardens, and natural history museums. *Conservation Biology* 18: 86–93. Eight tough questions are asked with hopes that these institutions can become more effective.

Mitchell, A. M., T. I. Wellicome, D. Brodie, and K. M. Cheng. 2011. Captive-reared burrowing owls show higher site-affinity, survival, and reproductive performance when reintroduced using a soft-release. *Biological Conservation* 144: 1382–1391. Different techniques for reintroduction are evaluated.

Molano-Flores, B. and T. J. Bell. 2012. Projected population dynamics for a federally endangered plant under different climate change emission scenarios. *Biological Conservation* 145: 130–138. Population viability analysis is used to evaluate the impact of climate change on an endangered species.

NatureServe. 2009. http://natureserve.org. This website organizes and presents data on biodiversity surveys from North America.

Schwartz, M. W. 2008. The performance of the Endangered Species Act. *Annual Review of Ecology, Evolution, and Systematics* 39: 279–299. Many listed species are recovering, but certain goals have not been achieved.

Sebastián-González, E., J. A. Sánchez-Zapata, F. Botella, J. Figuerola, F. Hiraldo, and B. A. 2011. Linking cost efficiency evaluation with population viability analysis to prioritize wetland bird conservation actions. *Biological Conservation* 144: 2354–2361. Different management approaches are evaluated for their cost effectiveness on bird populations in Spain.

Smith, M. J. and 11 others. 2011. Assessing the impacts of international trade on CITES-listed species: Current practices and opportunities for scientific research. *Biological Conservation* 144: 82–91. An expert panel identifies the types of practical information that is needed to determine how trade affects species.

Traill, L. W., B. W. Brook, R. R. Frankham, and C. J. A. Bradshaw. 2010. Pragmatic population viability targets in a rapidly changing world. *Biological Conservation* 143: 28–34. An analysis of published PVA studies shows that large numbers of individuals must be protected.

Vynne, C. and 10 others. 2011. Effectiveness of scat-detection dogs in determining species presence in a tropical savanna landscape. *Conservation Biology* 25: 154–162. The combination of trained dogs and DNA analysis is a powerful tool for detecting rare species and assessing population size.

Banff National Park, the oldest national park in Canada and a World Heritage Site, is known for its great natural beauty, such this view of Bow Lake and Crowfoot Mountain.

Chapter 7
Protected Areas

The most effective way to protect the species, ecosystems, and genetic variation that constitute biodiversity is to designate areas where human activities are regulated and, at times, even prohibited. A **protected area** is an area of land or sea dedicated by law or tradition to the protection of biodiversity and associated natural and cultural resources (www.iucn.org). Protecting areas that contain healthy, intact ecosystems is the most effective way to preserve overall biodiversity. One could argue that it is ultimately the only way to preserve species, since we have the resources and knowledge to maintain only a small minority of the world's species in captivity. Preserving ecosystems involves establishing single protected areas, creating networks of protected areas, managing those areas effectively, implementing conservation measures outside the protected areas, and restoring biological communities in degraded habitats (www.wri.org). The first three of these five topics are covered in this chapter; conservation outside of protected areas and restoration of degraded ecosystems are covered in Chapter 8.

Establishment and Classification of Protected Areas

Protected areas can be established in a variety of ways, but the most common mechanisms are these:

- Government action, usually at a national level, but often at regional or local levels as well
- Land purchases by private individuals and conservation organizations
- Actions of indigenous peoples and traditional societies
- Development of biological field stations (which combine biodiversity protection and research with conservation education) by universities and other research organizations

Although legislation and land purchases alone do not ensure habitat protection, they can lay the groundwork for it. Partnerships among governments of developing countries, international conservation organizations, multinational banks, research and educational organizations, and governments of developed countries represent another way to bring together funding, training, and scientific and management expertise to establish new protected areas.

Traditional societies also have established protected areas to maintain their ways of life or just to preserve their land. Many of these protected areas have been in existence for long periods and are linked to the religious and cultural beliefs of the people. National governments in many countries, including the United States, Canada, Colombia, Brazil, and Australia have recognized the rights of traditional societies to own and manage the land on which they live, hunt, and farm. However, in some cases, recognition of land rights results only after conflict in the courts, in the press, and on the land.

The International Union for Conservation of Nature (IUCN) has developed a six-category system of classifying protected areas that ranges from minimal (nature reserves, national parks, etc.) to intensive use of the habitat by humans, (**Table 7.1**). Of these categories, the first five can be defined as true protected areas, because their habitat is managed primarily for biological diversity. Areas in the sixth category, managed-resource protected areas, are administered for the production of natural resources, such as timber and cattle, but biodiversity conservation may be an important secondary priority. Managed-resource protected areas can be particularly significant for several reasons: (1) they are often much larger in area than other categories of protected areas, (2) they still may contain many or even most of their original species, and (3) they often adjoin or surround protected areas. Conservation in such places often

The IUCN has developed a classification system for protected areas, ranging from strict nature reserves to managed-resource protected areas, depending on the level of human impact and the needs of society for resources.

TABLE 7.1 IUCN Protected Area Designations I–VI

Category	Description
Ia Strict nature reserves	Managed mainly for scientific research and monitoring; areas of land and/or sea possessing some outstanding or representative ecosystems, geological or physiological features, and/or species
Ib Wilderness areas	Managed mainly for wilderness protection; large areas of unmodified or slightly modified land and/or sea retaining their natural character and influence, without permanent or significant habitation, which are protected and managed so as to preserve their natural condition
II National parks	Managed mainly for ecosystem protection and recreation; natural areas of land and/or sea designated to (1) protect the ecological integrity of one or more ecosystems for present and future generations; (2) exclude exploitation or occupation inimical to the purposes of designation of the area; and (3) provide a foundation for spiritual, scientific, educational, recreational, and visitor opportunities, all of which must be environmentally and culturally compatible
III Natural monuments	Managed mainly for conservation of specific natural features; areas containing one or more specific natural or natural/cultural features of outstanding or unique value because of inherent rarity, representative or aesthetic qualities, or cultural significance
IV Habitat/species management areas	Managed mainly for conservation through management intervention; areas of land and/or sea subject to active intervention for management purposes to ensure the maintenance of habitats and/or to meet the requirements of specific species
V Protected landscapes and seascapes	Managed mainly for landscape/seascape conservation and recreation; areas of land, with coast and sea as appropriate, where the interaction of people and nature over time has produced an area of distinct character with significant aesthetic, ecological, and/or cultural value, and often with high biological diversity
VI Managed-resource protected areas	Managed mainly for the sustainable use of natural ecosystems; areas containing predominantly unmodified natural systems, managed to ensure long-term protection and maintenance of biological diversity, while also providing a sustainable flow of natural products and services to meet community needs

Source: After www.iucn.org.

means finding a compromise between protecting biodiversity and ecosystem function, on the one hand, and satisfying immediate and long-term human needs for resources, on the other.

At least 180 countries—and perhaps more—currently have protected areas (Chape et al. 2008; www.iucn.org). While it could be argued that virtually all countries should have at least one national park, large countries with rich biotas and a variety of ecosystem types would obviously benefit from having many protected areas, especially if they are undergoing changes in

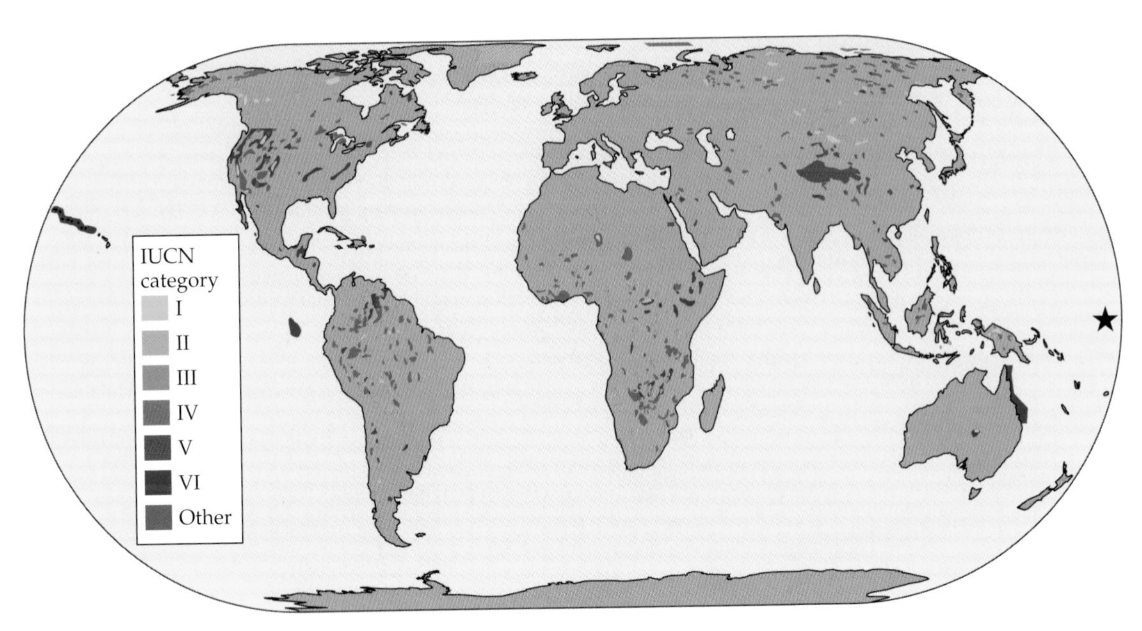

FIGURE 7.1 The world's terrestrial and marine protected areas. Although many small protected areas do not show up at this scale, all large areas in IUCN categories I–VI are indicated, as well as many areas that are protected in some manner (e.g., privately) but which do not have an official designation at the present time. Note the large protected areas in Greenland, the Hawaiian Islands, the Galápagos Islands, northern Alaska, northeastern Australia, and western China. The Phoenix Islands Protected Area is indicated by a star. (Based on World Database on Protected Areas 2005, www.wdpa.org.)

land use. More than 108,000 protected areas in IUCN categories I–VI have been designated worldwide, covering some 30 million km^2 on land and 2 million km^2 at sea (**Figure 7.1**).* This impressive total represents only about 13% of Earth's total land surface, which is about the same area as the land used to grow all of the world's crops. However, much of this protected land is not particularly valuable to people; the world's largest park is in Greenland on inhospitable terrain and covers 970,000 km^2, accounting for about 3% of the global area protected. About 6% of the Earth's surface is in categories I–IV, *strictly protected* in scientific reserves and national parks. Further increases in the amount of protected area will continue in coming years but are limited by the perceived need of governments and societies to manage land for natural resources and employment (McDonald and Boucher 2011).

*Uncertainty about the number and size of protected areas stems from the different standards used throughout the world, the degree of protection actually given to a designated area, and when the data were gathered.

The measurements of protected areas in individual countries and on continents are only approximate because sometimes the laws protecting national parks and wildlife sanctuaries are not strictly enforced; at the same time, there are sections of managed areas that, while not legally protected, are carefully protected in practice. Examples of this include the sections within U.S. national forests designated as wilderness areas. The coverage of strictly protected areas varies dramatically among countries: high proportions of land are protected in Germany (32%), Austria (36%), and the United Kingdom (15%), and surprisingly low proportions in Russia (8%), Greece (3%), and Turkey (3%). Even when a country has numerous protected areas, certain unique habitats of high economic value may remain unprotected (Dietz and Czech 2005; earthtrends.wri.org). In addition, protected areas may be reduced in size by the government, opened up for exploitation, or may even have their protected status removed (known as degazettement), particularly when they are found to contain valuable natural resources (Mascia and Pallier 2011).

Marine protected areas

Marine conservation has lagged behind terrestrial conservation efforts; however, rapid progress is now being made in establishing marine protected areas (Gaines et al. 2010). Only about 6% of the world's territorial seas near coastlines and about 1% of the total marine environment are included in protected areas, yet as much as 20% of the marine environment may need to be protected in order to manage declining commercial fishing stocks. Even more may be required to conserve the full range of coastal and marine biodiversity (www.iucn.org; Spalding et al. 2008). Over 5000 marine and coastal protected areas have been established worldwide, but most are small. The three largest marine protected areas (MPAs) account for about half of the total: the Great Barrier Reef Marine Park in Australia, the Northwestern Hawaiian Islands Marine National Monument, and the Phoenix Islands Protected Area (established by the nation of Kiribati in the South Pacific). The United States has 1700 marine protected areas, of which 13 are marine sanctuaries covering 46,548 km^2.

Urgent efforts are being made throughout the world to preserve marine biodiversity by establishing marine parks that seek to protect the nursery grounds of commercial species and to maintain water quality and both physical and biological features of the ecosystem. In the process, high-quality protected areas can also attract recreational activities such as swimming and diving and the economic benefits associated with tourism.

The effectiveness of protected areas

The value of protected areas in maintaining biodiversity is abundantly clear in many tropical countries (Nelson and Chomiz 2011). Inside the park boundaries there are trees and animal life, while outside the park the land

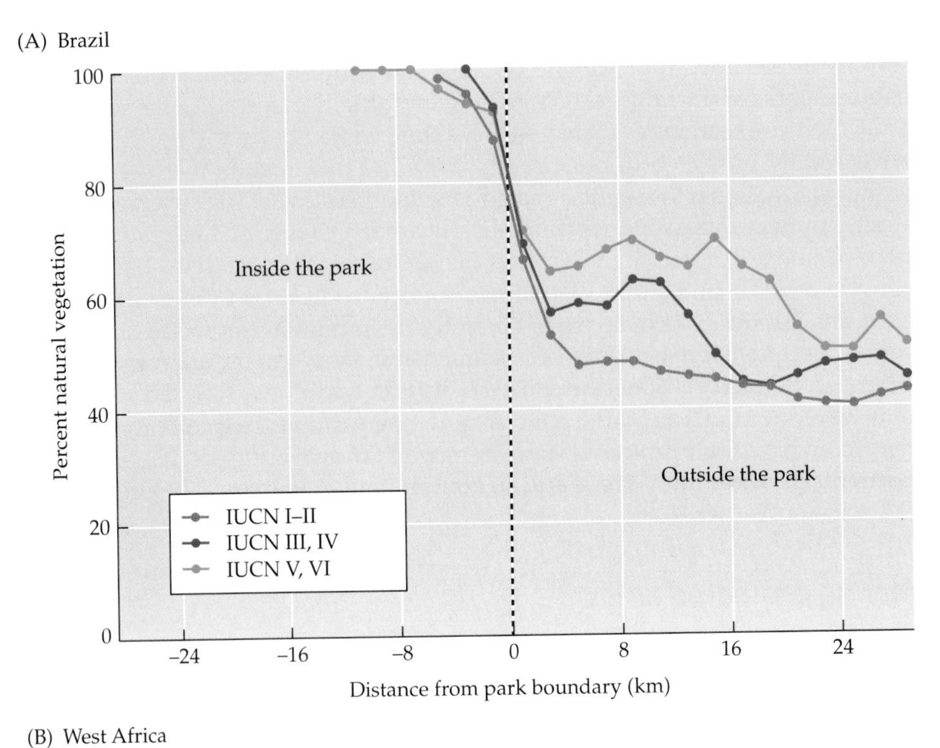
(A) Brazil
100
80
60
40
20
0
Percent natural vegetation
Inside the park
Outside the park
IUCN I–II
IUCN III, IV
IUCN V, VI
–24
–16
–8
0
8
16
24
Distance from park boundary (km)

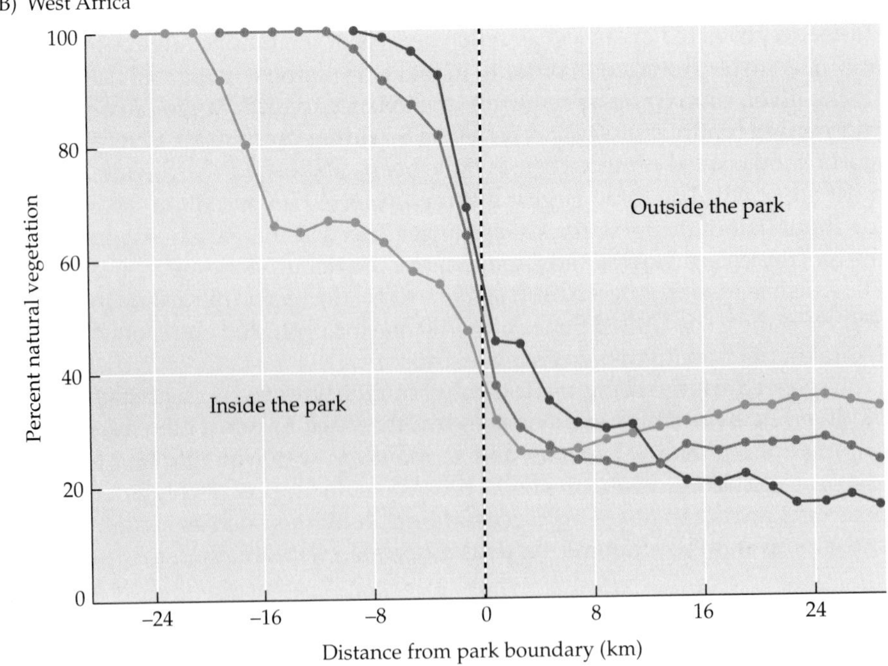
(B) West Africa
100
80
60
40
20
0
Percent natural vegetation
Outside the park
Inside the park
–24
–16
–8
0
8
16
24
Distance from park boundary (km)

◀ **FIGURE 7.2** National parks and other protected areas are able to prevent damage to the natural forests in (A) the Atlantic coast forests of Brazil, and (B) West Africa. In the Atlantic coast forest, there is a sharp boundary with intact forest inside the protected areas, and around 50% intact forest outside the protected area. For protected areas in West Africa, there is considerable forest degradation within 16 km of the park boundary, particularly for IUCN categories V and VI, which are mainly forest reserves. See Table 7.1 for a description of IUCN categories. (From Joppa et al. 2008.)

has been cleared, fires are common, and few animals are seen (**Figure 7.2**). These protected areas also continue to provide income and services to poor people living nearby (Andam et al. 2010). At the same time, these protected areas still need to be monitored and managed to protect them from logging, hunting, and other human activities.

If national parks can be established to include concentrations of species and diverse ecosystems, a large percentage of a country's species can be preserved. This explains why, in Mexico, 82% of mammal species are represented in protected areas that occupy only 4% of the country (Ceballos 2007). In Britain, 88% of plant species occur in the protected area system, and 26% of those species are found exclusively within protected areas (Jackson et al. 2009). China's nature reserves cover 15% of the total area and 81% of the country's vegetation types (Wu et al. 2011). However, the long-term future of many of these protected species and ecosystems remains in doubt because populations of many species and the area of ecosystems may be so reduced in size that their eventual fate is extinction.

Although the number of species living within a protected area is an important indicator of its potential to protect biodiversity, protected areas need to maintain healthy ecosystems and viable populations of important species.

Prioritization: What should be protected?

In a crowded world with limited natural resources and limited funding, it is crucial to establish priorities for conserving biodiversity. Although some conservationists would argue that no ecosystem or individual species should ever be lost, the reality is that numerous species are in danger of going extinct and there are too few resources available to save them all. The challenge to conservation efforts lies in finding ways to minimize the loss of biodiversity during a period of limited financial and human resources (Watson et al. 2011).

Conservation planners must address three interrelated questions: *What* needs to be protected? *Where* should it be protected? *How* should it be protected? Three criteria can be used to answer the first two of these questions and set conservation priorities:

Distinctiveness, endangerment, and utility are used to prioritize the protection of species and ecosystems.

1. *Distinctiveness (or irreplaceability)*. An ecosystem composed primarily of rare endemic species or with other unusual attributes (small area, scenic value, unique geological features) is given a high priority for conservation. A species is often given high conservation value if it is taxonomically distinctive—that is, it is the only species in its genus or family (Faith 2008). Similarly, a population of a species having unusual genetic characteristics that distinguish it from other populations of the species might be a high priority for conservation.
2. *Endangerment (or vulnerability)*. Species in danger of extinction are of greater concern than species that are not; thus, the whooping crane (*Grus americana*), with only about 382 individuals, requires more protection than the sandhill crane (*Grus canadensis*), with approximately 520,000 individuals. Ecosystems threatened with imminent destruction are also given priority, such as the rain forests of west Africa. Endangerment of an ecosystem can be estimated by the present size of its geographic range, the rate of decline in its range, and loss of its ecological functions, in a way comparable to the IUCN Red List criteria (Rodríguez et al. 2011).
3. *Utility*. Species that have present or potential value to people, such as wild relatives of wheat, are given high conservation priority. Species with major cultural significance, such as tigers in India and the bald eagle in the United States, are given high priority. Ecosystems of major economic value, such as coastal wetlands, are usually given greater priority for protection than less valuable communities. In coming years, forested protected areas may be increasingly valued for their ability to retain and absorb carbon (Soares-Filho et al. 2010).

The Komodo dragon (*Varanus komodoensis*) (**Figure 7.3**) of Indonesia is an example of a species that fits all three categories: it is the world's largest lizard (distinctiveness), it occurs on only a few small islands of a rapidly developing nation (endangerment), and it has major potential as a tourist attraction in addition to being of great scientific interest (utility). Appropriately, these Indonesian islands are now protected within the Komodo National Park and are designated as one of the United Nations Educational, Scientific and Cultural Organization (UNESCO) **World Heritage sites** (http://whc.unesco.org/en/35/; see p. 212). These specially designated protected areas include diverse sites of overwhelming natural and/or cultural significance that are deemed to be "of outstanding value to humanity" and "irreplaceable sources of life and inspiration," transcending national boundaries.

Using the above three criteria, several systems have been developed to prioritize areas for protection. These approaches are generally complementary; they differ more in their emphases than in fundamental principle. Such approaches are presently being reevaluated in relation to climate change, because in coming decades and centuries, species and ecosystems may not be able to survive where they presently occur and are protected.

FIGURE 7.3 The carnivorous Komodo dragon (*Varanus komodoensis*) of Indonesia is the largest living monitor lizard, with individuals reaching 3 m (10 feet) in length and 70 kg (150 pounds) in weight. Tourists flock to see these animals in the wild, and protecting this endangered species was an important reason for establishing the Komodo National Park. (Photograph © Corbis Flirt/Alamy.)

THE SPECIES APPROACH One approach to establishing conservation priorities involves protecting particular species—and in doing so, protecting an entire biological community and associated ecosystem processes (Branton and Richardson 2011). Protected areas are often established to protect individual species of special concern, such as rare species, endangered species, keystone species, and culturally significant species. Species such as these, that provide the impetus to protect an area and ecosystems, are known as **focal species**. One type of focal species is an **indicator species**, a species that is associated with an endangered biological community or set of unique ecosystem processes; for instance, the endangered northern spotted owl is a forest indicator species in the Pacific Northwest of the United States. Many national parks have also been created to protect **flagship species**, such as tigers and pandas, which capture public attention, have symbolic value, and are crucial to ecotourism. Flagship and indicator species are also known as **umbrella species**—protecting them automatically protects other species and aspects of biodiversity.

The species approach follows from creating survival plans for individual species, which are developed by governments and private conservation organizations. In the Americas, the Natural Heritage Programs and the NatureServe network are using information on rare and endangered species to target new localities for conservation—areas where there are concentrations of endangered species or where the last populations of a declining species exist (www.natureserve.org). In the IUCN Species Survival Commission, over 100 specialist groups provide action plans for endangered animals and plants that may include recommendations for creating new protected areas.

THE ECOSYSTEM APPROACH A number of conservationists have argued that ecosystems, and the biological communities that they contain, rather than species, should be targeted for conservation (Tallis and Polasky 2009). They claim that spending, say, $1 million on habitat protection and management of a self-maintaining ecosystem might preserve more species and provide more value to people in the long run than spending the same amount of money on an intensive effort to save just one conspicuous species (**Figure 7.4**). It often is easy to demonstrate an ecosystem's economic value to policy makers and the public in terms of flood control, clean water, and recreation, whereas arguing for a particular species may be more difficult. Combining species and ecosystem approaches may be a good conservation strategy in many circumstances.

(A)

(B)

FIGURE 7.4 (A) The Monterey Bay National Marine Sanctuary in California was established to protect a coastal marine environment that includes marine mammals, seabirds, and ocean bottom species. (B) Feather worms and various coral species living on the ocean bottom at the sanctuary. (Photographs courtesy of NOAA.)

When using this ecosystem-based approach, conservation planners should try to ensure that representative sites of as many types of ecosystems as possible are protected. A **representative site** includes the species and environmental conditions characteristic of the ecosystem. Although no site is perfectly representative, biologists working in the field can often identify suitable sites for protection.

Where immediate decisions must be made to determine park boundaries and which species and ecosystems to protect, biologists are being trained to make **rapid biodiversity assessments**, also known as RAPs (for *r*apid *a*ssessment *p*rograms). RAPs involve mapping vegetation, making lists of species, checking for species of special concern, estimating the total number of species, and looking out for new species and features of special interest.

THE WILDERNESS APPROACH Wilderness areas are a related priority for establishing new protected areas. Large blocks of land that have been minimally affected by human activity, that have a low human population density, and that are not likely to be developed in the near future are perhaps the only places in the world where large mammals can survive in the wild. These wilderness areas can serve as "controls," showing what natural ecosystems are like with minimal human influence. It is worth emphasizing that even these so-called wilderness areas have had a long history of human activity, and people have often affected the structure of the biological communities.

THE HOTSPOT APPROACH Certain organisms can be used as **biodiversity indicators** to highlight new areas where concentrations of species can be protected. For example, a site with a high diversity of flowering plants often, but not always, has a high diversity of mosses, spiders, and fungi. Further, areas with high diversity often have a high percentage of endemism, with species occurring there and nowhere else (Joppa et al. 2011). The IUCN Plant Conservation office in England has documented about 250 global centers of plant diversity with large concentrations of species (Hoffmann et al. 2008). BirdLife International has identified over 200 Important Bird Areas (IBAs) with more than 2400 restricted-range bird species; many of these localities are in urgent need of protection (www.birdlife.org).

Using a similar approach, Conservation International, World Wildlife Fund, and others have designated **hotspots** that have great biological diversity and high levels of endemism and that are under immediate threat of species extinctions and habitat destruction (**Figure 7.5**) (www.biodiversityhotspots.org). Using these criteria, 34 global hotspots have been targeted for new protected areas. These hotspots together encompass the entire ranges of 12,066 endemic species of terrestrial vertebrates (42% of the world's total) and at least part of the ranges of an additional 35% of the remaining terrestrial vertebrate species—all on only 2.3% of Earth's total land surface (**Table 7.2**).

Many protected areas have been established to protect large, well-known species, unique ecosystems, and hotspots where there are concentrations of species.

(A)

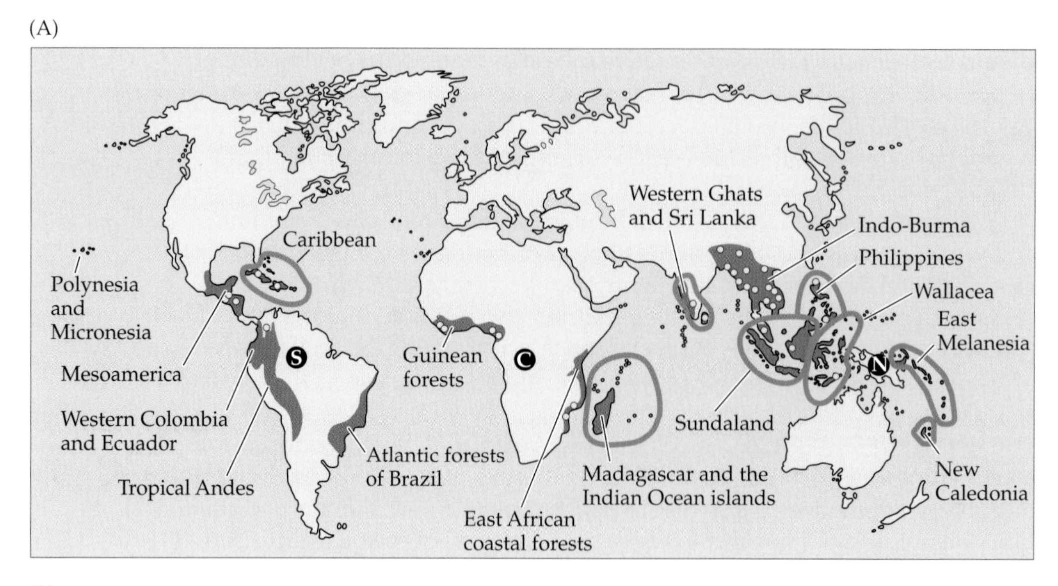

(B)

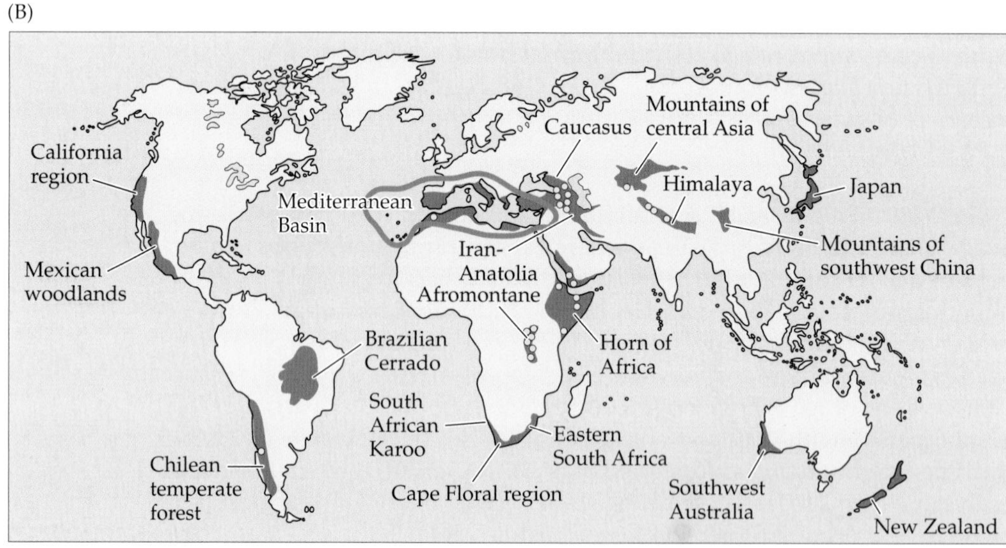

FIGURE 7.5 Hotspots are targets for protection because of their high biodiversity, endemism, and significant threat of imminent extinctions. (A) Sixteen tropical rain forest hotspots. Areas circled in green are island groups. The Polynesia/Micronesia region (far left) covers a large number of Pacific Ocean islands, including the Hawaiian Islands, Fiji, Samoa, French Polynesia, and the Marianas. Black-circled letters indicate the only three remaining undisturbed rain forest areas of any extent, in South America (S), the Congo basin of Africa (C), and the island of New Guinea (N). (B) Eighteen hotspots representing other ecosystems. Yellow dots denote areas that have experienced armed conflicts between 1950 and 2000 with over 1000 casualties. (After Mittermeier et al. 2005; Hansen et al. 2009.)

TABLE 7.2 A Comparison of 34 Global Hotspots

Location[a]	Original extent (× 1000 km²)	Undisturbed vegetation remaining (%)	Included in protected areas (%)[b]	Number of species		
				Plants	Birds	Mammals
The Americas						
Central Chile	397	30	11	3892	226	65
Tropical Andes	1543	25	8	30,000	1728	569
Western Colombia/Ecuador	275	24	7	11,000	892	283
Atlantic forest of Brazil	1234	8	2	20,000	936	263
Brazilian Cerrado	2032	22	1	10,000	605	195
Mexican pine-oak woodlands	461	20	2	5300	525	328
California region	294	25	10	3488	341	151
Mesoamerica	1130	20	6	17,000	1124	440
Caribbean islands	230	10	7	13,000	607	89
Africa						
Guinean forests of West Africa	620	15	3	9000	793	320
South African Karoo	103	29	2	6356	227	74
Cape region of South Africa	79	20	13	9000	324	90
Southeastern South Africa	274	24	7	8100	541	193
Madagascar and Indian Ocean islands	600	10	2	13,000	367	183
East African coastal forests	291	10	4	4000	639	198
East Afromontane	1018	10	6	7598	1325	490
Horn of Africa	1659	5	3	5000	704	219
Europe and Mideast						
Mediterranean Basin	2085	5	1	22,500	497	224
Caucasus Mountains region	863	20	1	6400	381	130
Iran-Anatolia	900	15	3	6000	364	141
Continental Asia						
Mountains of central Asia	863	20	7	5500	493	143
Himalaya	742	25	10	10,000	797	300
Western Ghats and Sri Lanka	190	23	11	5916	457	140
Indo-Burma	2373	5	6	13,500	1277	433
Mountains of southwest China	262	8	2	12,000	611	237

(Continued on next page)

TABLE 7.2 A Comparison of 34 Global Hotspots *(Continued)*

Location[a]	Original extent (× 1000 km²)	Undisturbed vegetation remaining (%)	Included in protected areas (%)[b]	Number of species		
				Plants	Birds	Mammals
Pacific Rim						
Sundaland island region	1501	7	6	25,000	771	381
Wallacea island region	338	15	6	10,000	650	222
Philippines	297	7	6	9253	535	167
Southwest Australia	357	30	11	5571	285	57
East Melanesian islands	99	30	0	8000	365	86
New Caledonia	19	27	3	3270	105	9
New Zealand	270	22	22	2300	198	4
Japan	373	20	6	5600	368	91
Micronesia/Polynesia (includes Hawaii)	47	21	4	5330	300	15

Source: Based on data from Mittermeier et al. 2005 and www.biodiversityhotspots.org.
[a]Tropical rain forest hotspots are shown in green; other hotspots encompass a variety of ecosystem types.
[b]Calculations based on protected areas in IUCN categories I–IV.

One major center of biodiversity is the tropical Andes, where 30,000 plant species, 1728 bird species, 569 mammal species, 610 reptile species, and 1155 amphibian species persist in tropical forests and high-altitude grasslands on about 0.3% of Earth's total land surface. The hotspot approach has generated a considerable amount of enthusiasm and funding during the last 10 years, and it will be worth watching to see how successful it is in advancing the goals of conservation in areas of intense human pressures, some of which are sites of armed conflicts (Hanson et al. 2009).

The hotspot approach can also be applied to individual countries and regions. In the United States, hotspots for rare and endangered species occur in the Hawaiian Islands, the southern Appalachians, the Florida panhandle, the Death Valley region, the San Francisco Bay area, and coastal and interior Southern California (Venevsky and Venevskaia 2005). Comparable hotspot analyses have been carried out for marine and freshwater ecosystems (Roberts et al. 2002), highlighting the great richness of endemic species in the western Pacific Ocean. The Coral Triangle Initiative on Coral Reefs, Fisheries and Food Security was established to coordinate conservation efforts among different countries, conservation organizations, and funding sources in that region (www.coraltriangleinitiative.org).

Measuring effectiveness: Gap analysis

Comparing biodiversity priorities with existing and proposed protected areas can identify gaps in preservation that need to be filled with new protected areas (Watson et al. 2011). In the past, this was done informally by establishing national parks in different regions that possess distinctive species and ecosystems. At the present time, a more systematic conservation planning process, known as **gap analysis**, is sometimes used (Tognelli et al. 2009). The following steps are often taken:

1. Data, including distribution, are compiled on the species, ecosystems, and physical features of the region, which are sometimes referred to as **conservation units**. Information on human densities and economic factors may also be included.
2. Conservation goals are identified, such as the amount of area to be protected for each ecosystem or the number of individuals of rare species to be protected.
3. Existing protected areas are reviewed to determine which conservation units are protected already and which are not (a process known as "identifying gaps in coverage").
4. Additional areas are identified for protection to increase the number of protected conservation units ("filling the gaps").
5. These additional areas are protected in some way, often by being directly purchased or designated as national parks, and management plans are developed and implemented.
6. The new protected areas are monitored to determine whether they are meeting their stated goals for conservation units. If not, the management plans may be changed or additional areas may be acquired to meet the goals.

On an international scale, scientists are comparing the distribution of endangered species and protected areas. The Global Gap Analysis project is helping to determine how effectively protected areas include populations of the world's vertebrates (Maiorano et al. 2008). This project compared the distribution of 11,633 species of mammals, birds, amphibians, turtles, and tortoises with the distribution of protected areas throughout the world to identify 1424 **gap species**: species not protected in any part of their range. The distressing result is that hundreds of these gap species are threatened with extinction. In the United States, 82% of endangered fish species are not found in the national parks; to increase effectiveness, it would be a priority for the government to acquire additional areas to protect these unprotected species (Lawrence et al. 2011).

Geographic information systems (GIS) represent the latest development in gap analysis technology—using computers to integrate the wealth of data on the natural environment with information on protected areas and biodiversity (Murray-Smith et al. 2009). The basic GIS approach involves storing, displaying, and manipulating many types of mapped data such as vegetation types, climate, soils, topography, geology, hydrology, species distributions, existing protected areas,

GIS is an effective tool for gap analysis, using a wide variety of information to pinpoint critical areas and species that are priorities for protection.

human settlements, and resource use (**Figure 7.6**). This approach can point out correlations among the biological and nonbiological (elevation, rock types, water sources, etc.) elements of the landscape, help plan parks that include a diversity of ecosystems, and even suggest sites to search for rare and protected species. Aerial photographs and satellite imagery are additional sources of data for GIS analysis, and they can highlight

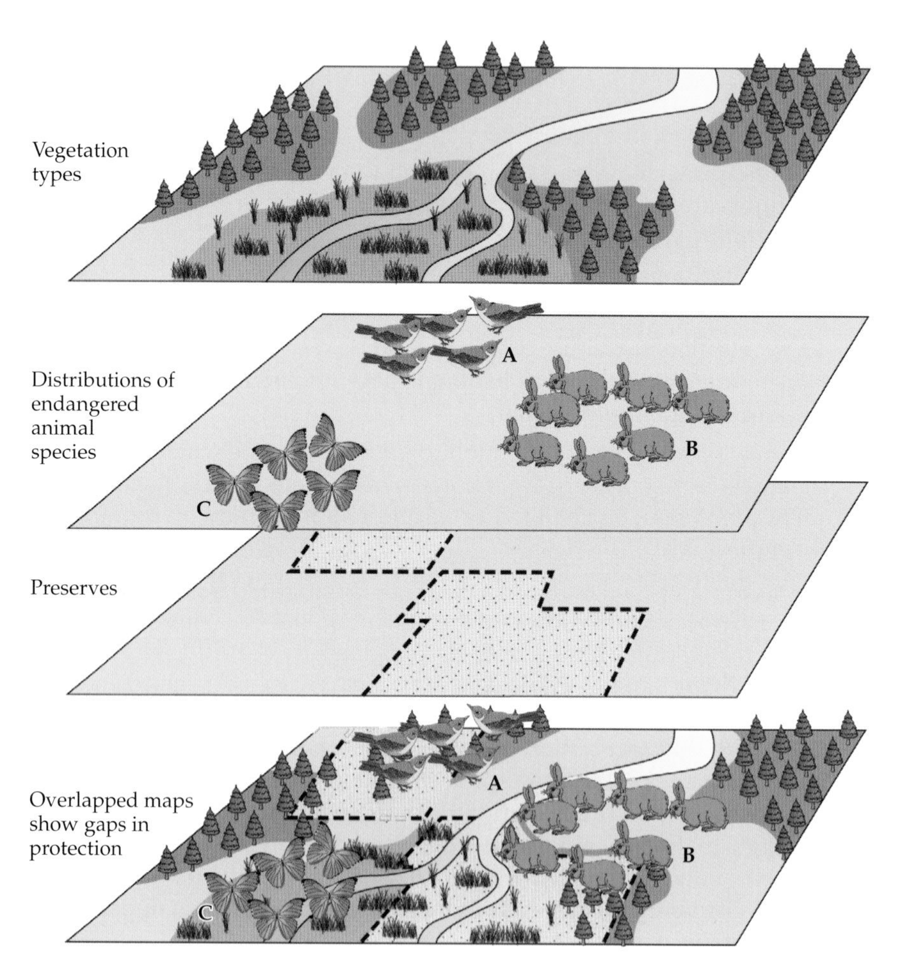

FIGURE 7.6 Geographic information systems (GIS) provide a method for integrating a wide variety of data for analysis and display on maps. In this example, vegetation types, distributions of endangered animal species, and preserved areas are overlapped to highlight areas that need additional protection. The overlapped maps show that the distribution of Species A is predominantly in a preserve, Species B is only protected to a limited extent, and Species C is found entirely outside of the preserves. Establishing a new protected area to include the range of Species C would be the highest priority. (After Scott et al. 1991.)

patterns of vegetation structure and distribution over local and regional scales (Ranganathan et al. 2007). In particular, a series of images taken over time can reveal patterns of habitat fragmentation and destruction that need prompt attention. These GIS images can dramatically illustrate when current conservation policies are not working and need to be changed.

Designing Protected Areas

The size and placement of protected areas throughout the world are often determined by the distribution of people, potential land values, the political efforts of conservation-minded citizens, and historical factors. In other cases, certain parcels of land may be purchased to protect a critical water supply or charismatic species. Sometimes lands are set aside for conservation protection because they have no immediate commercial value—they are "lands that nobody wants." Small reserves are acquired—sometimes at great cost—in urban areas. At present protected areas are often created in a haphazard fashion, dependent on the availability of money, land, and political influence, but a considerable body of ecological literature is now developing to address the most efficient way to design networks of conservation areas that more adequately protect the full range of biodiversity (Margules and Sarkar 2007; Adams et al. 2011).

Issues of reserve design have proved to be of great interest to governments, corporations, and private landowners, who are being urged—and mandated—to manage their properties for both the commercial production of natural resources and the protection of biological diversity. However, consideration of such issues does not necessarily produce universal design guidelines. Conservation biologists need to be cautious about using simplistic, overly general guidelines for designing protected areas, because every conservation situation requires special consideration (Cawardine et al. 2009). Also, academic scientists need to be better at communicating these principles of nature reserve design to the managers, planners, and policy makers who are actually creating specific new reserves (Sewall et al. 2011).

Conservation biologists often start the process of designing networks of protected areas by considering "the four *R*s":

- *Representation*. The protected areas should contain as many aspects of biodiversity (species, ecosystems, genetic variation, etc.) as possible.
- *Resiliency*. Protected areas must be sufficiently large to maintain all aspects of biodiversity in a healthy condition for the foreseeable future, including the predicted impacts of climate change.
- *Redundancy*. Protected areas must include enough examples of each aspect of biodiversity to ensure their long-term existence.
- *Reality*. There must be sufficient funds and political will, not only to acquire and protect lands but also to subsequently regulate and manage the protected areas.

The following more specific questions about reserve design are also useful for discussing how best to construct networks of protected areas:

- How large must a nature reserve be to effectively protect biodiversity that cannot survive outside the reserve?
- Is it better to have a single large protected area or multiple smaller reserves?
- What is the best shape for a nature reserve?
- When a network of protected areas is created, should the areas be close together or far apart? Should they be isolated from one another or linked together in some way?

Principles of design have been developed to guide land managers in establishing and maintaining networks of protected areas.

Some of these issues are being explored using the island biogeography model of MacArthur and Wilson (1967), described in Chapter 5. Many of them also have originated from the insights of wildlife and park managers (Roux et al. 2008). The island biogeography approach makes the significant assumption, which is often invalid, that parks are habitat islands completely isolated by an unprotected matrix of inhospitable terrain. In fact, many species are capable of living in and dispersing through this habitat matrix. Researchers working with island biogeography models and data from protected areas have proposed various principles of reserve design, but they are still being debated (**Figure 7.7**).

Also, all of these issues have been viewed mainly with land vertebrates, higher plants, and large invertebrates in mind. The applicability of these ideas to freshwater and marine nature reserves, where dispersal mechanisms are largely unknown, is an area of active investigation (Christie et al. 2010; Olds et al. 2012).

Protected area size and characteristics

An early debate in conservation biology occurred over whether species richness is maximized in one large protected area or in several smaller ones of an equal total area (Soulé and Simberloff 1986; McCarthy et al. 2006), known in the literature as the **SLOSS debate** (*s*ingle *l*arge *o*r *s*everal *s*mall). Is it better, for example, to establish one protected area of 10,000 ha or four protected areas of 2500 ha each? The proponents of large nature reserves argue that only a large protected area has sufficient numbers of large,

FIGURE 7.7 Principles of reserve design that are based in part on theories of island biogeography. Imagine that the reserves are "islands" of the original biological community surrounded by land that has been made uninhabitable for the original species by human activities such as farming, ranching, or industrial development. The practical application of these principles is still being studied and debated, but in general, the designs shown on the right are considered preferable to those shown on the left. (After Shafer 1997.) ▶

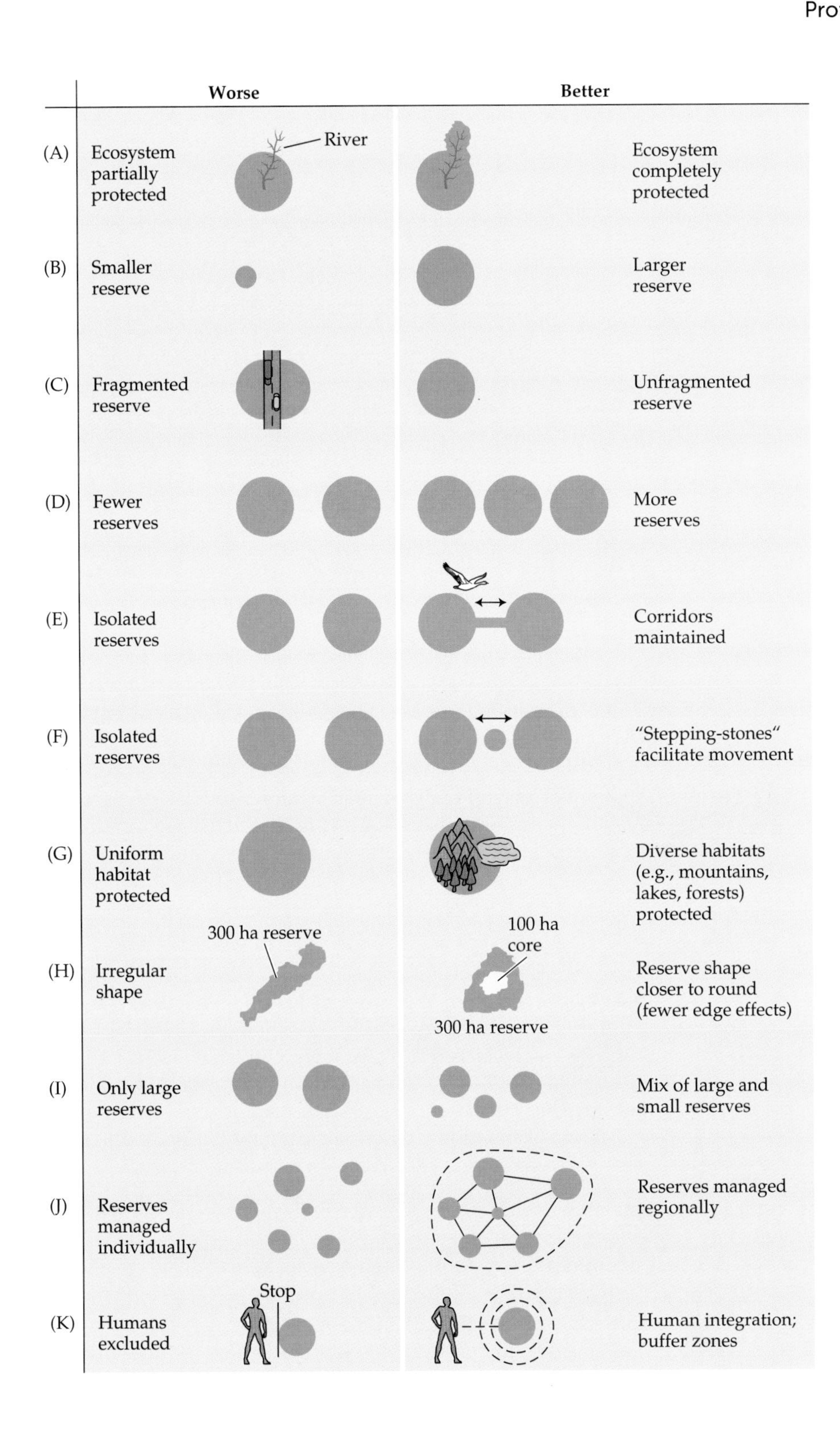
Worse
Better
(A)
Ecosystem partially protected
River
Ecosystem completely protected
(B)
Smaller reserve
Larger reserve
(C)
Fragmented reserve
Unfragmented reserve
(D)
Fewer reserves
More reserves
(E)
Isolated reserves
Corridors maintained
(F)
Isolated reserves
"Stepping-stones" facilitate movement
(G)
Uniform habitat protected
Diverse habitats (e.g., mountains, lakes, forests) protected
(H)
Irregular shape
300 ha reserve
100 ha core
300 ha reserve
Reserve shape closer to round (fewer edge effects)
(I)
Only large reserves
Mix of large and small reserves
(J)
Reserves managed individually
Reserves managed regionally
(K)
Humans excluded
Stop
Human integration; buffer zones

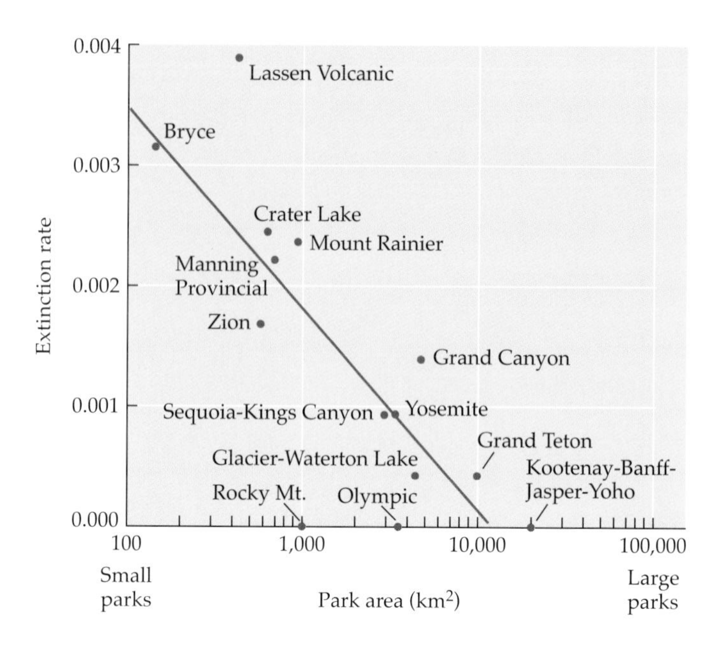

FIGURE 7.8 Mammals have higher extinction rates in smaller parks than in larger parks. If these same trends continue for the next 100 years, small parks such as Bryce are predicted to lose more than 30% of their mammal species, whereas large parks such as Yosemite will only lose 5% and Kootenay-Banff-Jasper-Yoho is not expected to lose any species. Each dot represents the actual extinction rate of mammal populations (expressed as the proportion of species that have gone extinct per year) for a particular U.S. national park, Canadian national park, or two or more adjacent parks. (After Newmark 1995.)

wide-ranging, low-density species (such as large carnivores) to maintain long-term populations. Large nature reserves also minimize the ratio of edge habitat to total habitat, encompass more species, and can have greater habitat diversity than small reserves.

The advantage of large parks is effectively demonstrated by an analysis of 299 mammal populations in 14 national parks in western North America (**Figure 7.8**). Local extinction rates have been very low or zero in parks with area over 1000 km^2 and have been much higher in parks smaller than 1000 km^2. It is also true that human population densities tend to be lower on the edges of large reserves than on the edges of small reserves, and this could contribute to the higher extinction rates in small parks (Wiersma et al. 2004). On the other hand, once a park reaches a certain size, the number of new species added with each increase in area starts to decline. At that point, creating a second large park, and/or another park some distance away, may be a more effective strategy for preserving additional species than simply adding on to the existing park.

Large reserves are generally better able to maintain many species because such reserves support larger population sizes and greater variety of habitats. However, small reserves are important in protecting particular species and ecosystems.

The research on extinction rates of populations in different-size parks has three practical implications:

1. When a new park is being established, it should be made as large as possible—to preserve as many species as possible, contain large populations of each species, and provide a diversity of habitats and natural resources. Keystone resources should be included, in addition to habitat features that promote biodiversity, such as elevation gradients.

2. Whenever possible, land adjacent to protected areas should be acquired in order to reduce external threats to existing parks and to maintain critical buffer zones. For example, terrestrial habitats adjacent to wetlands are often necessary for semiaquatic species such as snakes, toads, and turtles. The best protection may be provided when natural ecological units, such as entire watersheds or mountains, are encompassed within reserve borders as a means of reducing external threats.
3. The present and predicted effects of climate change will change ecosystems within existing protected areas. The result will often be a reduction in the area of habitat available for a species and a consequent decline in population size and increased probability of extinction. Protected areas may need to be enlarged in anticipation of these changes.

Often there is no choice but to accept the challenge of managing species and biological communities in small reserves. The 10,000 protected areas in Britain, for instance, have an average area of only 3 km^2 (**Figure 7.9**) (Jackson et al. 2009). Dense human populations and highly modified habitat often surround such small reserves. Small reserves, even ones less than a hectare, may be effective at protecting isolated populations of rare species, particularly when they encompass a unique habitat type found nowhere else (Jarosïk et al. 2011). Numerous countries have many more small protected areas (less than 100 ha) than medium and large ones, yet the combined area of these small reserves is only a tiny percentage of the total area under protection. This is particularly true in places that have been intensively cultivated for centuries, such as Europe, China, and Southeast Asia. Bukit Timah Nature Reserve in

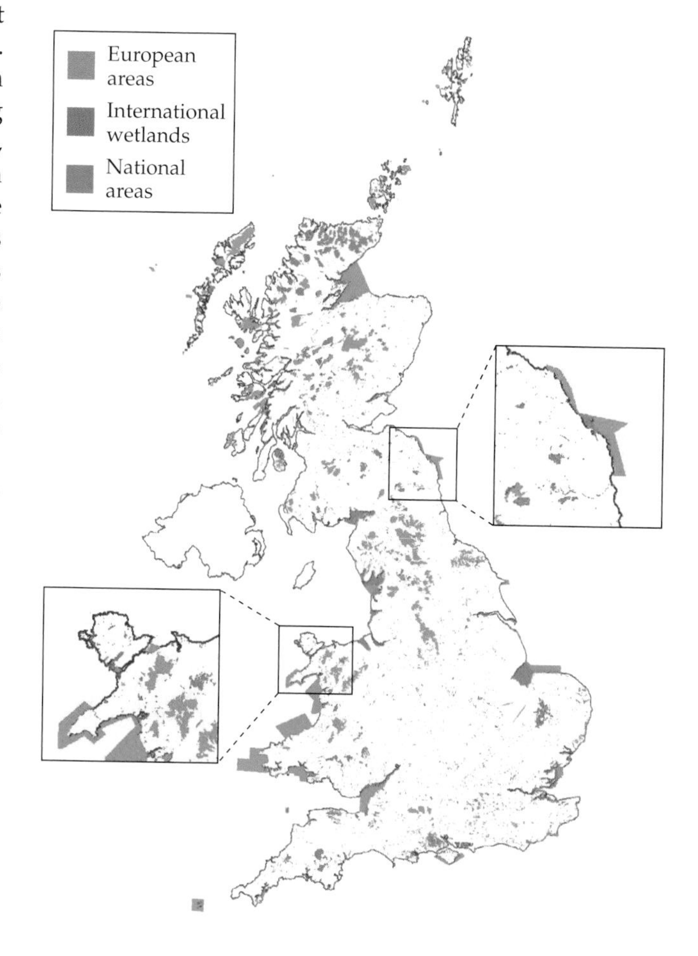

FIGURE 7.9 The geographic locations of protected areas in Britain managed for biodiversity conservation. Note their large number, varied sizes and shapes, and scattered distribution. At the scale of this map, most of the small protected areas cannot be seen. Many of the protected areas are covered by two or more designations. There are other areas managed for other purposes, not shown on this map. (Courtesy of Sarah Little.)

Singapore is an excellent example of a small reserve that provides long-term protection for numerous species. This 164 ha forest reserve represents 0.2% of the original forested area on Singapore and has been isolated from other forests since 1860, yet it still protects 74% of the original flora, 72% of the original bird species, and 56% of the fish (Corlett and Turner 1996). In addition, small reserves located near populated areas make excellent conservation education and nature study centers that further the long-range goals of conservation biology by developing public awareness of important issues. By 2030, over 60% of the world's population will live in urban areas; thus, there is a need to develop such reserves for public use and education.

It is generally agreed that protected areas should be designed to minimize harmful edge effects. Conservation areas that are rounded in shape minimize the edge-to-area ratio, and the center is farther from the edge than in other park shapes. Long, linear parks, such as those that protect the edges of streams and rivers, have the most edge, and all points in the park are close to it (Yamaura et al. 2008). Consequently, for parks with four straight sides, a square park is a better design than an elongated rectangle of the same area. Unfortunately, these ideas are rarely, if ever, implemented. Most parks have irregular shapes because land acquisition actions and the features they are protecting (such as rivers) are typically a matter of opportunity rather than a matter of design.

As discussed in Chapter 4, internal fragmentation of protected areas by roads, fences, farming, logging, and other human activities should be avoided as much as possible, because fragmentation creates barriers to dispersal and often divides a large population into two or more smaller populations, each of which is more vulnerable to extinction than the large population. Fragmentation also provides entry points for invasive species that may harm native species and creates more undesirable edge effects.

Networks of protected areas

Strategies exist for aggregating small and large protected areas into larger conservation networks (Bode et al. 2011). Such networks are needed because, as seen in the preceding discussions, it often requires a larger area to protect species and other aspects of biodiversity.

Cooperation among public and private landowners to create conservation networks is particularly important in developed metropolitan areas, where there are often many small, isolated nature reserves under the control of a variety of different government agencies and private organizations. An excellent example of such cooperation is the Chicago Wilderness project, which consists of more than 250 organizations collaborating to preserve and manage a network of conservation areas consisting of more than 145,000 ha (360,000 acres) of tallgrass prairies, woodlands, and rivers, streams, and other wetlands in the metropolitan Chicago area (**Figure 7.10**) (www.chicagowilderness.org).

(A)

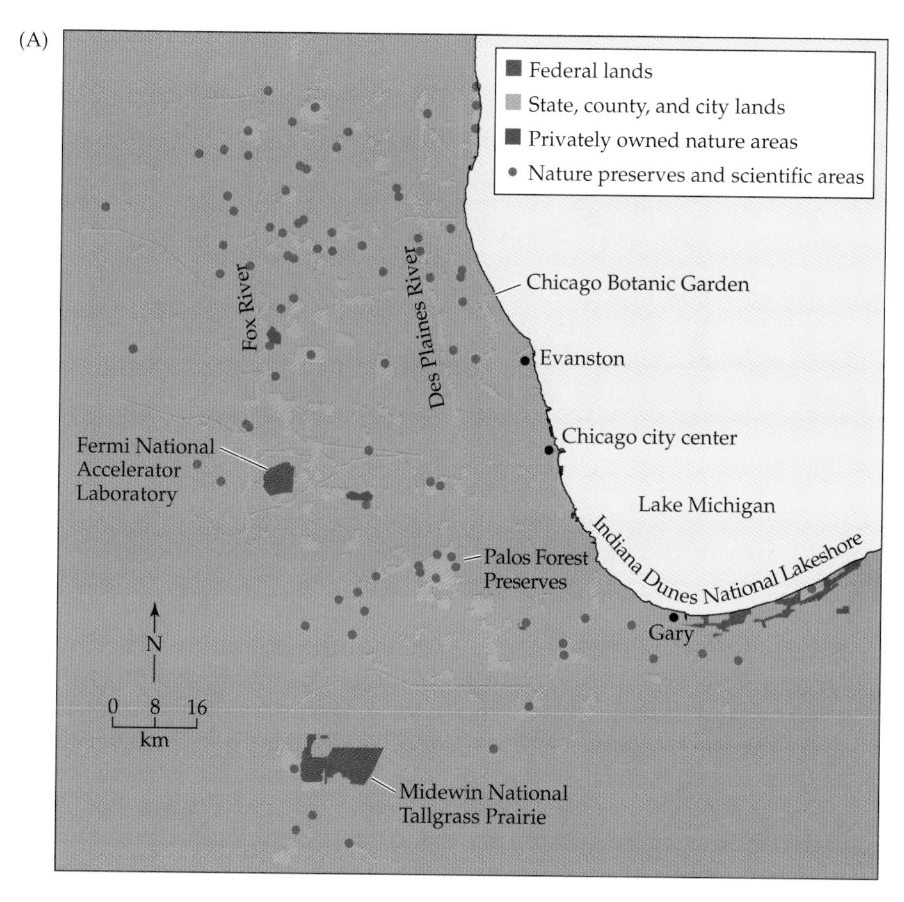

(B)

FIGURE 7.10 (A) The Chicago Wilderness project involves more than 250 organizations working to preserve biodiversity in a densely settled urban area surrounded by agriculture. Many of the linear areas are trails and the banks of rivers. (B) Children use hand lenses to examine milkweed pods as a part of an educational activity at the Chicago Botanic Garden. (Courtesy Chicago Regional Biodiversity Council and www.chicagowildernessmag.org; photograph © Chicago Botanic Garden.)

FIGURE 7.11 Tina Buijs, a park guard supervisor and operations manager with The Nature Conservancy (TNC), talks with Juan Antillanca, a farmer belonging to the Huiro indigenous community that borders TNC's Valdivian Coastal Reserve in Chile. The reserve is a 61,000 ha site comprising temperate rain forest and 36 km of Pacific coastline. Keeping in close contact with their neighbors helps TNC officials realize their conservation goals. (Photograph © Mark Godfrey/The Nature Conservancy.)

MIXED-USE PROTECTED AREAS Often a protected area is embedded in a larger matrix of habitat managed for human uses such as timber forest, grazing land, and farmland (**Figure 7.11**). If conservation biologists can make management for the protection of biodiversity a secondary priority of these areas, then larger habitat areas can be included in conservation plans, and the effects of fragmentation can be reduced (Hansen et al. 2011). Habitat managed for resource extraction can sometimes also be designated as important secondary sites for wildlife and as dispersal corridors between isolated protected areas. Using a network approach, populations of rare species can be managed as large metapopulations to facilitate gene flow and migration among populations (Nol et al. 2005).

HABITAT CORRIDORS One valuable conservation strategy has been to link isolated protected areas into one large system through the use of **habitat corridors**: strips of land running between the reserves (Beier 2011). Such habitat corridors, also known as **conservation corridors** or **movement corridors**, allow plants and animals to disperse from one protected area to another, facilitating gene flow and colonization of suitable sites.

Establishing habitat corridors can potentially transform a set of isolated protected areas into a linked network within which populations can interact as a metapopulation.

Corridors are clearly needed to preserve animals that must migrate seasonally among a series of different habitats to obtain food and water, such as the large grazing mammals of the African savanna; if these animals were confined to a single reserve, they could starve (Wilcove and Wikelski 2008). Observations of Brazilian arboreal mammals suggest that corridors 30 to 40 m wide may be adequate for migration of most species and that a corridor width of 200 m of primary forest will be adequate for all species (Laurance and Laurance 1999). Corridors only 4 m wide are adequate for many species (Rocha et al. 2011). In agricultural landscapes, increasing the connectivity of remnant fragments of natural habitat allows native species to persist at higher densities (Hilty and Merenlender 2004).

Some park managers have enthusiastically embraced the idea of corridors as a strategy for managing wide-ranging species. In Florida, millions of dollars have been spent to establish corridors between tracts of land occupied by the endangered Florida panther (*Felis concolor coryi*). In many areas, culverts, tunnels, and overpasses create passages under and over roads and railways that allow dispersal between habitats for lizards, amphibians, and mammals (Corlatti et al. 2009). An added benefit of these passageways is that collisions between animals and vehicles are reduced, which saves lives and money. In Canada's Banff National Park, road collisions involving deer, elk, and other large mammals declined by 96% after fences, overpasses, and underpasses were installed along a highway (**Figure 7.12A**) (Ford et al. 2009).

(A)

(B)

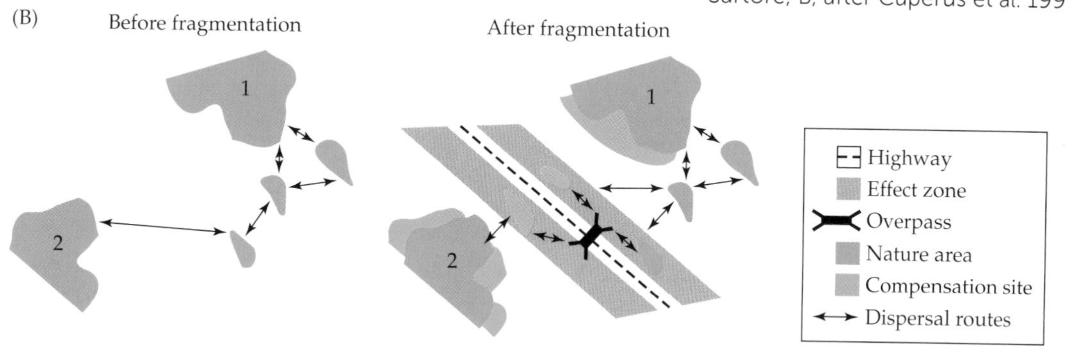

FIGURE 7.12 (A) An overpass above a fenced-off divided highway allows animals to migrate safely between two forested areas. (B) Individuals of a species naturally disperse between two large protected areas (areas 1 and 2, left) by using smaller protected areas as stepping-stones. The right-hand panel shows that habitat destruction and a large edge effect zone caused by a new road have blocked a migration route. To offset the effects of the road, compensation sites (orange) have been added to the system of protected areas, and an overpass has been built over the highway to allow dispersal. (A, photograph © Joel Sartore; B, after Cuperus et al. 1999.)

Some conservation biologists have started to plan habitat corridors on a truly huge scale. Wildlands Network has a detailed plan called the Spine of the Continent Initiative that would link all large protected areas in the western United States and Canada by habitat corridors, creating a system that would allow large and currently declining mammals to coexist with human society.

As the global climate continues to change in the coming decades, many species will need to migrate to higher elevations and to higher latitudes. Creating corridors to protect expected migration routes—such as north–south river valleys, ridges, and coastlines—would be a useful precaution (Hannah 2010). Corridors that cross gradients of elevation, rainfall, and soils will also allow local migration of species to more favorable sites.

Although the idea of corridors is intuitively appealing, there are some possible drawbacks. Corridors may facilitate the movement of pest species and disease; thus, a single infestation could quickly spread to all of the connected protected areas and cause the extinction of all populations of a rare species. Also, animals dispersing along corridors may be exposed to greater risks of predation, because human hunters as well as animal predators tend to concentrate on routes used by wildlife.

Some studies published to date support the conservation value of corridors, while other studies do not show any effect (Pardini et al. 2005). In general, maintaining existing corridors is probably worthwhile because many of them are along watercourses that may be biologically important habitats themselves. When new protected areas are carved out of large blocks of undeveloped land, preserving corridors between protected areas would be beneficial. There would also be value to leaving small clumps of original habitat between large conservation areas, to facilitate movement in a stepping-stone pattern (**Figure 7.12B**). The protected areas that birds use as stopping points along the flyways of their annual migration routes could be considered examples of such stepping-stones.

Landscape Ecology

The interaction of actual land-use patterns, conservation theory, and park design is evident in the discipline of **landscape ecology**, which investigates repeating patterns of habitat types on a regional scale and analyzes their influence on species distribution and ecosystem processes (Schwenk and Donovan 2011).

In some cases, long-term traditional human use has created landscape patterns that preserve and even increase biodiversity.

Landscape ecology has been more intensively studied in Europe and Asia, where long-term practices of traditional agriculture and forest management have determined the landscape pattern. In contrast, research in North America has emphasized single habitat types that are considered (often erroneously) to be minimally affected by people. In the European countryside, cultivated fields, pastures, woodlots, and hedges alternate to create a mosaic that affects the distribution of wild

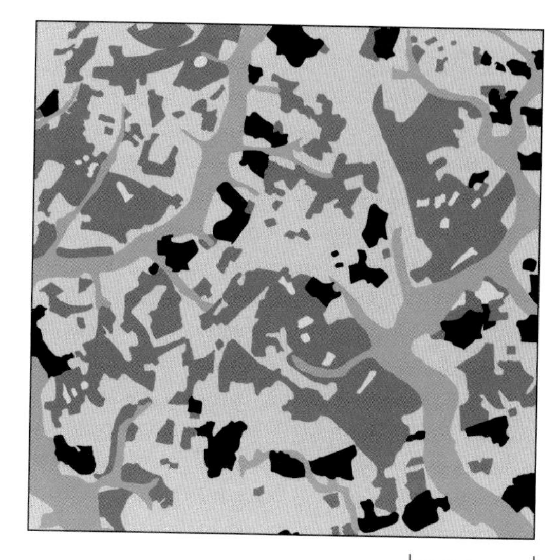

FIGURE 7.13 Traditional rural landscape near Tokyo, Japan, with an alternating pattern of villages (black), secondary forest (dark green), wet rice fields ("rice paddies"; light green), and hay fields (beige). Such landscapes were common in the past but are becoming more rare because of the increasing mechanization of Japanese agriculture, the movement of the population away from farms, and the urbanization of the Tokyo area. (After Yamaoko et al. 1977.)

species. In the traditional Japanese landscape known as *satoyama*, flooded rice fields, hay fields, villages, and forests provide a rich diversity of habitat for wetland species, such as dragonflies, amphibians, and waterfowl (**Figure 7.13**) (Kobori and Primack 2003; Kadoya et al. 2009). In many areas of Europe and Asia, traditional patterns of farming, grazing, and forestry are now being abandoned. In some places, rural people leave the land completely and migrate to urban areas, or their farming practices become more intensive, involving more machinery and inputs of fertilizer. To protect rare species and biological communities in such cases, conservation organizations and governmental agencies have to adopt strategies to maintain the traditional landscapes, sometimes by subsidizing traditional farming practices or having volunteers manage the land.

To increase the number and diversity of animals in nature reserves, wildlife managers sometimes seek to create the greatest amount of landscape variation possible within the confines of their game management unit. Fields and meadows are created and maintained, small thickets are encouraged, groups of fruit trees and crops are planted, patches of forests are periodically cut, little ponds and dams are developed, and numerous trails and dirt roads meander across and along all of the patches. Such landscaping is often appealing to the public, who are the main visitors and financial contributors to the park. However, such intensively managed reserves may lack many rare interior species that survive only in large blocks of undisturbed habitat. Large animals, such as bears, mountain lions, and tigers, will need to be managed on a regional landscape level, in which the sizes of the landscape units more closely correlate to the natural population sizes and migration patterns of the species (Wikramanayake et al. 2011).

Managing Protected Areas

A common misconception is that once protected areas are legally established, the work of conservation is largely complete, and nothing more needs to be done. Although some people believe that "nature knows best" and that biodiversity is best served when humans do not intervene, the reality is often very different. In many cases, humans have already modified the environment so much that the remaining species and ecosystems need human monitoring and intervention in order to survive. The world is littered with "paper parks" created by government decree but left to flounder without any management or funding (Joppa et al. 2008). These protected areas have gradually—or sometimes rapidly—lost species as their habitat quality has been degraded. In some countries, people readily farm, log, mine, hunt, and fish in protected areas because they feel that government land is owned by "everyone," "anybody" can take whatever they want, and "nobody" is willing to intervene. The crucial point is that often parks must be actively managed according to a carefully thought-out **management plan** to prevent deterioration. Important considerations in such a management plan include protecting biological diversity, maintaining ecosystem services and health, maintaining a historical landscape, and providing resources and experiences of value to local people and visitors (Hobbs et al. 2010). Part of the management plan involves making the public aware of which activities are encouraged (for example, wildlife photography) and which are prohibited (for example, hunting), and then enforcing the rules (**Figure 7.14**).

In many European, Asian, and African countries with long traditions of cultivation, ranching, and grazing, hundreds and even thousands of years of human activity have shaped habitats such as woodlands, meadows, and hedges. These habitats support high species diversity as a result of traditional land-management practices, which must be maintained if the species are to persist (Jacquemyn et al. 2011). Similarly, grasslands that have been grazed in the past by large wild animals or domesticated animals, such as cattle, still need to be grazed. If protected areas that include these types of habitats

FIGURE 7.14 Management of protected areas may involve deciding which types of activities are encouraged and which are prohibited, as indicated by this sign at the entrance to a national park in Romania. (Photograph by Richard B. Primack.)

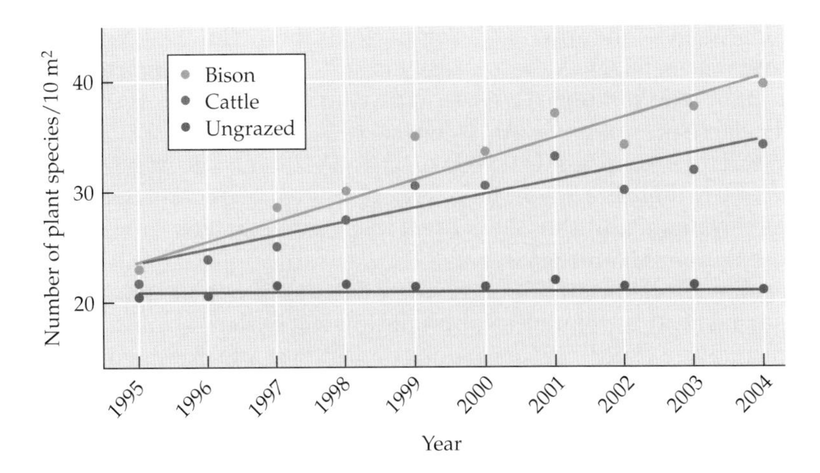

FIGURE 7.15 Large herbivores originally grazed the tallgrass prairies of the midwestern United States. The loss of these herbivores has altered the ecology of this ecosystem, with a resulting loss of plant species. Management involving grazing by cattle and bison resulted in a gradual increase in plant species in prairie research plots over a 10-year period, compared with ungrazed control plots. (After Towne et al. 2005; photograph courtesy of Jim Peaco/U.S. National Park Service.)

are not managed, they will undergo succession over a few years or decades, and many of their characteristic species will disappear (**Figure 7.15**).

It is also true, however, that management practices are sometimes ineffective or even detrimental. For example, active management to promote the abundance of a game species such as deer has frequently involved eliminating top predators, such as wolves and cougars; without predators to control them, game populations (and, incidentally, rodents) may increase far beyond expectations. The result is overgrazing, habitat degradation, and a collapse of animal and plant communities. Overenthusiastic park managers who remove hollow trees, dead trees, rotting logs, and underbrush to "improve" a park's appearance may unwittingly remove a critical keystone resource needed by certain animal species for nesting and overwintering, by rare plants for seed germination, and by all species as an integral part of nutrient cycling (Keeton et al. 2007). In many protected areas, fire is part of the natural ecology of the area, and attempts to suppress fire are expensive and waste scarce management resources. Suppressing the normal fire cycle may eventually lead to the loss of fire-dependent plant species, as well as to uncontrollable fires of unnatural intensity.

Many examples of successful park management come from the United Kingdom, where there is a history of scientists and volunteers successfully monitoring and managing small reserves such as the Monks Wood and Castle Hill National Nature Reserves (www.naturalengland.org.uk). At these sites, the effects of different grazing methods (sheep versus cattle, light versus heavy) on populations of wildflowers, butterflies, and birds are closely followed using an experimental approach in which treated areas are compared with control areas. These experiments demonstrate that parks often must be actively managed to prevent deterioration. The most effective parks are usually those whose managers have the benefit of information provided by research and monitoring programs and have funds available to implement the management plans.

In some protected areas, difficult choices may have to be made; for example, when protected sea lions are eating endangered yellow-eyed penguins in New Zealand, park managers may need to prioritize one species over the other in determining management practices (Lalas et al. 2007).

Management plans are needed that articulate conservation goals and practical methods for achieving them. Management activities can include controlled burns, enforcement of restrictions on human use, and maintenance of keystone resources, especially water.

Managing sites

Park managers sometimes must actively manage sites to ensure that all successional stages are present so that species characteristic of each stage will have a place to persist and thrive, as discussed in Chapter 2. Two common ways to do this is to set localized, controlled fires periodically or to cut trees in a small area to reinitiate vegetation succession (Davis et al. 2009). Obviously, any habitat management burning must be done in a legal and carefully controlled manner to prevent damage to nearby property (**Figure 7.16A**). Also, prior to burning or tree cutting, land managers need to develop a program of public education to explain to local residents the role of habitat management in maintaining the balance of nature. In other situations, however, parts of protected areas must be managed carefully to *minimize* fire and other disturbance, providing the conditions required by old-growth species (**Figure 7.16B**). In urban areas, if managers want to maintain healthy frog populations, they may need to actively remove predatory fish, improve water quality, and remove excessive vegetation (Hamer and Parris 2011).

WETLANDS Wetlands management is a particularly crucial issue. Maintaining healthy wetlands is necessary for populations of waterbirds, fish, amphibians, aquatic plants, and a host of other species (Jähnig et al. 2011). Yet, protected areas may end up directly competing for water resources

(A)

(B)

FIGURE 7.16 Conservation management: intervention versus leave-it-alone. (A) Heathland in protected areas of Cape Cod, Massachusetts, is burned on a regular basis in order to maintain the open vegetation habitat and to protect wildflowers and other rare species. (B) Sometimes management involves keeping human disturbance to an absolute minimum. Muir Woods National Monument is a forest of old-growth coast redwoods, protected in the midst of the heavily urbanized San Francisco Bay area. (A, photograph by Elise Smith, U.S. Fish and Wildlife Service; B, photograph courtesy of U.S. National Park Service.)

with irrigation projects, demands for residential and industrial water supplies, flood control schemes, and hydroelectric dams. Wetlands are often interconnected, so a decision that affects water levels and quality in one place has ramifications for other areas. Water quality monitoring can help to document alterations in quality and quantity of water in ecosystems and provide information needed to convince government officials and the public of the seriousness of the problem. One strategy for maintaining wetlands is to include an entire watershed within the protected area. In any case, park managers need to be effective advocates for maintaining the quantity and quality of water needed for lands under their control.

KEYSTONE RESOURCES In many parks, it may be necessary to preserve, maintain, and supplement **keystone resources** on which many species depend. These resources include trees that supply fruit when little or no other food is available, pools of water during a dry season, exposed mineral licks, and so forth. Keystone resources can be expanded in managed conservation areas to increase the size and number of populations of species of conservation concern whose numbers have declined. By planting native fruit trees and building artificial ponds, for example, it may be possible to maintain vertebrate species in a smaller conservation area, and at higher densities, than would be predicted based on studies of species distribution in undisturbed habitat. For example, small artificial ponds not only provide needed habitat for attractive insects such as dragonflies; they also serve as important centers of public education in urban areas (Kobori and Primack 2003). Providing nesting boxes or drilling nesting holes in trees allows certain bird populations to reproduce and grow when there are few dead trees with nesting cavities.

In each case a balance must be struck between, at one extreme, establishing nature reserves free from human influence and, at the other extreme, creating seminatural gardens in which the plants and animals are dependent on people.

Monitoring sites

An important aspect of protected areas management involves monitoring components that are crucial for biodiversity, such as the water level of ponds and streams; the number of individuals of rare and endangered species; and the density of herbs, shrubs, and trees (Lindenmayer and Likens 2010). Methods for monitoring these components can also include recording standard observations, carrying out surveys, and taking photographs from fixed points. Monitoring an area's biodiversity is being combined with the monitoring of its social and economic characteristics in recognition of the linkages between people and conservation. In particular, the amount and value of plant and animal materials people obtain from nearby ecosystems are important features to monitor. The exact types of information gathered depend on the goals of park management. Not only does monitoring allow managers to determine the health of the park, it can also suggest which management practices are working and which are not (de Bello et al. 2010). The effectiveness of park management practices can sometimes be investigated with carefully designed experiments involving comparisons with control areas or prior baseline data. Managers must continually refine the information they need on conditions inside, or sometimes outside, protected areas and be ready to adjust park management practices in an adaptive manner to achieve conservation objectives, sometimes referred to as **adaptive management** (**Figure 7.17**).

> Parks must be monitored to determine whether their goals are being met. Management plans may need to be adjusted based on new information from monitoring.

One important problem for monitoring programs is that they are often discontinued after a few years, because of a lack of funding or lack of interest. The fact that environmental effects such as acid rain and climate change may lag behind their initial causes for many years creates a challenge to understanding change in ecosystems. To remedy this situation, many scientific research organizations and government departments have begun to implement programs for monitoring ecological change over the course of decades and centuries. One program is the system of 26 **Long-Term Ecological Research** (**LTER**) sites established by the U.S. National Science Foundation. Another is UNESCO's Man and the Biosphere system of biosphere reserves.

The scale and methods of monitoring have to be appropriate for management needs. For large parks in remote areas, remote sensing using satellites and airplanes may be an effective method for monitoring logging, shifting cultivation, mining, and other activities, both authorized and unauthorized (Morgan et al. 2010). In some cases, local people can be trained to carry out the required monitoring at a local scale. Often these local people have extensive and useful knowledge of a protected area that they are willing to share as part of the monitoring process (Anadón et al. 2009).

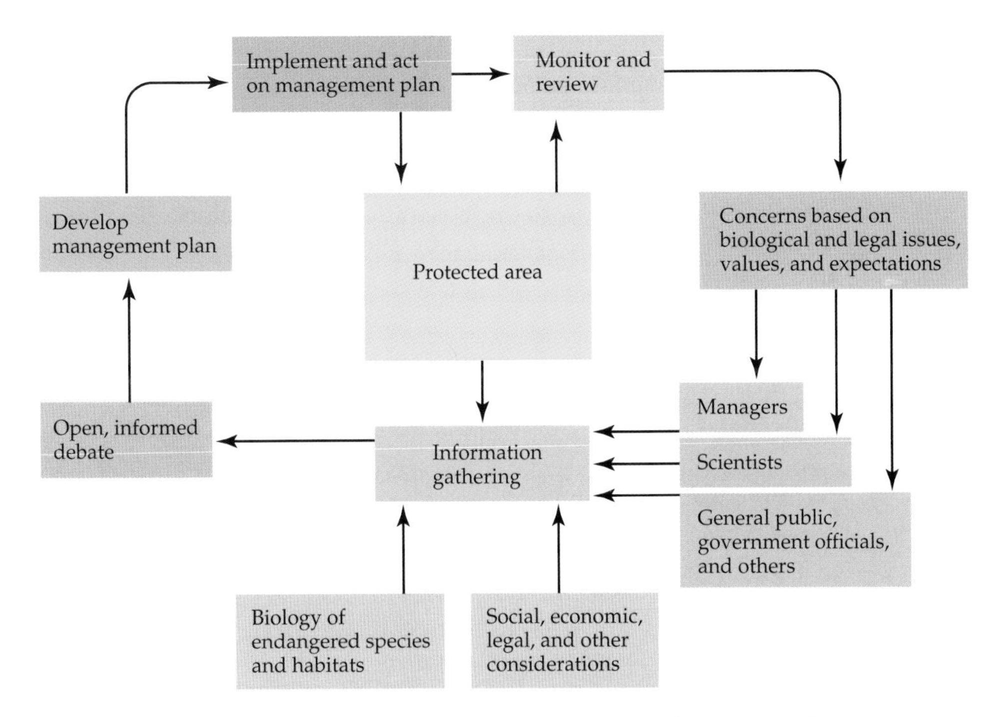

FIGURE 7.17 Model of an adaptive management process for protected areas, emphasizing the decision-making stages. Input is solicited from many sources, and then the plan is developed, implemented, and monitored. (After Cork et al. 2000.)

> The involvement of local people is often the crucial element missing from conservation strategies. Local people need to be involved in conservation programs, as participants, employees, and leaders.

Management and people

In both developed and developing countries, a central part of any park's management plan must be a policy on the use of the park by local people and outside visitors (Grumbine and Xu 2011). People who have traditionally long used a protected area and are suddenly not allowed to enter the area and use the resources will be understandably angry and frustrated, and people in such a position are unlikely to be strong supporters of conservation (Mascia and Claus 2009). Many parks flourish or are destroyed depending on the degree of support, neglect, hostility, or exploitation they receive from the people in the surrounding area who use them. If the purpose of a protected area is explained to local residents, and if most residents agree with the objectives of the park, they may become strong supporters of what they see as common land. In the most positive scenario, local people become involved in park management and planning, are trained and employed by the park authority, and receive economic benefits from the protection of biodiversity and regulation of activity within the park (Ferro et al. 2011).

At the other extreme, if there is a history of bad relations and mistrust between local people and the government, or if the purpose of the park is not explained adequately, the local people may reject the park concept and ignore park regulations. In this case, the local people will come into conflict with park personnel, to the detriment of the park. Park personnel and even armed soldiers may have to patrol constantly to prevent illegal activity. In some protected areas local people have been arrested and jailed for activities that they regard as necessary for their livelihood. Escalating cycles of conflict can lead to outright violence by local people, and park personnel can be threatened, injured, and even killed.

Unfortunately, it is sometimes necessary to exclude local people from protected areas when resources are being overharvested, either legally or illegally, to the point that the health of the ecosystem and the existence of endangered species are threatened (Singh and Gibson 2011). Such degradation and loss of biodiversity can result from overgrazing by cattle, excessive collection of fuelwood, or hunting with guns. In such cases the only solution may be a strategy of "fences and fines."

Zoning as a solution to conflicting demands

A possible way to deal with conflicting demands on a protected area is **zoning**, which considers the overall management objectives for a park and sets aside designated areas that permit or give priority to certain activities (Eigenbrod et al. 2009). Different areas may be designated for timber production, hunting, wildlife protection, nature trails, or watershed maintenance. Other zones may be established for the recovery of endangered species, restoration of degraded communities, and scientific research.

The challenge in zoning is to find a compromise that people are willing to accept that provides for the long-term, sustainable use of natural resources. Managers often need to spend considerable effort informing the public about what activities are acceptable in particular areas of a park, and then enforcing park regulations (Andersson et al. 2007).

Zoned marine reserves, also known as **marine protected areas (MPAs)**, marine parks, and no-fishing zones, have proven an effective way to rebuild and maintain populations of fish and other marine organisms (**Figure 7.18**) (McCook et al. 2010). In comparison with nearby unprotected sites, marine parks often have greater total weight of commercially important fish, greater numbers of individual fish, and greater coral

Zoning allows the separation of mutually incompatible activities. Marine protected areas are often zoned with no-fishing areas where fish and other marine organisms can recover from harvesting.

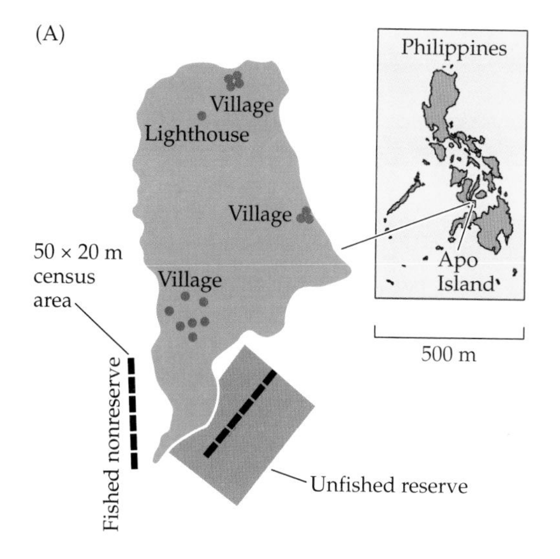

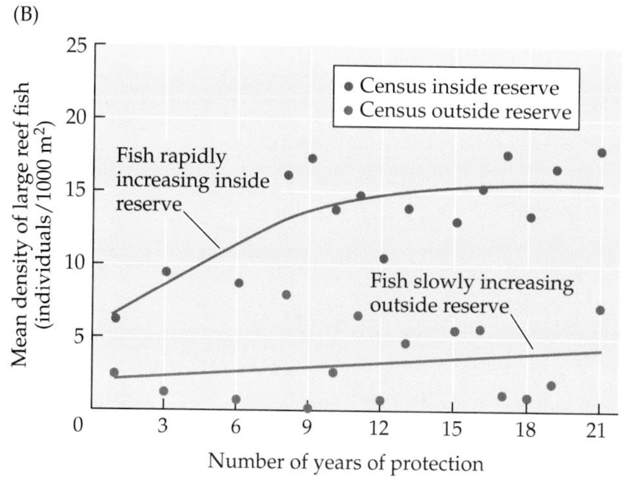

FIGURE 7.18 On Apo Island in the Philippines, large reef fish had been overharvested to the point where they were rarely seen. (A) In response to overharvesting, a reserve was set up (blue area) on the eastern side of the island, while fishing continued as before at a specified nonreserve area on the western side. A censusing study measured the number of large reef fish at each site (six underwater census areas are shown as black rectangles for each site). (B) Resulting data show that after the marine reserve was established, the number of fish observed in the unfished reserve increased substantially. The number of fish in the unprotected area did not increase initially, because the fish were still being intensively harvested; after about 8 years, however, an increase became detectable, originating from the spillover of fish from the reserve area. (After Abesamis and Russ 2005.)

FIGURE 7.19 (A) The general pattern of a biosphere reserve: a core protected ▶ area is surrounded by a buffer zone, where human activities are monitored and managed and where research is carried out; this, in turn, is surrounded by a transition zone, where sustainable development and experimental research take place. (B) Kuna people still practice traditional methods of catching fish in Kuna Yala Indigenous Reserve. (Photograph © Andoni Canela/AGE Fotostock.)

reef cover. Evidence shows that fish from marine reserves spill over into adjacent unprotected areas, where they can help rebuild populations and also be caught by fishermen. Enforcement of zoning is often a major challenge in marine reserves because fishermen will tend to move toward and into the fishing-exclusion zones, because those areas are where the fishing is best, and that leads to overfishing at the margins of the marine reserve. Conservation and zoning in freshwater environments share many of the same concerns (Nogueira et al. 2010). A combination of local involvement, publicity, education, clear posting of warning signs, and visible enforcement significantly increases the success of any zoning plan, especially in the marine environment (Aburto-Oropeza et al. 2011; Fox et al. 2012).

BIOSPHERE RESERVES UNESCO has pioneered one such zoning approach with its **Biosphere Reserves** Program, which integrates human activities, research, protection of the natural environment, and sometimes tourism at a single location. These locations often have well-established human settlements, traditional land-use patterns, and scenic landscapes. A desirable feature of biosphere reserves is a system in which there are zones delineating varying levels of use (**Figure 7.19A**). At the center is a core area in which ecosystems are strictly protected. This is surrounded by a buffer zone in which traditional human activities—such as collection of edible plants and small fuelwood—are monitored and nondestructive research is conducted. Surrounding the buffer zone is a transition zone in which some forms of sustainable development (such as small-scale farming) are allowed, along with some extraction of natural resources (such as selective logging) and experimental research. This general strategy of surrounding core conservation areas with buffer and transitional zones can encourage local people to support the goals of the protected area.

One instructive example of a biosphere reserve is the Kuna Yala Indigenous Reserve on the northeast coast of Panama. In this protected area comprising 60,000 ha of tropical forest and coral islands live 50,000 Kuna people, in 60 villages, who practice traditional medicine, agriculture, fishing, and forestry while documentation and research is undertaken by scientists from outside institutions (**Figure 7.19B**). At present Kuna conservation beliefs and practices are gradually changing because of outside influences, and younger Kuna often question the need to rigidly protect the reserve. Further, rising sea levels and declining marine resources are forcing village leaders to consider other options for their future (Posey and Balick 2006).

(A)

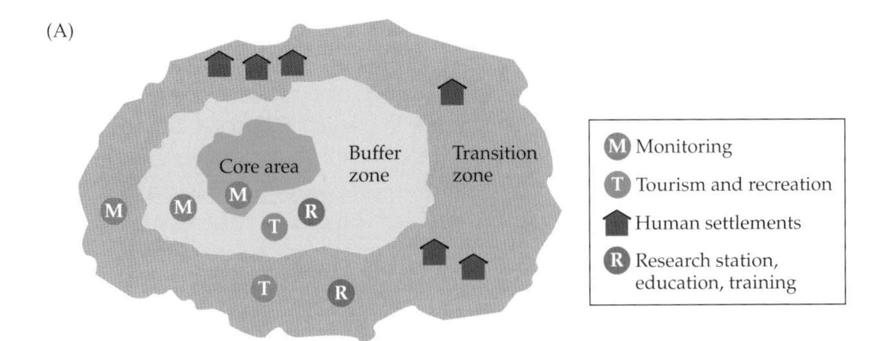

(B)

Challenges to Park Management

Human populations will continue to increase dramatically in the coming decades, and resources such as fuelwood, medicinal plants, and bushmeat will become harder to find. Managers of protected areas in the developing world need to anticipate ever-greater demand for use of the remaining patches of natural habitat. Conflict is inevitable as more people live and farm closer to high concentrations of wildlife that, when food is scarce, have nowhere to go but out of the park and into nearby agricultural fields and residential areas. Elephants, primates, and flocks of birds that live inside protected areas can all be significant crop raiders, while predators such as tigers and bears can kill domestic animals and pose a threat to people. Similarly, people who are poor and hungry will enter the nearby protected areas to take what they need to live, whether or not they have permission.

For park management to be effective, there must be adequate funding for a sufficient number of well-equipped, properly trained, and motivated park personnel who are willing to carry out park policy (Armsworth et al. 2011; Tranquilli et al. 2012). Buildings, communications equipment, and other appropriate elements of infrastructure are necessary to manage a park. In many areas of the world, particularly in developing but also in developed countries, protected areas are understaffed and lack the equipment to patrol remote areas of the reserve. Without enough radios and vehicles, the park staff may be restricted to the vicinity of headquarters, unaware of what is happening in their own park. The importance of sufficient personnel and equipment should not be underestimated; in Panama's protected areas, for instance, the abundance of large mammals and the seed dispersal services they provide are directly related to the frequency of antipoaching patrols by park guards (that is, parks with more patrols have more large mammals) (Brodie et al. 2009). International conservation organizations and governmental agencies often assist in providing funds for managing protected areas in developing countries, but the funding is still often not sufficient.

It is ironic that at the same time that zoos, aquariums, botanical gardens, and museums in the developed countries spend vast sums on visitor centers and research laboratories, the biologically rich national parks of developing countries often languish for lack of resources. Increasing funding for the management of protected areas needs to be a priority for government agencies and conservation organizations. And at the end of the day, conservation biologists need to account for whether their management of protected areas achieved the stated goals and whether money was spent effectively.

Summary

- Protecting habitat is the most effective method of preserving biodiversity. Thirteen percent of the Earth's land surface is included in about 108,000 protected areas, but the percentage may not increase much further, because of the needs of human societies for natural resources.
- Government agencies and conservation organizations have set priorities for establishing new protected areas based on the relative distinctiveness, endangerment, and utility of the species and ecosystems that occur in particular locations. Many protected areas are established to safeguard focal species of special concern, unique ecosystems, wilderness areas, and concentrations of rare and endangered species. Gap analysis is used to identify areas and elements of biodiversity in need of additional protection.
- Conservation biologists are developing guidelines for designing protected areas: protected areas should be large whenever possible, they should not be fragmented, and managers should create networks of conservation areas for maximum protection.
- Habitat corridors connecting protected areas may allow species dispersal to take place and may be particularly important in maintaining known migration routes.
- Protected areas often must be actively managed in order to maintain their biodiversity. Monitoring provides information needed to evaluate whether management activities are achieving their intended objectives or need to be adjusted.
- Management might involve zoning to establish areas where certain activities are allowed or prohibited. Managing interactions with local people are critical to the success of protected areas and should be part of a management plan.
- Adequate staffing and funding are necessary for park management.

For Discussion

1. Obtain a map of a town, state, or nation that shows protected areas (such as nature reserves and parks) and multiple-use managed areas. Who is responsible for each parcel of land, and what is the goal in managing it? Consider the same issues for aquatic habitats (ponds, lakes, rivers, coastal zones, etc.).
2. If you could protect additional areas on the map, where would they be and why? Show their exact locations, sizes, and shapes, and justify your choices.
3. Think about a national park or nature reserve you have visited. In what ways was it well run or poorly run? What were the goals of this protected area, and how could they be achieved through better management?
4. Can you think of special challenges in the management of aquatic preserves such as coastal estuaries, islands, or freshwater lakes that would not be faced by managers of terrestrial protected areas?

Suggested Readings

Chape, S., M. D. Spalding, and M. D. Jenkins (eds.). 2008. *The World's Protected Areas: Status, Values, and Prospects in the Twenty-First Century*. University of California Press, Berkeley. Massive source of information on protected areas.

Ferro, P. J., M. M. Hanauer, and K. R. E. Sims. 2011. Conditions associated with protected area success in conservation and poverty reduction. *PNAS* 108: 13,913–13,918. Protected areas provide many benefits for poor people living nearby.

Fox, H. E. and 14 others. 2012. Reexamining the science of marine protected areas: Linking knowledge to action. *Conservation Letters* 5: 1–10. Biological, social, and policy issues are all important in effective marine protected areas.

Grumbine, R. E. and J. Xu. 2011. Creating a "Conservation with Chinese Characteristics." *Biological Conservation* 144: 1347–1355. Characteristics of Chinese society may result in a distinctive approach to conservation issues.

Hannah, L. 2010. A global conservation system for climate-change adaptation. *Conservation Biology* 24: 70–77. The world's network of protected areas needs to be planned with climate change in mind.

Hobbs, R. J. and 15 others. 2010. Guiding concepts for park and wilderness stewardship in an era of global environmental change. *Frontiers in Ecology and the Environment* 8: 483–490. Excellent statement of the need for guiding principles in park management.

Mascia, M. B. and S. Pallier. 2011. Protected areas downgrading, downsizing, and degazettement (PADD) and its conservation implications. *Conservation Letters* 4: 9–20. In many areas of the world, the official protection of national parks and other conservation areas is being withdrawn.

McCook, L. J. and 20 others. 2010. Adaptive management of Great Barrier Reef: A globally significant demonstration of the benefits of networks of marine reserves. *PNAS* 107: 18,278–18,285. This premier protected area provides great value to Australia and important lessons to other countries.

Olds, A. D., R. M. Connolly, K. A. Pitt, and P. S. Maxwell. 2012. Habitat connectivity improves reserve performance. *Conservation Letters* 5: 56–63. Networks of marine reserves benefit from connectivity.

Spalding, M. D., L. Fish, and L. J. Wood. 2008. Towards representative protection of the world's coasts and oceans—Progress, gaps, and opportunities. *Conservation Letters* 1: 217–226. Marine protected areas are now being established throughout the world, but efforts still lag far behind terrestrial conservation.

Tranquilli, S. and 40 others. 2012. Lack of conservation effort rapidly increases African great ape extinction risk. *Conservation Letters* 5: 48–55. Patrols by park guards and the actions of conservation organizations are the key to protecting wildlife.

Farmers and nursery workers plant tree seedlings on degraded land in Costa Rica.

Chapter 8
Conservation Outside Protected Areas

Establishing protected areas with intact ecosystems is essential for species conservation (Gibson et al. 2011). It is, however, shortsighted to rely solely on protected areas to preserve biodiversity. Such reliance can create a paradoxical situation in which species and ecosystems inside the protected areas are preserved while the same species and ecosystems outside are allowed to be damaged, which in turn results in the decline of biodiversity *within* the protected areas (Newmark 2008). This decline is due in part to the fact that many species must migrate across protected area boundaries to access resources that the protected area itself cannot provide. In India, for example, tigers leave their protected sanctuaries to hunt in the surrounding human-dominated landscape.

In general, the smaller the protected area, the more it depends on unprotected neighboring lands for the long-term maintenance of biodiversity. Unprotected areas, including those immediately *outside* protected areas, are thus crucial to an overall conservation strategy (Hansen et al. 2011). Developing strategies for unprotected areas is essential because, according to even the most optimistic predictions, more than 80% of the world's land will remain outside protected areas. Such unprotected areas include government and private lands managed primarily for grazing, logging, and other resource extraction; privately owned farms; highly modified urban areas; and oceans, lakes, rivers, and other aquatic systems where food is harvested.

Many endangered species and unique ecosystems are found partly or entirely on unprotected lands. Consequently, the conservation of biodiversity in these places has to be considered.

Unprotected Public and Private Lands

Human use of ecosystems varies greatly in unprotected lands, but significant portions are not used intensively by humans and still harbor some of their original biota (**Figure 8.1**). Strategies for reconciling human needs and conservation interests in unprotected areas are critical to the success of conservation plans (Koh et al. 2010; Cox and Underwood 2011). In almost every country, numerous rare species and ecosystems exist primarily or exclusively on unprotected public lands or on lands that are privately owned. In the United States, 60% of species that are globally rare or listed under the U.S. Endangered Species Act occur on private forested lands (Robles et al. 2008). Even when endangered species occur on public land, it is often not land managed for biodiversity but rather land managed primarily for timber harvesting, grazing, mining, or other economic uses. For example, 75% of the remaining orangutans in Indonesia live outside of protected forests, often in logged forests and tree plantations (Meijard et al. 2010).

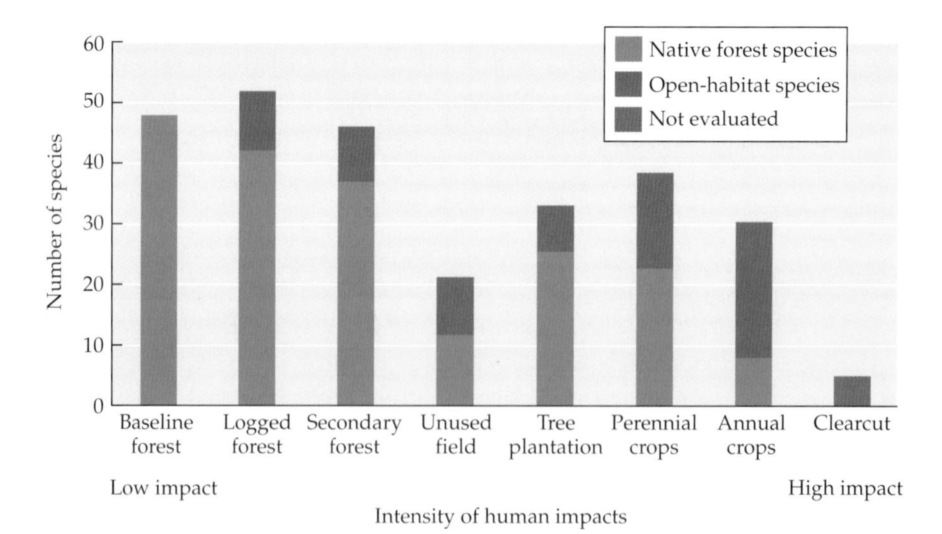

FIGURE 8.1 For a range of land uses in West Africa, the average number of vertebrate native forest species declines with increasing intensity of human impacts, and the number of open-habitat species increases. Some native forest species are still present even with intensive land use, such as tree plantations, but the overall number of species and the proportion of native species are much lower. (After Norris et al. 2010.)

Native species can continue to live in unprotected areas when those areas are set aside or managed for some purpose that is not harmful to the ecosystem. Security zones surrounding government installations are some of the most outstanding natural areas in the world. In the United States, excellent examples of natural habitat occur on the sites of military bases such as Fort Bragg in North Carolina, nuclear processing facilities such as the Savannah River site in South Carolina, and watersheds adjacent to metropolitan water supplies. More than 200 threatened and endangered species of plants and animals occur on Department of Defense managed land.

> Even ecosystems that are managed primarily for the production of natural resources can retain considerable biodiversity, and they have importance in conservation efforts.

In estuaries and seas managed for commercial fisheries, many of the native species remain because commercial and noncommercial species alike require an undamaged chemical and physical environment. Other areas that are not protected by law may retain species because the human population density and degree of utilization are very low. Border areas such as the demilitarized zone between North and South Korea often have an abundance of wildlife because they remain undeveloped and depopulated. Mountain areas, often too steep and inaccessible for agriculture and development, are frequently managed by governments as valuable watersheds that produce a steady supply of water and prevent flash flooding and erosion; they also harbor important ecosystems.

In many parts of the world, wealthy individuals have acquired large tracts of land for personal estates and for private hunting. These estates are frequently used at very low intensity, often in a deliberate attempt by the landowners to maintain large wildlife populations. Some estates in Europe preserve unique old-growth forests that have been owned and protected for hundreds of years by royal families. Such privately owned lands, whether owned by individuals, families, corporations, or tribal groups, often contain important aspects of biodiversity. Strategies that encourage private landowners and government land managers to protect rare species and ecosystems are obviously essential to the long-term conservation of biodiversity. In this chapter and Chapter 9, these strategies will be presented.

Human-dominated landscapes

Most of the world's landscapes have been affected in some way by human activity. The extent of human dominance over the natural landscape varies greatly, however (**Figure 8.2**). Considerable biological diversity can be maintained in well-managed and low-intensity traditional agricultural systems, grazing lands, hunting preserves, forest plantations, and recreational lands (Mendenhall et al. 2011; Wright 2012). Bird species are often abundant in traditional agricultural landscapes, which are characterized by a mixture of small fields, hedges, and woodlands. In comparison with lands subjected to intensive "modern" agricultural practices, traditionally managed agricultural landscapes experience less exposure to herbicides, fertilizers,

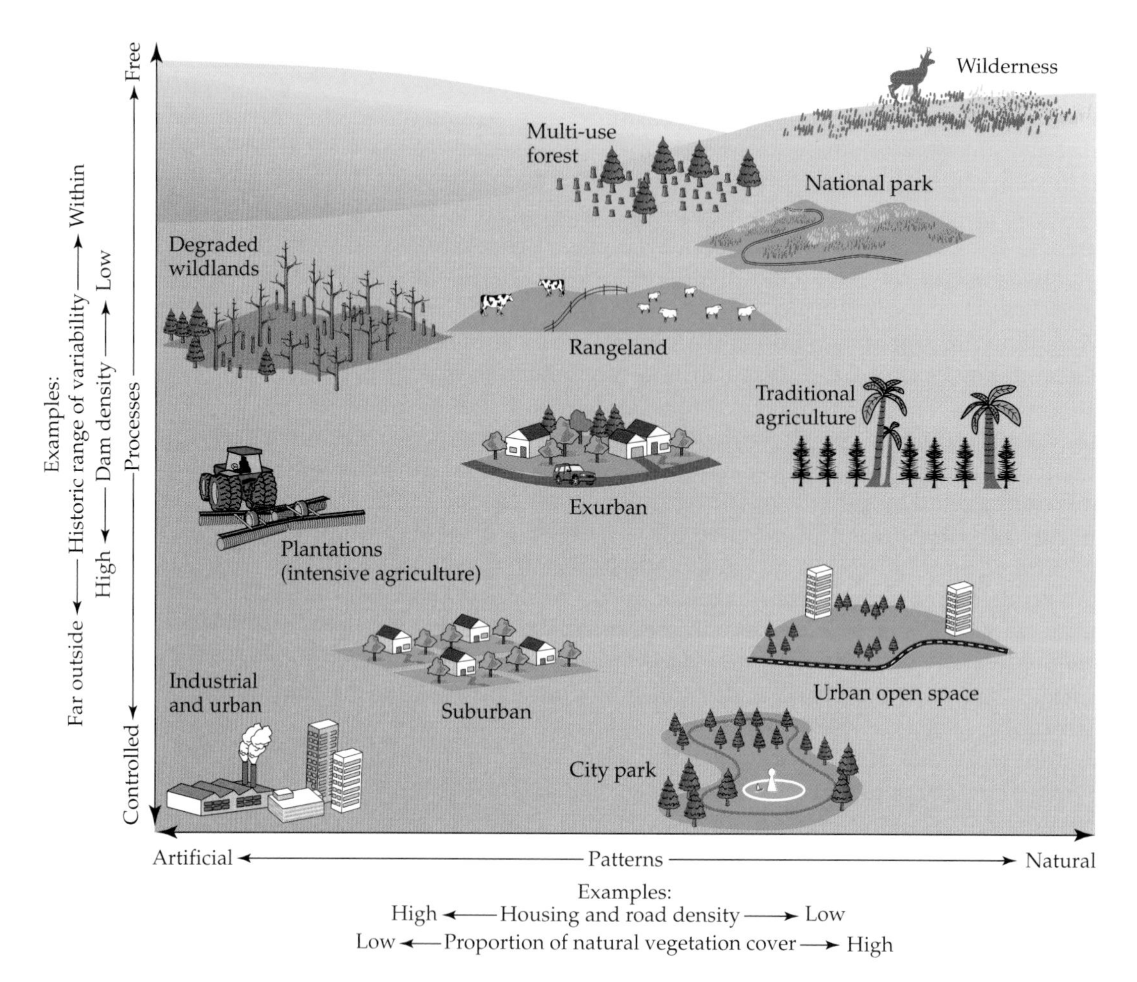

FIGURE 8.2 Landscapes vary in the extent to which humans have altered the patterns of species composition and natural vegetation through activities such as agriculture, road construction, and housing; ecosystem processes (water flow, nutrient cycling, etc.) also vary because of fire control activities, dam construction, and other activities that alter plant cover. Wilderness areas retain most of their original patterns and processes, urban areas retain the least, and other landscapes retain various intermediate amounts. (After Theobald 2004.)

and pesticides and have more habitat heterogeneity. Similarly, farmlands worked using organic methods support more birds than farmlands worked using nonorganic methods, because organic farms have more insects for the birds to eat. In many areas of the world, however, the best agricultural lands are being ever more intensively used, while less optimal lands are abandoned as people leave for urban areas (West and Brockington 2006).

One notable example of preserving biodiversity in an agricultural setting comes from tropical countries and their traditional shade coffee plantations, in which coffee is grown under a wide variety of shade trees, often as many as 40 tree species per farm (**Figure 8.3**). In northern Latin America alone, shade coffee plantations cover 2.7 million ha. These plantations have structural complexity, created by multiple vegetation layers and a diversity of birds and insects, that is sometimes comparable to that in adjacent natural forest (Vandermeer et al. 2010). "Environmentally friendly," shade-grown coffee is sold at a premium price, but let the buyer beware; there are no uniform standards for shade coffee. Thus some coffee marketed as "environmentally friendly, shade coffee" may actually be grown as sun coffee with only a few small, interspersed trees.

Government programs that subsidize traditional agriculture, and include the preservation of wildlife as a major goal, are being developed in countries throughout the world. In Japan, for example, the government provides financial incentives to farmers who maintain traditional rice fields that are flooded during the winter; these traditional farming practices support much greater densities of winter bird populations than intensively managed farms that are not flooded.

Forests that are recovering from selective logging, clear-cutting, or the abandonment of agriculture may still contain a considerable percentage of their original biota and maintain most of their ecosystem services. This is particularly true when species can migrate from nearby undisturbed lands and colonize the sites. In African tropical forests, gorillas, chimpanzees, and elephants can tolerate selective logging and other land uses that involve low levels of disturbance, though only when hunting levels are controlled by active antipoaching patrols (Stokes et al. 2010).

In many countries, large parcels of government-owned land are designated as **multiple-use habitat**; that is, they are managed to provide a variety of goods and services. An emerging and important research area involves the development of innovative ways to reconcile competing claims on land use. This will require careful analyses and considerations of the trade-offs of pursuing alternative

(A)

(B)

FIGURE 8.3 Two types of coffee management systems. (A) Shade coffee is grown under a diverse canopy of trees, providing a forest structure in which birds, insects, and other animals can live. (B) Sun coffee is grown as a monoculture, without shade trees. Animal life is greatly reduced. (A, photograph © John Warburton-Lee Photography/Alamy; B, photograph © Elder Vieira Salles/shutterstock.)

development options in regard to both environmental and socioeconomic priorities (Koh et al. 2010). A different approach is to use the legal system to halt government-approved activities on public lands if these activities threaten the survival of endangered species.

In the United States, the Bureau of Land Management oversees more than 110 million ha of multiple-use land, including 83% of the state of Nevada and large areas of Utah, Wyoming, Oregon, Idaho, and other western states. In the past, these lands were primarily managed for logging, mining, grazing, wildlife, and recreation. Increasingly, however, these multiple-use lands are being valued and managed for their ability to protect species, biological communities, and ecosystem services (Minteer and Miller 2011). The protection of biodiversity is incorporated into "ecological forestry" or "green-tree retention," a logging method that has been developed for the Pacific Northwest (Zarin et al. 2007). This method involves leaving patches of intact forest, and some dispersed living trees, standing dead trees, and fallen trees during logging operations, to provide structural complexity and to serve as habitat for animal species during forest regrowth. If logging near streams is avoided, water quality and other ecosystem services can also be protected. Many other methods of selective logging, including "low-impact logging" and "light-touch logging," are being developed throughout the world.

The Forest Stewardship Council has been one of the leading organizations to promote the certification of timber produced from sustainably managed forests. Certification of forests is increasing rapidly, with demand, especially in Europe, exceeding supply. For certification to be granted by the Forest Stewardship Council and other comparable organizations, the forests need to be managed and monitored for their long-term environmental benefits, and the rights and well-being of local people and workers need to be recognized. At the same time, major industrial organizations representing such industries as logging, mining, and agriculture are lobbying for their own alternative certification programs, which generally have lower requirements for monitoring and weaker standards for judging practices to be sustainable.

Expanding the objectives of multiple-use lands to include protection of biodiversity extends even to the military. In some countries, military commanders include wildlife protection as a management goal for the lands under their control. For example, personnel at the Barksdale Airforce Base in Shreveport, Louisiana, have reflooded wetlands along the Red River, restoring wetlands for wading birds. It is also true, of course, that during times of political upheaval, war, or national emergency, such regulations are ignored.

Ecosystem management links private and public landowners, businesses, and conservation organizations in a planning framework that facilitates acting together on a larger scale.

Ecosystem Management

Resource managers around the world increasingly are being urged by their governments and conservation organizations to expand their traditional emphasis on the maximum production of goods (such as volume of timber harvested) and services (such as the number of visitors to parks) and take a broader perspective that includes the conservation of biodiversity and the protection of ecosystem processes (Altman et al. 2011; Chapin et al. 2010). This viewpoint is encompassed in the concept of **ecosystem management**, a system of management involving multiple stakeholders, the primary goal of which is preserving ecosystem components and processes for the long term while still satisfying the current needs of society (**Figure 8.4**). Rather than each government agency, private conservation organization, business, or landowner acting in isolation for its own narrow interests, all would cooperate to achieve the common objectives of ecosystem management (Koontz and Bodine 2008). For example, in a large forested

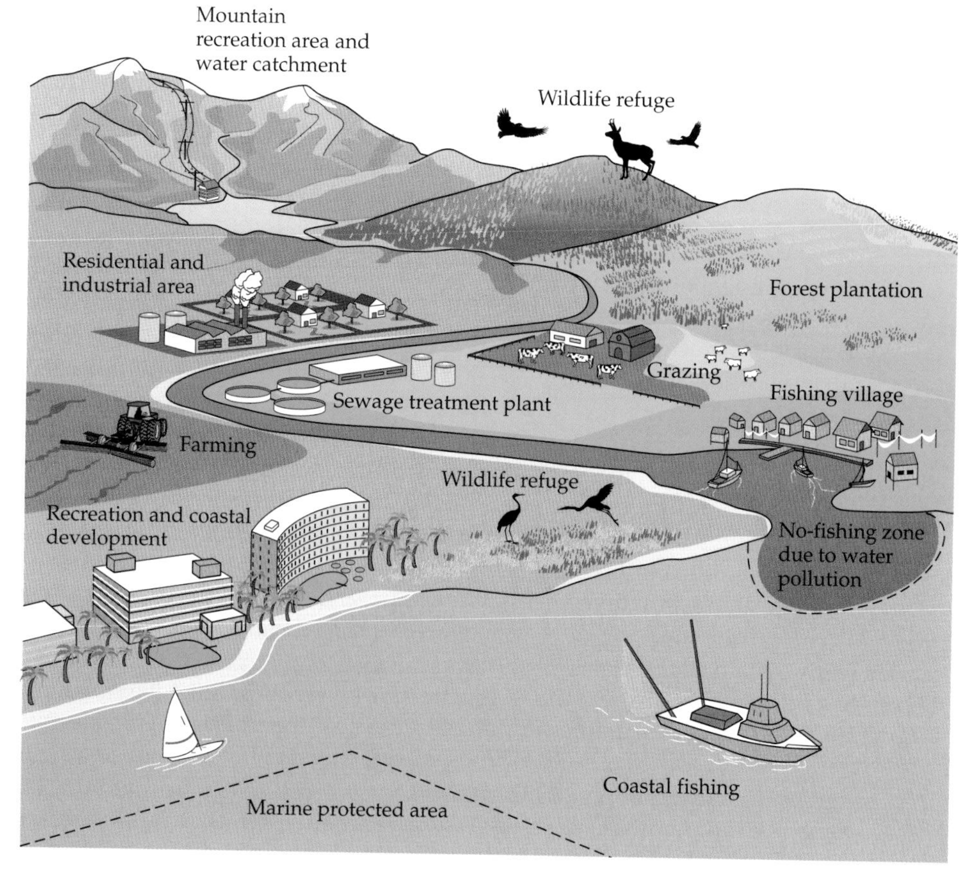

FIGURE 8.4 Ecosystem management involves bringing together all of the stakeholders who affect a large ecosystem and receive benefits from it. In this case, a watershed needs to be managed for a wide variety of purposes, many of which influence one another. (After Miller 1996.)

watershed along a coast, ecosystem management would link all owners and users, from the tops of the hills to the seashore, including foresters, farmers, business groups, townspeople, and the fishing industry.

Important themes in ecosystem management include the following:

- Using the best science available to develop a coordinated plan for the area that is sustainable; includes biological, economic, and social components; and is shared by all levels of government, business interests, conservation organizations, and private citizens
- Ensuring viable populations of all species, representative examples of all ecosystems and successional stages, and healthy ecosystem functions
- Seeking and understanding connections between all levels and scales in the ecosystem hierarchy—from the individual organism to the species, the biological community, the ecosystem, and even to regional and global scales
- Monitoring significant components of the ecosystem (numbers of individuals of significant species, vegetation cover, water quality, etc.), gathering the needed data, and then using the results to adjust management in an adaptive manner—a process sometimes referred to as **adaptive management** (see Figure 7.17)

One example of ecosystem management is the work of the Malpai Borderlands Group, a cooperative enterprise of ranchers and landowners who promote collaboration between private landowners, government agencies, and conservation organizations such as The Nature Conservancy (www.malpaiborderlandsgroup.org). The group is working to develop a network of cooperation across nearly 400,000 ha of unique, rugged mountain and desert habitat along the Arizona and New Mexico border, where many rare species live. The Malpai Borderlands Group is using controlled burning as a range management tool, reintroducing native grasses, applying innovative approaches to cattle grazing, incorporating scientific research into management plans, and taking action to avoid habitat fragmentation by using conservation easements (agreements not to develop land; see Chapter 9) to regulate residential development.

A logical extension of ecosystem management is **bioregional management**, which focuses on a single large ecosystem, such as the Caribbean Sea or the Great Barrier Reef of Australia, or a series of linked ecosystems such as the protected areas of Central America. A bioregional approach is particularly appropriate where there is a single, continuous, large ecosystem that crosses international boundaries or when activity in one country or region will directly affect an ecosystem in another country. For example, 21 countries in the European Union participate in the Mediterranean Action Plan to maintain the health of the enclosed Mediterranean Sea. The cooperation of each of these countries is needed to prevent water pollution and overfishing, because they all share the same body of water.

Working with Local People

In many parts of the world, areas with high biodiversity are inhabited by indigenous people with long-standing systems for resource protection and use. These people are important, and possibly essential, to conservation efforts in those areas.

Even remote regions that are considered "wilderness" by governments and the general public often have small, sparse human populations. Societies that practice a traditional way of life in rural areas, with relatively little outside influence in terms of modern technology, are variously referred to as "tribal people," "indigenous people," "native people," or "traditional people" (Timmer and Juma 2005; www.iwgia.org). These people may be the original inhabitants of the region and are often organized at the community or village level. It is necessary to distinguish these established indigenous people from more recent settlers, who may not be as concerned with the health of surrounding biological communities. In many countries, such as India and Mexico, there is a striking correlation between areas occupied by indigenous people and areas of high conservation value and intact forest. Indigenous people often have established traditional systems of rights to natural resources, which sometimes are recognized by their governments; local people therefore are potentially important partners in conservation efforts (Nepstad et al. 2006).

Worldwide, there are approximately 400 million indigenous people occupying between 12% and 19% of the Earth's total land surface (Redford and Mansour 1996; www.indigenouspeople.net). Rather than being threats to the "pristine" environments in which they live, many traditional peoples have been integral parts of these environments for thousands of years (Borghesio 2009). The present mixture and relative densities of plants and animals in many biological communities may reflect the historical activities—such as fishing, selective hunting of game animals, clearing of forests for fields, and planting of useful plant species—of people in those areas. These activities often do not degrade the environment as long as human population density is low and there are abundant land and resources.

Most indigenous societies have come into contact with the modern world, resulting in changing belief systems (particularly among younger individuals) and greater use of outside manufactured goods. It is important to consider that this shift can lead to a weakening of ties to the land and conservation ethics. However, many traditional societies still have strong conservation ethics. These ethics may be more subtle and less clearly stated than Western conservation beliefs, but they tend to affect people's actions in their day-to-day lives more than Western beliefs do (Abensperg-Traun 2009). In such societies, people use their traditional ecological knowledge to create management practices that are linked to belief systems and enforced by village consent and the authority of leaders. One example of such a conservation perspective is that of the Tukano Indians of northwest Brazil (Andrew-Essien and Bisong 2009), who live on a diet of root crops and river fish. They have strong religious and cultural prohibitions against cutting the forest along the Upper Río Negro, which they recognize as important to the maintenance of fish populations. The Tukano believe that these forests belong to the fish and cannot be cut down by people. They have also designated extensive refuges for fish and permit fishing along less than 40% of the river margin.

In Papua New Guinea, the establishment and linking of multiple protected areas in the TransFly Ecoregion of wetlands, grasslands, and tropical rain forest has resulted in over 2 million ha of protected wild lands. Over 60 different groups of indigenous people live in or have cultural ties to this region, and most have joined the World Wildlife Fund in supporting the protection of forests and wildlife (wwf.panda.org). The TransFly region is a biodiversity hotspot (see Chapter 7), being home to many endemic species, including the beautiful and intriguing birds of paradise (**Figure 8.5A**). New Guinea tribesmen have long hunted birds of paradise and other native species for the males' fabulous feathers, which are used in headdresses and other regalia (**Figure 8.5B**). Now that many species of these birds are declining in abundance, the people are eager to learn about and support efforts to maintain their populations, including limiting harvesting of feathers and eggs.

Local people who support conservation and the protection of their local natural resources are often inspired to take the lead in protecting biodiversity. Empowering them by helping them to obtain **legal title**—the right to ownership of the land that is recognized by the government—to their traditional lands is often an important component of efforts to establish locally managed protected areas in developing countries (Bhagwat and Rutte 2006). Today indigenous communities own 97% of the land in Papua New Guinea and over 100 million ha (22%) of incredibly diverse habitat in the Amazon basin; together these two regions encompass a huge percentage of the world's biodiversity (**Figure 8.6**). The Inuit people govern one-fifth of Canada. In Australia, tribal people control 90 million ha, including many of the most important areas for conservation.

(A)

(B)

FIGURE 8.5 (A) Many bird of paradise species, such as Goldie's bird of paradise (*Paradisaea decora*) shown here, are endemic to New Guinea. (B) Payakona and other New Guinea tribesmen use bird of paradise feathers in ceremonial costumes. Local people are cooperating with international conservation organizations to create a huge international reserve that will protect these birds and other wildlife. (Photographs © Tim Laman.)

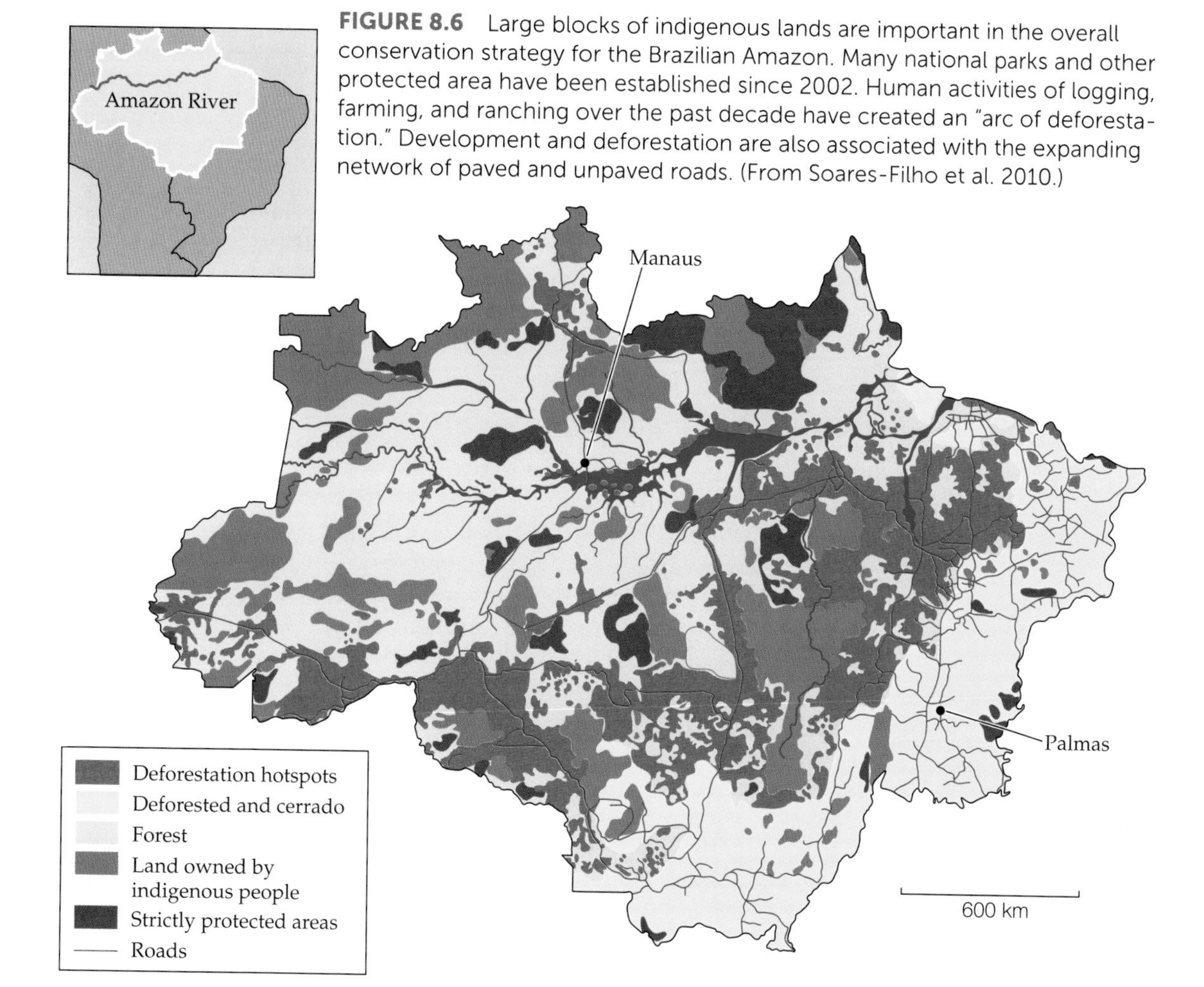

FIGURE 8.6 Large blocks of indigenous lands are important in the overall conservation strategy for the Brazilian Amazon. Many national parks and other protected area have been established since 2002. Human activities of logging, farming, and ranching over the past decade have created an "arc of deforestation." Development and deforestation are also associated with the expanding network of paved and unpaved roads. (From Soares-Filho et al. 2010.)

The challenge, then, is to develop strategies for incorporating these local peoples in conservation programs and policy development (McShane et al. 2011). The partnership of traditional peoples, government agencies, and conservation organizations working together in protected areas has been termed **co-management** (Borrini-Feyerabend et al. 2004). Co-management involves sharing in decision making and the consequences of management decisions. Such new approaches have been developed in an effort to avoid **ecocolonialism**, the common practice by some governments and conservation organizations of disregarding the traditional rights and practices of local people during the establishment of new conservation areas. This practice is called ecocolonialism because of its similarity to the historical abuses of native rights by colonial powers of past eras (Cox and Elmqvist 1997).

Integrated conservation and development projects (ICDPs) involve local people in sustainable activities that combine biodiversity conservation and economic development.

The economic needs, goals, and opinions of indigenous people are now often included in conservation management plans. For instance, the Biosphere Reserves Program (see Chapter 7) attempts to combine the protection of biodiversity and the customs of traditional societies with aspects of economic development, including job creation, improved health, and food security. Numerous such projects, known as **integrated conservation and development projects (ICDPs)**, have been started in the last decades. In practice, however, they are often problematic to manage (Linkie et al. 2008). The following sections illustrate some of the ICDP types that are currently in place.

In situ agricultural conservation

The long-term health of modern agriculture depends on the preservation of the genetic variability found in local varieties of crops maintained by traditional farmers (Bisht et al. 2007). However, in many areas of the world traditional crop varieties are being abandoned in favor of a few high-yielding modern cultivars. To maintain the genetic resources needed for crop improvement, the Chinese government supports a program that pays for farmers to interplant high-quality traditional and high-yielding hybrid rice varieties (Zhu et al. 2003).

A different approach linking traditional agriculture and genetic conservation is being used in arid regions of the American Southwest, with a focus on dryland crops with drought tolerance (www.nativeseeds.org). A private organization, Native Seeds/SEARCH, collects the seeds of traditional crop cultivars for long-term preservation. The organization also encourages a network of 4600 farmers and other members to grow traditional crops, provides them with the seeds of traditional cultivars, and buys their unsold production.

Countries have also established special reserves to conserve areas that contain wild relatives and traditional strains of commercial crops. Species reserves protect the wild relatives of wheat, oats, and barley in Israel and of citrus in India.

Extractive reserves

In many areas of the world, indigenous people have extracted products from natural communities for decades and even centuries. The sale or barter of these natural products is a major part of people's livelihoods. Understandably, local people are very concerned about retaining their rights to continue collecting natural products from newly created protected areas. A type of protected area known as an **extractive reserve** may present a sustainable solution to this problem. In such areas, the level of extraction is monitored and regulated to prevent overexploitation.

In Brazil, the government allows local citizens to collect natural materials such as wood products, edible seeds, rubber, resins, and Brazil nuts from

FIGURE 8.7 Extractive reserves established in Brazil provide a reason to maintain forests. The trunks of wild rubber trees are cut for their latex which flows down the grooves into the cup. Later the latex will be processed and used to make natural rubber products. (Photograph © Edward Parker/Alamy.)

extractive reserves in ways that minimize damage to the forest ecosystem (**Figure 8.7**) (Duchelle et al. 2012). Such extractive areas in Brazil, which comprise about 3 million ha, permit local people to continue their way of life and guard against the possible conversion of the land to cattle ranching and farming. However, populations of large animals in extractive reserves are often substantially reduced by subsistence hunting by local people, and the density of Brazil nut seedlings is reduced by the intense collection of mature nuts.

Many countries in eastern and southern Africa, with financial support from Western countries, are applying sustainable harvesting strategies in their efforts to preserve wildlife populations. The government of Namibia in southwestern Africa, in particular, has developed innovative programs in which wildlife management provides clear financial benefits to local people (Lindsey et al. 2007). One example is the Community Based Natural Resource Management program, in which local communities working with the government sell sport-hunting rights of high-value trophy species, such as lions and elephants, to safari companies. Revenue is also generated through operating tourist facilities. To maintain the needed densities of wildlife, the village community must work together with government officials to prevent illegal hunting. Despite the apparent success of this approach, animal rights organizations from Western countries question why their governments are subsidizing the killing of big game animals by wealthy safari hunters.

Community-based initiatives

In many cases, local people already protect biological communities, forests, wildlife, rivers, and coastal waters in the vicinity of their homes. Protection of such areas, sometimes called **community conserved areas**, is often enforced by village elders because maintaining natural resources such as food supplies and drinking water provides clear benefit to the local people. Protection is also sometimes justified on the basis of religious and traditional beliefs (Borrini-Feyerabend et al. 2004). Governments and conservation organizations assist local conservation initiatives by providing legal title to traditional lands, access to scientific expertise, and financial assistance to develop needed infrastructure. One example of such an initiative is the Community Baboon Sanctuary in eastern Belize, created by a collective agreement among several villages to maintain the forest habitat required by the local population of black howler monkeys (*Alouatta pigra*), locally known as baboons (Waters and Ulloa 2007). Ecotourists visiting the sanctuary pay a fee to the village organization, and additional payments are required if they stay overnight and eat meals with local families. Conservation biologists working at the site have provided training for local nature guides, a body of scientific information on the local wildlife, funds for a local natural history museum, and business training for the village leaders.

In the Pacific islands of Samoa, much of the rain forest land is under customary ownership—meaning that it is owned by communities of indigenous people. The local people have a strong desire to preserve the land because of the forest's religious and cultural significance, as well as its value for medicinal plants and other products, but they are under increasing pressure to sell logs from their forests to pay for their children's schools, roads, medicines, and other necessities. A variety of solutions is being developed to meet these conflicting needs. One solution was applied in American (or Eastern) Samoa by the U.S. government in 1988 when it leased forest and coastal land from the villages to establish a new national park. The villages gained needed income and retained ownership of the land and traditional hunting and collecting rights (www.nps.gov).

Payments for ecosystem services

A new creative strategy involves direct payments to individual landowners and local communities that protect critical ecosystems and the services that they provide, in effect paying the community to be good land stewards (Honey-Roses et al. 2011). These types of programs are sometimes referred to as **payments for ecosystem services (PES)** and are becoming increasingly popular (**Figure 8.8**). Such an approach has the advantage of greater simplicity than programs that attempt to link conservation with economic development. Governments, nongovernmental conservation organizations, and businesses develop markets in which local landowners can participate by protecting and restoring ecosystems. For example, owners of a forest may receive direct payments from a city government for the ecosystem services provided by the forest, such as flood control and providing drinking water. Local landowners and farmers can be paid for allowing large predators such as wolves, bears, tigers, and mountain lions to be on their land, with additional payments if their livestock is attacked (Dickman et al. 2011).

> New markets are being developed in which local people and landowners are paid for providing ecosystem services such as protecting forests to maintain water supplies and planting trees to absorb carbon dioxide. Programs that address climate change issues are predicted to become more common in coming years.

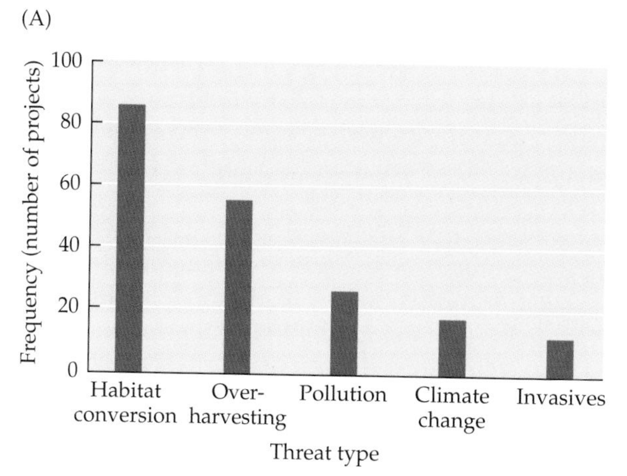

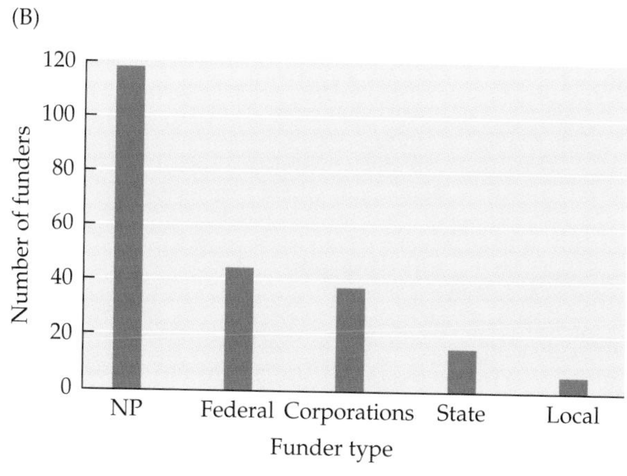

FIGURE 8.8 Patterns of payments for 103 ecosystem services (PES) projects from 37 countries. (A) Number of projects addressing different types of threat. Most projects address issues of habitat conversion (from forest to agricultural land) and overharvesting of trees. (B) Funding sources for the projects are primarily nonprofit (NP) conservation organizations but also include government agencies at national, state, and local levels, as well as corporations. (After Tallis et al. 2009.)

Rural people can also be drawn into newly developing international markets for ecosystem services (www.ecosystemmarketplace.com). In the Scolel Té project in Chiapas, Mexico, begun in 1996, farmers agree to maintain their existing forest land and restore degraded land in order to increase natural carbon sequestration. Farmers participating in the plan receive payments from a European car racing association, the World Bank, and religious groups seeking carbon credits to offset their own carbon dioxide emissions. These programs for Reducing Emissions from Deforestation and Forest Degradation (REDD+) are likely to expand greatly in coming years and may provide substantial funds for land protection (Venter and Koh 2012). However, at present such programs are sometimes unable to pay enough money to prevent landowners from selling logging rights to timber companies (Butler et al. 2009).

Evaluating conservation initiatives

Unfortunately, when external funding ends and if the projected income stream fails to develop, many of these integrated conservation and development projects end abruptly. Even for projects that appear successful, there is often no monitoring of ecological and social parameters to determine whether project goals are being achieved. It is essential that any conservation program design include mechanisms for evaluating the progress and success of measures taken (Kapos et al. 2008).

A key element in the success of many of the projects discussed in the preceding sections is the opportunity for conservation biologists to work with stable, flexible, local communities that have effective leaders and competent government agencies (Persha et al. 2011; Baker et al. 2012). However, in many cases a local community may have internal conflicts and poor leadership, making it incapable of administering a successful conservation program. Traditional practices may change or disappear, economic pressures for exploitation may increase, and programs may sometimes be mismanaged. Further, government agencies at all levels may be ineffective or even corrupt. Although working with local people may be a desirable goal, in some cases it simply is not possible. Some scientists have argued that the only way to preserve biodiversity is to exclude people from clearly zoned, core protected areas and rigorously patrol their boundaries (Minteer and Miller 2011).

Restoring Damaged Ecosystems

Ecosystems can be damaged by natural phenomena such as hurricanes or fires triggered by lightning, but they typically recover their original community structures and even similar species compositions through the process of ecological succession (see Chapter 2). However, some ecosystems destroyed by intensive human activities such as mining, ranching, and logging may have lost much of their natural ability to rebound and will not recover on their own without human intervention. For example, the original plant species will not be able to grow at a site if the soil has been washed away by erosion. Recovery is also often unlikely when the damaging agent is still present in the ecosystem (Christian-Smith and Merenlender 2010). Restoration of degraded savanna woodlands in the western United States, for instance, is not possible as long as the land continues to be overgrazed by introduced cattle; reduction of the grazing pressure is obviously the key starting point in restoration efforts there (**Figure 8.9**). When the original plant and animal species have been eliminated from an area by human activities, they may have to be reintroduced into the site, as described in Chapter 6. In some cases, reestablishing populations of keystone species, such as wolves, may be necessary for ecosystem restoration (Licht et al. 2010).

> Some ecosystems have been so severely degraded by human activity that their ability to recover on their own is severely limited. Ecological restoration reestablishes functioning ecosystems, with some or all of the original species or sometimes a different group of species.

Ecological restoration is the practice of restoring the species and ecosystems that occupied a site at some point in the past (www.ser.org; Clewell and Aronson 2006). **Restoration ecology** is the science of restoration—the research and scientific study

(A)

(B)

FIGURE 8.9 (A) Trout stream habitat that has been degraded by human activities. (B) Trout stream habitat that has been restored by installing fencing to exclude cattle, planting native species, and reinforcing stream banks with rocks. (From Hobbs et al. 2010; photographs courtesy of K. Matthews.)

of restored populations and ecosystems. Rebuilding damaged ecosystems has great potential for enlarging and connecting the current system of protected areas. In addition, restoration projects provide opportunities to completely reassemble biological communities in different ways, to see how well they function, and to test ecological ideas on a larger scale than would be possible otherwise (**Figure 8.10**). They can also provide local employment and opportunities for volunteers and conservation education.

FIGURE 8.10 An experiment to test the effects of different treatments on restoration at the Friendship Marsh in Tijuana Estuary, California. The marsh is divided into six experimental units (dashed lines), three with tidal creeks and three without creeks, to test the effects of drainage. Within each unit, restoration treatments (green blocks) involve different species, different planting densities, and different soil additions. In the largest blocks (light green rectangles), salt marsh grass has been planted with or without kelp compost. The impacts of these treatments on plants, fish, invertebrates, and algae are being evaluated. (After Zedler 2005.)

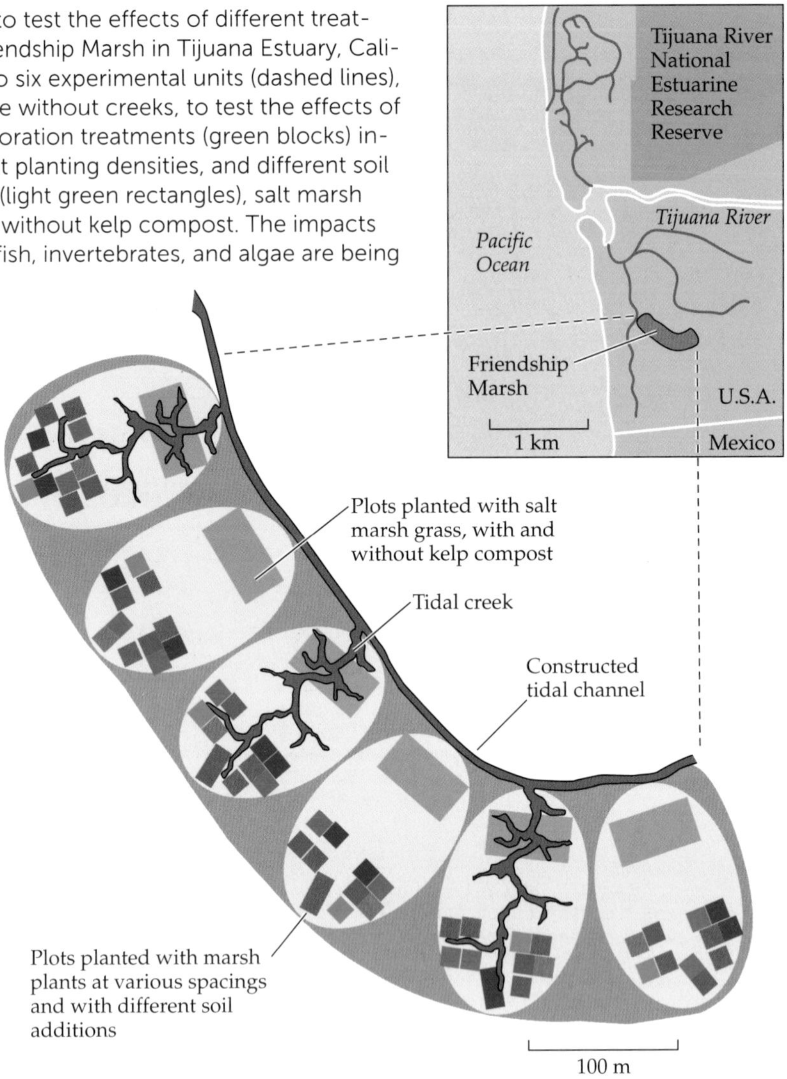

Ecological restoration has its origins in older applied technologies that attempted to restore ecosystem functions or species of known economic value, such as wetland creation (to prevent flooding), mine site reclamation (to prevent soil erosion), range management of overgrazed lands (to increase production of grasses), and tree planting on cleared land (for timber, recreational, and ecosystem values). However, these technologies often produced only simplified ecosystems that lasted only a few years. As concern for biodiversity has grown, restoration plans have included the permanent reestablishment of original species assemblages and ecosystems as a major goal.

The 2005 destruction of New Orleans and other Gulf Coast cities by Hurricane Katrina (and, to a lesser extent, by Hurricane Rita, which closely followed Katrina) was in part a result of the loss and overdevelopment of the region's wetlands that left the coastline vulnerable to storm damage. The ensuing natural disaster has become a classic example of the importance of such ecosystem services to biological and human communities alike (see Chapter 3). Ironically, the damage that followed these hurricanes had been predicted seven years earlier in an assessment of coastal wetlands by the Louisiana Coastal Wetlands Conservation and Restoration Task Force (1998), which stressed the urgent need for immediate action to restore lost wetlands. Restoration projects have begun but if they are not adequately funded and large enough in scope, New Orleans will remain vulnerable to another destructive flood.

Restoration efforts are sometimes part of **compensatory mitigation**, in which a new site is created or rehabilitated as a substitute for a site that has been destroyed by development (Clewell and Aronson 2006). At other times ecosystem processes, rather than ecosystems, are the goal of restoration efforts. For example, annual floods, disrupted by the construction of dams and levees, and natural fires, stopped by fire suppression efforts, may need to be reintroduced if the absence of these processes proves harmful to local and regional ecosystems and the species that live there.

Many restoration efforts are supported and even initiated by local conservation groups because they can see the direct connection between a healthy environment and their own personal and economic well-being. People can understand that planting trees produces firewood, timber, and food; prevents soil from washing away; and cools off the surrounding area in hot weather. An excellent example of a restoration effort with strong local support is the Green Belt Movement, a grassroots effort involving mainly rural women in Kenya that has planted over 30 million trees in degraded sites. The movement also organizes rural people, especially poor women, to have a voice in the political process, to maintain access to public forests, and to resist illegal logging.

Often the goal of restoration efforts is to create biological communities that are comparable in ecosystem functions or species composition to existing **reference sites** (Hopfensperger et al. 2006). Reference sites provide explicit goals for restoration and supply quantitative measures of the success of a project. Indeed, reference sites act as control sites and are central to the very concept of restoration. To determine whether the goals of restoration projects are being achieved, both the restoration and reference sites need to be monitored for years, even decades, to determine how well management goals are being achieved and whether further intervention is required. This approach is known as **adaptive restoration** (Zedler 2005). In particular, native species may have to be introduced again if they did not survive, and invasive species may have to be removed again if they are still abundant.

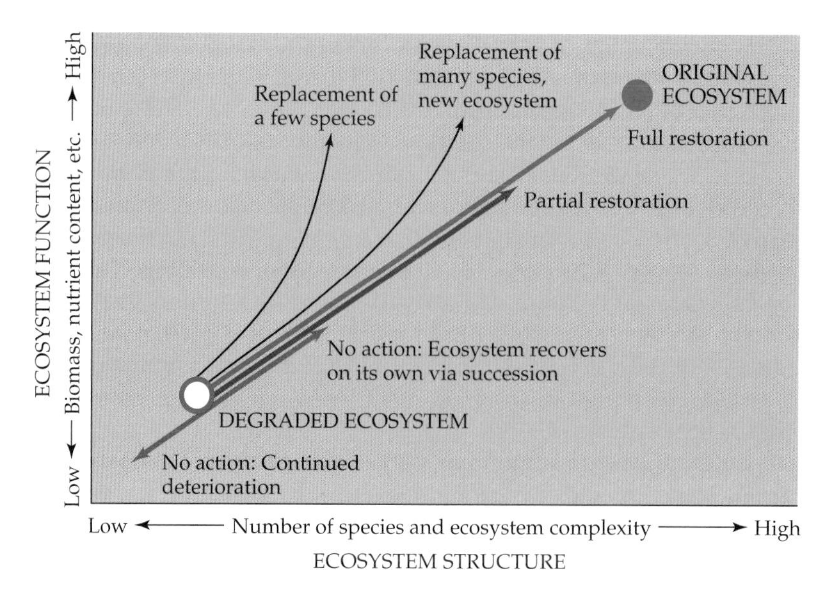

FIGURE 8.11 Decisions must be made about whether the best course of action is to restore a degraded site completely, partially restore it, replace the original species with different species, or take no action. (After Bradshaw 1990.)

Approaches to ecosystem restoration

Figure 8.11 outlines the main approaches to restoring biological communities and ecosystems. These include the following:

1. *No action.* Restoration is deemed too expensive, previous attempts have failed, or experience has shown that the ecosystem will recover on its own. Such passive restoration occurs when abandoned agricultural fields return to forest.
2. *Rehabilitation.* A degraded ecosystem is replaced with a different but productive ecosystem type (for example, a degraded forest might be replaced with a tree plantation). Just a few species may be replaced, or a larger-scale replacement of many species may be attempted.
3. *Partial restoration.* At least some of the ecosystem functions and some of the original dominant species are restored. An example is replanting a degraded grassland with a few species that can survive and are critical to ecosystem function, while delaying action on the rare species that are part of a complete restoration program.
4. *Complete restoration.* The area is restored to its original species composition, ecosystem structure, and ecosystem processes with an active program of elimination (or reduction) of damaging agents, site modification, and reintroduction of original species.

To be practical, ecological restoration must consider the speed of restoration, the cost, the reliability of results, and the ability of the final biological community to persist with little or no further maintenance (Falk et al. 2006). Considerations such as how much seeds cost and how available they are, when to water plants, how much fertilizer to add, how to remove invasive species, and how to prepare the surface soil may become paramount in reaching a successful outcome.

Restoration ecology will play an increasingly important role in the conservation of biological communities if degraded lands and aquatic communities can be restored to their original species compositions and added to the limited existing area under protection. While conservation of existing natural ecosystems is critical, it is often only through restoration that we can increase the area of ecosystems dominated by native species. Because degraded areas are unproductive and of little economic value, governments may be willing to restore them to increase their productive and conservation value.

Targets of major restoration efforts

Many efforts to restore ecosystems have focused on wetlands, lakes, and prairies. These environments have suffered severe alteration from human activities and are good candidates for restoration work. In addition, restoration projects in urban areas can enhance the quality of life for people living in the area.

WETLANDS Because of the recognized importance of wetlands in providing flood control and other ecosystem services, large development projects that damage wetlands are often required by governments to repair them or create new wetlands to compensate for those damaged beyond repair (Robertson 2006). The focus of these efforts has been on re-creating the original hydrology of the area and then planting native species (Brinson and Eckles 2011). Strategies to restore the biodiversity of rivers include the complete removal of dams and other structures and controlled releases of water from dams (Helfield et al. 2007).

> Highly visible restoration efforts are taking place in many urban areas to reduce the intense human impact on ecosystems and enhance the quality of life for city dwellers. Other restoration projects are improving polluted lakes, damaged wetlands, and abandoned farmlands.

Experience has shown that such efforts to restore wetlands frequently do not closely match the species composition or hydrologic characteristics of natural sites. The subtleties of species composition, water movement, and soils, as well as the site history, are often impossible to match, and after a few years, the restored wetlands are dominated by exotic, invasive species. However, the restored wetlands can have some of the original wetland plant species, often provide ecosystem services, and can have a degree of functional equivalency to the reference sites (Meyer et al. 2010).

An informative example of wetlands restoration comes from Japan, where parents, teachers, and children have built hundreds of small dragonfly ponds next to schools and in public parks to provide habitat for dragonflies and other native aquatic species (Primack et al. 2000). Dragonflies, an important symbol in Japanese culture, are a useful starting point for teaching children about science. The ponds are planted with aquatic plants; many dragonflies colonize them on their own, and additional species are "imported" as nymphs from other ponds. The children are responsible for the regular weeding and maintenance of these "living laboratories," which helps them to feel ownership in the project and to develop environmental awareness.

LAKES One of the most common types of damage to lakes and ponds is **cultural eutrophication**, which occurs when there are excess mineral nutrients in the water resulting from human activity. Signs of eutrophication include increases in the algae population (particularly surface scums of blue-green algae and floating plants), lowered water clarity, lowered oxygen content in the water, and fish kills.

In many lakes, the eutrophication process can be reversed and the ecosystem at least partially restored by reducing amounts of mineral nutrients entering the water, through better sewage treatment or by diverting polluted water. One of the most dramatic and expensive examples of lake restoration has been the effort to restore Lake Erie (Sponberg 2009). Lake Erie was the most polluted of the Great Lakes in the 1950s and 1960s, characterized by deteriorating water quality and declining indigenous fish populations. To address this problem, the governments of the United States and Canada have invested billions of dollars since 1972 in wastewater treatment facilities, dramatically reducing the annual discharge of phosphorus into the lake. Once water quality began to improve, populations of native fish began to increase on their own. Even though altered water chemistry and the presence of numerous exotic fish and invertebrate species may prevent the lake from ever returning to its historical condition, the investment of billions of dollars has resulted in a significant degree of restoration in this large, highly managed ecosystem.

PRAIRIES AND FARMLANDS Many parcels of former agricultural land in North America have been restored as prairies. Because they are species rich, have many beautiful wildflowers, and can be established within a few years, prairies are ideal subjects for restoration work (Foster et al. 2009). Also, the techniques used for prairie restoration are similar to common gardening and agriculture techniques and are well suited to volunteer labor. The basic method in prairie restoration involves site preparation by shallow plowing, burning, and raking, if prairie species are present, or the elimination of all vegetation by plowing or applying herbicides, if only exotics are present. Native plant species are then established by transplanting them in prairie sods obtained elsewhere, planting individuals grown from seed, or scattering prairie seed collected from the wild or from cultivated plants (**Figure 8.12**).

One of the most ambitious proposed grassland restorations involves re-creating a shortgrass prairie ecosystem, or "buffalo commons," on about

FIGURE 8.12 (A) In the late 1930s, members of the Civilian Conservation Corps (one of the organizations created by President Franklin Roosevelt in order to boost employment during the Great Depression) participated in a University of Wisconsin project to restore the wild species of a midwestern prairie. (B) The prairie as it looks today. (A, photograph courtesy of the University of Wisconsin Arboretum and Archives; B, photograph courtesy of Molly Field Murray.)

380,000 km^2 of the Great Plains states, from the Dakotas to Texas and from Wyoming to Nebraska (Adams 2006). Some of this land is currently used for environmentally damaging and often unprofitable agriculture and grazing that are supported by government subsidies. The human population of this region is declining as farmers and townspeople go out of business and young people move away. From the ecological, sociological, and even economic perspectives, the best long-term use of much of the region might be as a restored prairie ecosystem. The human population of the region would be supported by less damaging core industries such as tourism, wildlife management, and low-level grazing by cattle and bison, leaving only the best lands in agriculture.

A further thought-provoking proposal argues for releasing large game animals from Africa and Asia, such as elephants, cheetahs, camels, and even lions, into this area in an attempt to re-create the types of ecological interactions that occurred in North America more than 12,000 years ago, before humans arrived on the continent (Hayward 2009). Both of these proposed projects are controversial because many of the farmers and ranchers in the region want to continue their present way of life without alteration, and they tend to be highly resentful of unwanted advice and/or interference from scientists or the government.

URBAN AREAS Local citizen groups often welcome the opportunity to work with government agencies and conservation groups to restore degraded urban areas. Unattractive drainage canals in concrete culverts can be replaced with winding streams bordered with large rocks and planted with native wetland species. Vacant lots and neglected lands can be replanted with native shrubs, trees, and wildflowers. Gravel pits can be packed with soil and restored as ponds. These efforts have the additional benefits of fostering neighborhood pride, creating a sense of community, and enhancing property values. Developing places in which people and biodiversity can coexist has been termed **reconciliation ecology**, and this approach will increase in importance as urban areas expand (Rosenzweig 2003).

Restoring native communities on huge urban landfills presents one of the most unusual opportunities for conservation biologists. In the United States, 100 million tons of trash is buried in over 1800 active landfills each year. These eyesores can be the focus of conservation efforts. When they have reached their maximum capacity, landfills can be capped by sheets of plastic and layers of clay to prevent, or at least minimize, toxic chemicals and pollutants from seeping into nearby streams, marshes, and other wetlands. Planting native shrubs and trees in a layer of topsoil above the clay will attract birds and mammals that will bring in and disperse the seeds of a wide range of native species. One such example is the Freshkills landfill, covering 1000 ha on Staten Island in New York City, which is being restored through site contouring and planting of native trees and shrubs (www.nycgovparks.org). In the process, a new public park will be created, with abundant wildlife and many recreational, cultural, and educational amenities (**Figure 8.13**).

The future of restoration ecology

Restoration ecology is a major growth area in conservation biology, and it has its own scientific society (the Society for Ecological Restoration) and journals (*Restoration Ecology* and *Ecological Restoration*). Scientists are increasingly able to make use of the growing range of published studies and suggest improvements in how to carry out restoration projects. At its

(A)

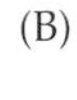

(B)

FIGURE 8.13 (A) The Freshkills landfill on Staten Island while active dumping was still occurring. Note the large number of gulls. (B) The future planned restoration of the site, based on an artist's viewpoint. The restoration will include recreational sites, natural areas, and rebuilt wetlands. (A, photograph from *Infrastructure: A Field Guide to the Industrial Landscape*, © Brian Hayes; B, image courtesy of NYC Parks and Recreation.)

best, restored land can provide new opportunities for protecting biodiversity. However, conservation biologists in this field must take care to ensure that restoration efforts are legitimate, rather than just a public relations cover by environment-damaging industrial corporations only interested in continuing business as usual. A 2 ha "demonstration" or "best practices" project in a highly visible location is not compensation for thousands of hectares damaged elsewhere and should not be accepted as such by conservation biologists or the general public. The best long-term strategy remains the protection and management of species and ecosystems where they are found naturally; only in these places can we be sure that the requirements for the long-term survival of all species and ecosystems are available. In addition, we need to consider restoring ecosystems in anticipation of the impacts of climate change.

Summary

- Considerable biodiversity exists outside of protected areas, particularly in habitat managed for multiple-use resource extraction. Traditional farmland, selectively logged forests, and lightly grazed grasslands often retain many of their original species and other elements of biodiversity; such lands outside of protected areas need to be considered in conservation management.
- Working with local people is often crucial to the success of conservation projects. Integrated conservation and development projects (ICDPs) link the protection of biodiversity with providing economic development opportunities to local people. One promising idea is payments for ecosystem services, involving direct payments to landowners for good land management practices, especially when these involve maintaining intact forests.
- Government agencies, private conservation organizations, businesses, and private landowners can cooperate in large-scale ecosystem management projects to achieve conservation objectives and to use natural resources sustainably.
- Restoration ecology provides methods for reestablishing species, whole biological communities, and ecosystem functions in degraded habitat. Restoration projects often enhance the quality of life for people living nearby. Restoration is sometimes carried out as a part of a mitigation effort to compensate for habitat destruction elsewhere.

For Discussion

1. Imagine that you are informed by the government that the endangered Florida panther lives on a piece of land that you own and have been planning to develop as a golf course. Are you happy, angry, confused, or proud? What are your options? What would be a fair compromise that would protect your rights, the rights of the public, and the rights of the panther?
2. Describe the diverging perspectives of participants in a large and complicated environmental activity, such as cleaning up after an oil spill or establishing a new habitat conservation plan. How can the needs of these participants be fairly considered in a compromise? And what are the alternatives to a negotiated compromise?
3. What methods and techniques could you use to monitor and evaluate the success of a restoration project? What time scale would you suggest using?

Suggested Readings

Baker, J., E. J. Milner-Gulland, and N. Leader-Williams. 2012. Park gazettement and integrated conservation and development as factors in community conflict at Bwindi Impenetrable Forest, Uganda. *Conservation Biology* 26: 160–170. The authors suggest management changes to avoid conflict between conservation and rural development.

Cox, R. L. and E. C. Underwood. 2011. The importance of conserving biodiversity outside of protected areas in Mediterranean ecosystems. *PLoS ONE* 6(1): e14508. Unprotected lands have the potential to contribute to an overall conservation strategy.

Falk, D. A., M. A. Palmer, and J. B. Zedler (eds.). 2006. *Foundations of Restoration Ecology: The Science and Practice of Ecological Restoration*. Island Press, Washington, DC. A good source for more information about this rapidly developing field.

Koontz, T. M. and J. Bodine. 2008. Implementing ecosystem management in public agencies: Lessons from the U.S. Bureau of Land Management and the Forest Service. *Conservation Biology* 22: 60–69. Often the barriers to effective management are political, cultural, and legal.

Minteer, B. A. and T. R. Miller. 2011. The new conservation debate: Ethical foundations, strategic trade-offs, and policy opportunities. *Biological Conservation* 144: 945–947. A special issue of the journal devoted to the balance between protecting biodiversity and providing opportunities for rural people.

Norris, K. and 7 others. 2010. Biodiversity in a forest-agriculture mosaic—The changing face of West African rainforests. *Biological Conservation* 143: 2341–2350. A diverse community of animals can adapt to range of human activities as long as some forest structure remains.

Vandermeer, J., I. Perfecto, and S. Philpott. 2010. Ecological complexity and pest control in organic coffee production: uncovering an autonomous ecosystem service. *BioScience* 60: 527–537. Organic farms can provide significant pest control, reducing the need for pesticides.

Venter, O. and L. P. Koh. 2012. Reducing emissions from deforestation and forest degradation (REDD+): Game changer or just another quick fix? *Annals of the New York Academy of Sciences* 1249: 137–150. This new program has great conservation potential for forest conservation, but may have unintended consequences.

Wright, H. L., I. R. Lake, and P. M. Dolman. 2012. Agriculture—A key element for conservation in the developing world. *Conservation Letters* 5: 11–19. Open farmland habitat is important habitat for many threatened bird species in developing countries, and needs to be part of conservation policy.

An extensive solar power station at the foot of an undeveloped mountain region.

Chapter 9
The Challenge of Sustainable Development

Efforts to preserve biological diversity sometimes conflict with both real and perceived human needs (**Figure 9.1**). Increasingly, many conservation biologists, policy makers, and land managers are recognizing the need for **sustainable development**—economic development that satisfies both present and future needs for resources and employment while minimizing the impact on biological diversity (Holden and Linnerud 2007). Sustainable development can be contrasted with more typical development that is *unsustainable*, meaning it cannot continue without causing irreparable harm to both natural and human communities. As defined by some environmental economists, **economic development** implies improvements in efficiency and organization *but not necessarily increases in resource consumption*. Economic development is clearly distinguished from **economic growth**, which is defined as material increases in the amount of resources used. Sustainable development is a useful and important concept in conservation biology because it emphasizes *improving current economic development and limiting unsustainable economic growth*.

FIGURE 9.1 Sustainable development seeks to address the conflict and heal the rift between development to meet human needs and the preservation of the natural world. (Top photograph © FloridaStock/shutterstock; bottom photograph © kavram/shutterstock.)

By this definition, investing in national park infrastructure to improve protection of biological diversity and provide revenue opportunities for local communities would be an example of movement toward sustainable development, as would implementation of less destructive logging and fishing practices. Unfortunately, the term "sustainable development" has become overused and is often misappropriated. Few politicians or businesses are willing to proclaim themselves to be against sustainable development. Thus, many large corporations, and the policy organizations that they fund, misuse the notion of sustainable development to "greenwash" their industrial activities, with only limited change in practice.

The goal of sustainable economic development is to provide for the current and future needs of human society while at the same time protecting species, ecosystems, and other aspects of biodiversity.

For instance, a plan to establish a huge mining complex in the middle of a forest wilderness cannot justifiably be called sustainable development simply because a small percentage of the land area is set aside as a park. Similarly, building huge houses filled with "energy-efficient" appliances and oversized SUVs that boast the latest energy-saving technology cannot really be called sustainable development or "green technology" when the net result is increased energy use. Alternatively, some people champion the opposite extreme, claiming that sustainable development means that vast areas of the world must be kept off-limits to all human interference and should remain as, or be allowed to return to, wilderness. As with all such disputes, informed scientists and citizens must study the issues carefully, identify which groups are advocating which positions and why, and then make careful decisions that best meet seemingly contradictory demands—the needs of human society and the protection of biological diversity. Addressing both demands necessitates compromise, and in most cases compromises form the basis of government policy and laws, with conflicts resolved by government agencies and in the courts.

Sustainable Development at the Local Level

Most efforts to find the right balance between the preservation of biodiversity and the needs of society rely on initiatives from concerned citizens, conservation organizations, and government officials. The results of these initiatives are often codified into environmental regulations or laws and backed by the powers of the government (or sometimes citizens themselves using the courts). These efforts may take many forms, but they begin with individual and group decisions to prevent the destruction of habitats and species in order to preserve something of perceived economic, cultural, biological, scientific, or recreational value. One of the most significant developments of recent decades has been the rise of **nongovernmental organizations** (**NGOs**), many of which mobilize people to protect the environment and promote the welfare of citizens. Many NGOs have a local focus, but there are already over 40,000 international NGOs (**Figure 9.2**). Among the best known are the World Wildlife Fund, The Nature Conservancy, Conservation International, the Sierra Club, the Audubon Society and the National Trust in Britain.

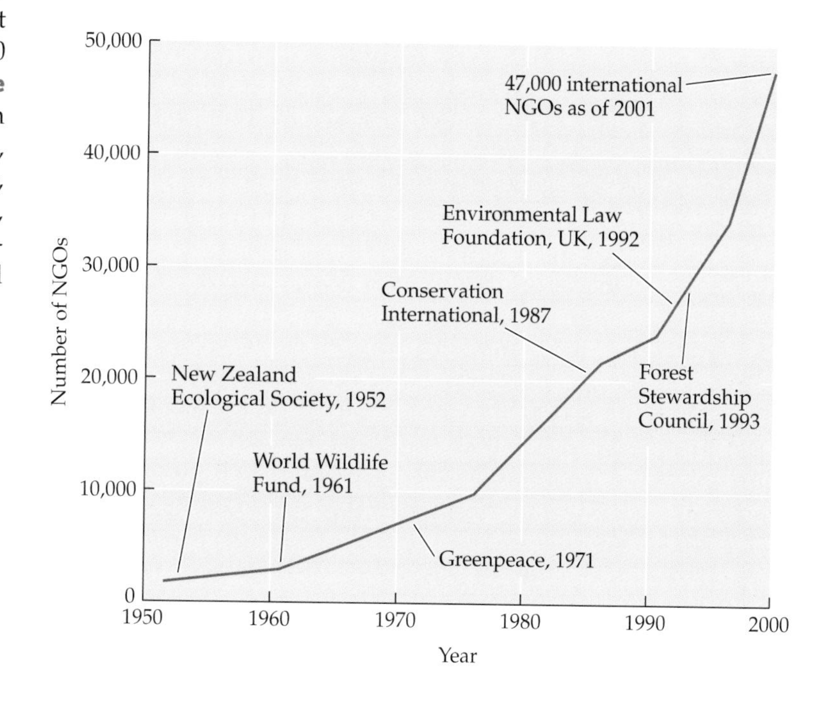

FIGURE 9.2 There has been enormous growth in the number of international nongovernmental organizations (NGOs) since 1950; many of these organizations protect the environment and promote the welfare of people. (After WRI 2003.)

Local and regional conservation regulations

In modern societies, local (city and town) and regional (county, state, and provincial) governments pass laws to provide effective protection for species and ecosystems while at the same time allowing development for the continued needs of society. Conservation laws regulate activities that directly affect species and ecosystems. The most prominent of these laws govern when and where hunting and fishing can occur; the size, number, and species of animals that can be taken; and the types of weapons, traps, and other equipment that can be used. Restrictions are enforced through licensing requirements and patrols by game wardens and police. In some densely settled areas and protected areas, hunting and fishing are banned entirely. Similar laws affect the harvesting of plants, seaweed, and shellfish. Certification of origin of biological products may be required to ensure that wild populations are not depleted by illegal collection or harvest. These restrictions have long applied to animals, such as trout and deer, and to plants of horticultural interest, such as orchids, azaleas, and cacti. They are increasingly being applied to timber products, coffee, and other widely used commodities to certify that environmental regulations have been followed.

Laws that control the ways in which land is used are another means of protecting biodiversity. For example, vehicles and even people on foot may be restricted from habitats and resources that are sensitive to damage, such as bird-nesting areas, bogs, sand dunes, wildflower patches, and sources of drinking water. Uncontrolled fires may severely damage habitats, so practices (such as building campfires) that contribute to accidental fires are often rigidly controlled. Zoning laws sometimes prevent construction in sensitive areas such as barrier beaches and floodplains. Even where development is permitted, building permits are reviewed with increasing scrutiny to ensure that damage is not done to endangered species or ecosystems, particularly wetlands. For major regional and national projects, such as dams, mining operations, oil extraction, and highway construction, environmental impact statements must be prepared that describe a project's potential damage.

One of the most powerful strategies in protecting biodiversity at the local and regional level is the designation of intact biological communities as nature reserves, conservation land, and state and provincial parks and forests. Government bodies buy land and establish protected areas for various uses—local parks for recreation, conservation areas to maintain biodiversity, forests for timber production and other uses, and watersheds to protect water supplies. In some cases, land is purchased outright, but often land is donated to conservation organizations by public-minded citizens.

Land trusts

In many countries, private conservation organizations are among the leaders in acquiring land for conservation (Bode et al. 2011). In the Netherlands, about half of the protected areas are privately owned. In the United States, over 15 million ha of land is protected at the local level by some 1700 **land trusts**, which are private, nonprofit corporations established to protect land and natural resources (www.landtrustalliance.org). At the national level, major organizations such as The Nature Conservancy and the Audubon Society have protected an additional 10 million ha in the United States (**Figure 9.3**).

Land trusts are common in Europe. In Britain, the National Trust has more than 3.8 million members and 62,000 volunteers and owns about 250,000 ha of land, much of it farmland, including 57 National Nature Reserves, 466 Sites of Special Scientific Interest, 355 Areas of Outstanding Natural Beauty, and 40,000 archaeological sites. The Royal Society for the Protection of Birds has more than a million members and manages 200 reserves with an area of almost 130,000 ha. A major emphasis of many of these reserves is nature conservation and education, often linked to school programs. These private reserve networks are collectively referred to as CARTs—Conservation, Amenity, and Recreation Trusts, a name that reflects their varied objectives.

Land trusts are private conservation organizations that purchase and protect land. Conservation easements and limited development agreements are also used by land trusts to further increase the area that can protect biodiversity.

FIGURE 9.3 The Nature Conservancy, a land trust active in the United States and 33 other countries, manages this rare bog habitat in western Massachusetts in collaboration with area colleges. In 1997, volunteers built a 200-meter long boardwalk so that students, professors, and the public can access this "outdoor classroom" without damaging the fragile habitat and without having to walk through waist-deep water. (Photograph by David McIntyre.)

In addition to outright purchase of land, both local and regional governments and private conservation organizations protect land through **conservation easements**, in which landowners give up the right to develop, build on, or subdivide their property in exchange for a sum of money or for lower real estate taxes or some other tax benefit (Farmer et al. 2011). For many landowners, accepting a conservation easement is an attractive option: they receive a financial advantage while still owning their land and are able to feel that they are supporting conservation objectives. The offer of lower taxes or money is not always necessary; many landowners will voluntarily accept conservation restrictions without compensation.

Another strategy that land trusts and local and regional governments use is **limited development**, also known as **conservation development** (Milder et al. 2008). In these situations, a landowner, a property developer, and a government agency and/or conservation organization reach a compromise that allows part of the land to be commercially developed while the remainder is protected by a conservation easement. Limited development allows the construction of necessary buildings and other infrastructure for an expanding human society; the projects are often successful precisely because being adjacent to conservation land enhances the value of the developed land.

Governments and conservation organizations can further encourage conservation on private lands through other mechanisms, including compensating private landowners for refraining from some damaging activity and implementing some positive activity in its place. **Conservation leasing** involves providing payments to private landowners who actively manage their land for biodiversity protection.

A related idea is **conservation banking**, in which a landowner deliberately preserves an endangered species or a protected habitat type, such as wetlands, or even restores degraded habitat and creates new habitat (Dreschler and Watzold 2009). A developer can then pay the landowner or a conservation organization to protect this new habitat or endangered species in compensation for a similar habitat that is being destroyed elsewhere by a construction project. A similar approach, described in Chapter 8, is **payments for ecosystem services** (**PES**), in which a landowner is paid for providing specific conservation services such as maintaining water quality, preserving habitat for endangered wildlife, or planting trees for absorbing carbon (**Figure 9.4**). Utility companies may also gain carbon credits by paying for such habitat protection and restoration; these carbon credits are then used to offset the carbon emissions produced through the burning of fossil fuels.

Conservation concessions are a recent approach in which conservation organizations outbid extractive industries such as logging companies, not for ownership of the land but for the rights to use and protect the land. The government or large landowner receives the same annual income from a conservation organization instead of a logging company, and the animals and plants of the area are protected rather than destroyed.

(A)

(B)

FIGURE 9.4 Different regulations and management styles can have different outcomes for conservation. (A) An agriculturally improved pasture in south Florida, with primarily nonnative plants and inputs of fertilizer. (B) Ranchers can maintain native Florida prairie pasture with many native plants and minimal fertilizer if they are provided with conservation subsidies. (From Bohlen et al. 2009; photographs by Patrick Bohlen and Carlton Ward Jr.)

Enforcement and public benefits

The conservation measures described in this chapter must be continuously monitored to make sure that regulations and laws are enforced and that agreements are being carried out, particularly in cases where destruction

> Conservation lands require ongoing management and continuous vigilance, and they often provide extensive benefits to the local society and economy.

cannot be easily reversed (Rissman and Butsic 2011). In one scenario, a developer may be issued construction permits after agreeing to limit the amount of development and to conserve an area of forest but then ignore the agreement and clear all the trees. In such cases, legal action may be needed to force the developer to restore the site.

Public perception can also be a source of problems. Local efforts by land trusts to protect land are sometimes criticized as being elitist because they provide tax breaks only to those wealthy enough to take advantage of them, while they lower the revenue collected from land and property taxes. Others argue that land used in other ways, such as for agriculture or commercial activity, is more important because it creates jobs and products. Although land in trust may initially lower tax revenues, the loss is often offset by the increased value and consequent increased property taxes of houses and land adjacent to the conservation area. In addition, the employment, recreational activities, and tourist spending associated with nature reserves and other protected areas generate revenue throughout the local economy, which benefits local residents. Finally, by preserving important features of the landscape and natural communities, local nature reserves also preserve and enhance the quality of life and cultural heritage of the local society, considerations that must be valued for sustainable development to be achieved.

Conservation at the National Level

Throughout the modern world, national governments play a leading role in conservation activities (Zimmerer 2006). National governments can identify endangered species within their borders and take steps to conserve them, such as protecting and acquiring habitat for the species; controlling the harvest, sale, and use of the species; developing research programs; and implementing in situ and ex situ recovery plans. A revealing study from Europe demonstrates greater increases in bird populations in those countries that take conservation actions than in those countries that do not take such actions (Donald et al. 2007). Conservation biologists contribute to these efforts by providing government officials with key information on threats to biodiversity, and the resulting laws and regulations are used to protect biodiversity (**Figure 9.5**).

Similar to local and regional governments, national governments can use their revenues and authority to buy new land for conservation. In the United States, special funding mechanisms exist at the national level, such as the Lands Legacy Initiative and the Land and Water Conservation Fund, to purchase land for conservation purposes.

The establishment of national parks is a particularly important conservation strategy. National parks are the single largest source of protected lands in many countries. For example, Costa Rica's national parks protect about

FIGURE 9.5 A field biologist takes a U.S. congressman into a woodland to explain how habitat fragmentation and the loss of biodiversity contribute to increasing rodent populations and incidence of Lyme disease. (Photograph by J. Halpern.)

62,000 ha, or about 12% of the nation's land area (www.costarica-nationalparks.com). Outside the protected areas, deforestation is occurring rapidly, and soon national parks may represent the only undisturbed habitat and source of natural products, such as timber, in the whole country. The U.S. National Park Service protects about 8.4 million ha with 391 sites. The U.S. government also protects biodiversity in its 550 National Wildlife Refuges covering 62 million ha, the Bureau of Land Management's National Landscape Conservation System with 886 sites covering 11 million ha, and many of the National Forests.

National legislatures and governing agencies are the principal bodies for developing policies that regulate environmental pollution. Laws are passed by legislatures and then implemented in the form of regulations by government agencies. Laws and regulations affecting air emissions, sewage treatment, waste dumping, and the development of wetlands are often enacted to protect human health and property and natural resources such as drinking water, forests, and commercial and sport fisheries. The level of enforcement of these laws demonstrates a nation's determination to protect the health of its citizens and the integrity of its natural resources. At the same time, these laws protect biological communities that would otherwise be destroyed by pollution and other human activities. Air pollution that exacerbates human respiratory disease, for example, also damages commercial forests and grasslands, and the pollution that ruins drinking water also kills terrestrial and aquatic species such as turtles and fish.

National governments protect designated endangered species within their borders, establish national parks, and enforce legislation on environmental protection.

Finally, national governments can have a substantial effect on the protection of biodiversity through the control of their borders, ports, and commerce. To protect forests and regulate their use, governments can ban logging, as was done in Thailand following disastrous flooding or they can restrict the export of logs, as was done in Indonesia. Certain kinds of environmentally damaging mining can be banned. Methods of shipping oil and toxic chemicals can be regulated.

Despite the fact that many countries have enacted legislation to protect endangered species, forests, wetlands, and other aspects of biodiversity, it is also true that national governments are sometimes unresponsive to requests from conservation groups to protect the environment. Governments have even acted to remove the protected status of national parks, sacred forests, and other conservation areas in order to facilitate the extraction of natural resources (Mascia and Pallier 2011). Governments sometimes do this because they feel that the needs of the broader regional and national society for natural resources and economic development are more important than the needs of the local people. There are also many cases where such downgrading, downsizing, and degazettement of protected areas are linked to certain government officials who are personally profiting from their actions. This damage to protected areas sometimes occurs despite the vigorous protests and demonstrations of local people. In some cases, national governments recognize that local people are better able to protect ecosystems close to where they live and have relinquished control of these resources to local governments, village councils, and conservation organizations (WRI 2003).

International Approaches to Sustainable Development

The biological diversity needed for humanity's future well-being is concentrated in the tropical countries of the developing world, most of which are relatively poor and experiencing rapid rates of population growth, development, and habitat destruction. Developing countries may be willing to preserve biodiversity, but they are often unable to pay for the habitat preservation, research, and management required for the task. The developed countries of the world (including the United States, Canada, Japan, Australia, and many European nations) must work together with tropical countries to preserve the biodiversity needed by the world as a whole.

While the major legal and policing mechanisms that presently exist in the world are based within individual countries, international cooperation to protect biodiversity is an absolute requirement for several reasons:

International cooperation and agreements to protect biodiversity are needed for migratory species and when threats occur across countries.

- *Many species migrate across international borders.* Conservation efforts must protect species at all points in their ranges; efforts in one country will be ineffective if critical habitats are destroyed in another country to which an animal migrates. For example, efforts to protect migratory bird species in northern Europe will not work if the birds' overwintering habitat in Africa is destroyed.
- *International trade in biological products is commonplace.* A strong demand for a product in one country can result in the overexploitation of the species by a poor country or in illegal trade to supply this demand.
- *The benefits of biodiversity are of international importance.* All nations are helped by the species and varieties that can be used in agriculture, medicine, and industry; by the ecosystems that help regulate climate;

and by the national parks and biosphere reserves of international scientific and tourist value.

- *Many problems of environmental pollution are international in scope and require international cooperation.* Among threats that readily spread across borders are atmospheric pollution and acid rain; the pollution of lakes, rivers, and oceans; greenhouse gas production and global climate change; and ozone depletion (Srinivasan et al. 2008).

International conservation agreements

Countries can gain international recognition for protected areas through the Ramsar Convention, the World Heritage Convention, and the Biosphere Reserves Program. Transfrontier parks in border areas provide opportunities for both conservation and international cooperation.

To address the protection of biodiversity, countries of the world have signed a variety of international agreements to protect species, habitats, ecosystem processes, and genetic variation. Treaties are negotiated at international conferences and put into practice when they are ratified by a certain number of countries.

Habitat conventions at the international level complement the species conventions described in Chapter 6 by emphasizing unique biological communities and ecosystem features that need to be protected (and within these habitats, a multitude of individual species can be protected). Three of the most important international conventions are the **Ramsar Convention on Wetlands**, the **Convention Concerning the Protection of the World Cultural and Natural Heritage** (or the **World Heritage Convention**), and the **UNESCO Biosphere Reserves Program**. Countries designating protected areas under these conventions voluntarily agree to manage them under the terms detailed in the conventions; countries do not give up sovereignty over these areas to an international body but retain full control over them. Such conventions have been effective at protecting lands and meeting conservation goals.

The Ramsar Convention on Wetlands was established in 1971 to recognize the ecological, scientific, economic, cultural, and recreational values of wetlands (**Figure 9.6A**). The Ramsar Convention covers freshwater, estuarine, and coastal marine habitats and includes 1995 sites with a total area of more than 192 million ha. The 160 countries that have signed the Ramsar Convention agree to conserve and protect their wetland resources, particularly those that support migratory waterfowl, and to designate for conservation purposes at least one wetland site of international significance (www.ramsar.org).

The goal of the World Heritage Convention is the protection of cultural areas and natural areas of international significance (whc.unesco.org). This convention is distinctive because it emphasizes the cultural as well as the biological significance of natural areas and recognizes that the world community has an obligation to support the sites financially. The 936 World Heritage sites in 153 countries include some of the world's premier conservation

(A)

(B)

FIGURE 9.6 (A) Los Lípez is a Ramsar-listed site in Bolivia covering 1.5 million ha and noted for its populations of two flamingo species. (B) Iguaçu Falls in Iguaçu National Park, Brazil is a World Heritage Site. (A, photograph © J. Marshall-Tribaleye Images/Alamy; B, photograph © Joris Van Ostaeyen/istock.)

areas: Serengeti National Park in Tanzania, Iguaçu Falls in Brazil and Argentina, the Wet Tropics of Queensland in Australia, and Great Smoky Mountains National Park in the United States, to name a few (**Figure 9.6B**).

UNESCO's Biosphere Reserves Program began in 1971. Biosphere reserves are designed to be models that demonstrate the compatibility of conservation efforts and sustainable development for the benefit of local people, as described in Chapter 7 (see Figure 7.19). A total of 580 biosphere reserves have been created in 114 countries, covering more than 260 million ha, including 47 reserves in the United States, 40 in Russia, 16 in Bulgaria, 29 in China, 15 in Germany, and 40 in Mexico. The largest designated biosphere reserve, located in Greenland, is over 97 million ha in area.

These three conventions, which together support the provisions of the Convention on Biological Diversity (see the next section), establish an overarching consensus on appropriate conservation of protected areas and certain habitat types. More limited international agreements protect unique ecosystems and habitats in particular regions, including the Western Hemisphere, the Antarctic, the South Pacific, Africa, the Caribbean, and Europe. Other international agreements have been ratified to prevent or limit pollution that poses regional and international threats to the environment. For example, the Convention on Long-range Transboundary Air Pollution in the European region recognizes the role that long-range transport of air pollution plays in acid rain, lake acidification, and forest dieback; the Convention for the Protection of the Ozone Layer was signed in 1985 to regulate and phase out the use of chlorofluorocarbons; and the Convention on the Law of the Sea promotes the peaceful use and conservation of the world's oceans.

In many regions of the world, rugged, undeveloped border areas mark the boundaries between countries. These border regions are often carved into artificial units according to political boundaries rather than natural ecosystem boundaries. An alternative to this situation is to establish transfrontier parks that include larger areas. Park personnel from the countries involved can manage the park resources collectively and promote conservation on a larger scale. An early example of this collaboration was the decision to manage Glacier National Park in the United States and Waterton Lakes National Park in Canada, which are adjacent to one another, as the Waterton Glacier International Peace Park. Intensive efforts are underway to link national parks and protected areas in Zimbabwe, Mozambique, and South Africa into the Greater Limpopo Transfrontier Park and other large management units (**Figure 9.7**). This joint management would have the advantage of protecting the seasonal migratory routes of large animals. Transfrontier parks have many potential benefits, but they are difficult to establish where countries do not have a friendly relationship and a common language, and they require additional funding.

Marine pollution is another area of vital concern because extensive areas of the world's oceans constitute international waters, not under any one nation's control, and because pollutants released in one area can easily spread to another area. Agreements that cover marine pollution include the Convention on the Prevention of Marine Pollution by Dumping of Wastes and Other Matter, the Convention on the Law of the Sea, and the Regional Seas program of the

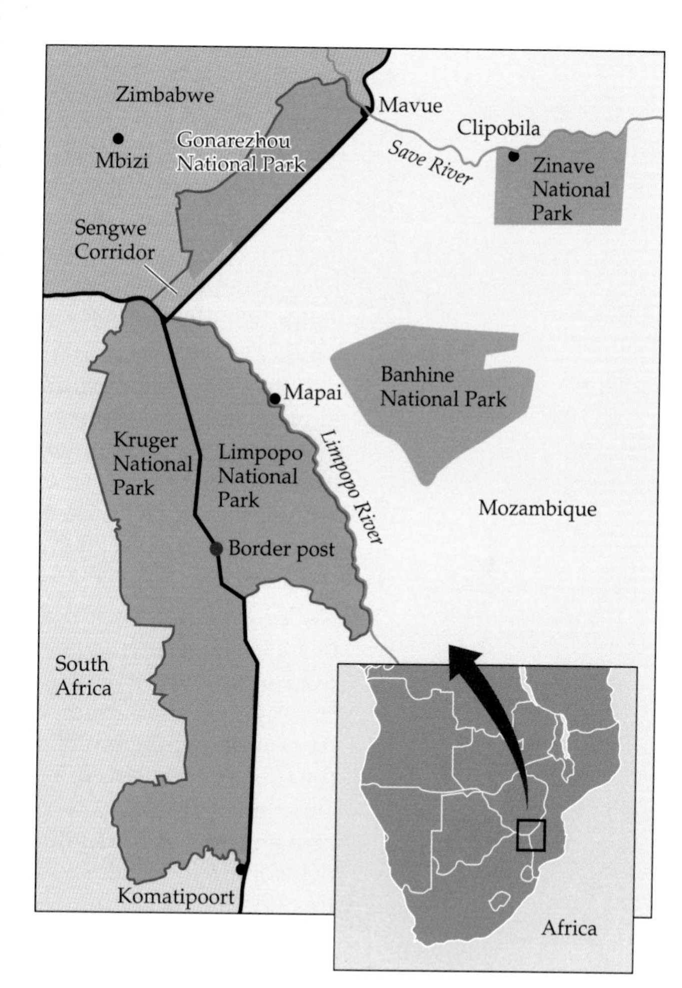

FIGURE 9.7 The Greater Limpopo Transfrontier Park has the potential to unite wildlife management activities in national parks and conservation areas of South Africa, Mozambique, and Zimbabwe. A larger conservation area in Mozambique will eventually connect all of the national parks and will include private game reserves, and private farms and ranches. (After www.sanpark.org.)

United Nations Environment Programme (UNEP). Vast areas of the open ocean, far from coastlines and the watchful eyes of governments, are still largely unexplored and unregulated at this point and are in urgent need of protection.

International summits

One of the most significant hallmarks of progress in international efforts to protect biodiversity was the international conference held in 1992 in Rio de Janeiro, Brazil. Known officially as the United Nations Conference on Environment and Development (UNCED), and unofficially as the **Earth Summit** or the **Rio Summit**, the conference brought together representatives from 178 countries, including heads of state, plus leaders of the United Nations and major nongovernmental and conservation organizations. Participants discussed ways of combining increased protection of the environment with more sustainable economic development in less wealthy countries.

The meeting produced a number of major documents, including the following:

- *The Rio Declaration on Environment and Development*. The Rio Declaration is a nonbinding declaration that recognizes the right of nations to utilize their own natural resources for economic and social development, as long as the environments of other nations are not harmed in the process. The declaration affirms the "polluter pays" principle, in which companies and governments take financial responsibility for the environmental damage that they cause.
- *United Nations Framework Convention on Climate Change*. This agreement requires industrialized countries to reduce their emissions of carbon dioxide and other greenhouse gases and to make regular reports on their progress. Although this convention has received broad support, the United States and other major producers of greenhouse gases have refused to ratify it. Subsequent summits at Kyoto in 1997, Bali in 2007, and Copenhagen in 2009 discussed this issue but failed to produce a comprehensive plan that all key countries would support (**Figure 9.8**).
- *United Nations Convention on Biological Diversity*. The **Convention on Biological Diversity**, signed by 180 countries, has three objectives: protecting biodiversity, using biodiversity in a sustainable manner, and sharing the benefits of new products made using biodiversity. Developing international intellectual property rights laws that fairly share the financial benefits of biodiversity among countries, biotechnology companies, and local people is proving to be a major challenge to the convention, with roadblocks still to overcome. In a number of cases, developing countries have refused or impeded permission to scientists to collect biological samples for their research because of concerns about how the material will be used (or misused).

FIGURE 9.8 In December 2009, many international leaders attended the United Nations Climate Change Conference in Copenhagen, Denmark, along with observers from intergovernmental and nongovernmental organizations. Development of the next comprehensive agreement to reduce greenhouse gas emissions—a successor to the Kyoto Protocol—dominated the conference discussions. (Photograph by Pete Souza.)

In August 2002, another major environment summit was held: The World Summit on Sustainable Development in Johannesburg, South Africa, with representatives of 191 countries and 20,000 participants (WRI 2003). Although the conference emphasized the need to reduce the rate of biodiversity loss, the main focus was on achieving the social and economic goals of sustainability. This shift in focus from the Rio Summit highlights a significant, ongoing debate over whether the emphasis in conservation should be to promote sustainable use of natural resources for the benefit of poor people or should be to protect natural areas and biodiversity (Naughton-Treves et al. 2005). More recently, countries agreed to Millennium Development Goals aimed at reducing extreme poverty by half by 2015 and lowering the rate of biodiversity loss.

Funding for Conservation

One of the most contentious issues resulting from these international conferences and treaties has been deciding how countries will fund the proposals, particularly the Convention on Biological Diversity and other programs related to sustainable development and conservation. Over the past 20 years, international funding for conservation by developed countries, foundations, and private donors has increased, though not as much as developing countries and conservation biologists had hoped. In the United States government, funding for international conservation programs is spread across many departments, including the Agency for International Development, the National Science Foundation, the Smithsonian Institution, and the Fish and Wildlife Service.

The World Bank and international NGOs

Much of the increase in conservation funding by developed countries has been channeled through the **World Bank** (www.worldbank.org) and associated **Global Environment Facility** (**GEF**) (www.thegef.org/gef). The World Bank is a multilateral development bank established to promote international trade and economic activity. The World Bank is governed mainly by developed countries, and only a small portion of its activities is related to conservation. Its partner organization, the GEF, was established specifically to channel money from developed countries to conservation and environmental projects in developing countries, with much of its funding distributed by the World Bank. The World Bank and the GEF have emerged as the leading sources of international conservation funding. From 1988 to the present, the World Bank and the GEF have provided grants and loans totaling over $8 billion to support more than 500 biodiversity projects in 185 countries, as well as 39 multicountry regional programs. Funded activities include establishing protected areas, protecting endangered species, restoring degraded habitats, training conservation staff, developing conservation infrastructure, addressing global climate change, and managing forest, freshwater, and marine resources. The World Bank and the GEF provide funding to national governments and conservation organizations to carry out such projects.

The scale of World Bank activities is illustrated by its joint Forest Alliance program with the World Wildlife Fund, which has already established 47 million ha of protected forest. Upcoming goals include establishing 25 million more ha of protected forest and improving the management of a further 300 million ha of forest through a combination of forest certification of good practices, forest restoration, and community forestry (www.worldwildlife.org/alliance). The World Bank is also one of the leaders in efforts to reduce carbon dioxide emissions caused by deforestation in tropical countries such as Indonesia. Through its Forest Carbon Partnership Facility, corporations and developed countries are able to offset their present production of greenhouse gases by purchasing carbon credits for maintaining these tropical forests.

NGOs have also become leading sources of conservation funding, raising funds from membership dues, donations from wealthy individuals, sponsorship from corporations, and grants from foundations, developed countries and international development banks, including the World Bank and the GEF (**Figure 9.9**). The big international conservation organizations (sometimes

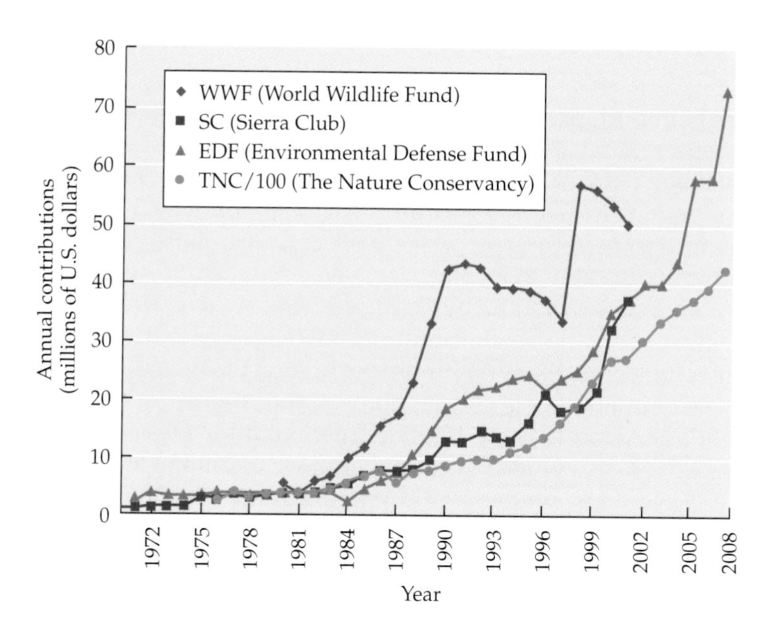

FIGURE 9.9 Over the past four decades, there has been a dramatic increase in the annual contributions of many conservation organizations, as illustrated by four large NGOs from the United States: The Nature Conservancy, World Wildlife Fund, Environmental Defense Fund, and Sierra Club. Note that the values for The Nature Conservancy should be multiplied by 100; for 2007, its contributions were approximately $4 billion. (After Zaradic et al. 2009.)

called BINGOs), such as The Nature Conservancy, BirdLife International, World Wildlife Fund, and Conservation International, use these funds for scientific research, training, land purchases, and running large-scale conservation projects around the world. While such large organizations have provided important benefits to conservation, they need to be monitored more carefully to make sure that they keep their focus on conservation and use their money in the most efficient way possible. In particular, there is often a tendency for such conservation organizations to compete with each other, duplicating their efforts rather than cooperating, and to spend a large percentage of their funds on maintaining an extensive headquarters staff rather than dealing with problems in the field (Bode et al. 2011).

The international NGOs are often active in establishing, strengthening, and funding both local NGOs and government agencies in the developing world that run conservation programs (see the Appendix for a list of major conservation NGOs). From the perspective of a BINGO, such as the World Wildlife Fund, working with local organizations in developing countries is an effective strategy because it relies on local knowledge, and it trains and supports groups of citizens within the country who can then be advocates for conservation for years to come. NGOs are sometimes perceived to be more effective at carrying out conservation projects than government departments, but programs initiated by NGOs may end after a few years when funding runs out, and often they do not achieve a lasting effect. Also, there is sometimes a mismatch between the shifting agendas of BINGOs and the long-term concerns of local NGOs.

Government, World Bank, and NGO funding for conservation projects has increased in recent decades. Environmental trust funds and debt-for-nature swaps provide additional mechanisms to support conservation activities.

Environmental trust funds

An increasingly important mechanism used to provide secure, long-term support for conservation activities in developing countries is the **national environmental fund** (**NEF**). NEFs are typically set up as conservation trust funds or foundations in which a board of trustees—composed of representatives of the host government, conservation organizations, and donor agencies—allocates the annual income from an endowment to support inadequately funded conservation activities by government departments and nongovernment conservation organizations. NEFs have been established in over 50 developing countries, with funds contributed by developed countries and by major international conservation organizations.

One important example is the Bhutan Trust Fund for Environmental Conservation, established in 1991 by the government of Bhutan in cooperation with the World Bank and the World Wildlife Fund. The Bhutan Trust Fund has received about $40 million (exceeding its goal of $20 million). It provides $1 million per year for conservation activities in this eastern Himalayan country. NEFs have proliferated in recent years, with the Latin American and Caribbean Network of Environmental Funds (RedLAC) alone comprising 13 countries and over 3000 projects supported by an annual budget of over $70 million (www.redlac.org).

Debt-for-nature swaps

Another approach, the **debt-for-nature swap**, has made innovative use of the huge international debt owed by developing countries to international financial institutions as an opportunity to protect biodiversity (Greiner and Lankester 2007). The commercial banks that hold these huge debts sometimes sell the debts at a steep discount on the international secondary debt market when there is a low expectation of repayment. Thus, an international conservation organization is able to buy a developing country's discounted debt from a bank. The debt is then canceled in exchange for the developing country's agreement to make annual payments, in its own currency, for conservation activities such as land acquisition, park management, and public education. In other "swaps," governments of developed countries owed money directly by developing countries may cancel the debts if the developing countries agree to contribute to national environmental trust funds or to fund other conservation activities.

Such programs have converted debt valued at $1.5 billion to conservation and sustainable development activities in Costa Rica, Colombia, Poland, the Philippines, Madagascar, and a dozen other countries. In certain specific places, debt-for-nature swaps have been an effective conservation tool. However, increasing spending on conservation programs by a developing country through debt-for-nature swaps may divert needed funding from social programs such as medical care, schools, and agricultural development.

How effective is conservation funding?

Evaluations of projects funded by the World Bank, the GEF, and other large sources have revealed a mixture of positive and negative outcomes (see Kapos et al. 2008 and various reports at www.thegef.org/gef). On the positive side, these international organizations have provided increased funding for conservation projects, planned national biodiversity strategies, identified and protected important ecosystems and habitats, and provided training. However, the lack of participation by community groups, local scientists, and government leaders; an overreliance on foreign consultants; and a lack of understanding of organizational objectives by people in the recipient countries are identified as major problems. An additional problem is the mismatch of funding over short periods with the long-term needs of poor countries.

It must be recognized that many environmental projects supported by international aid do not provide lasting solutions to the problems, because of failure to deal with the "4 Cs"—concern, contracts, capacity, and causes. Environmental aid will be effective only when applied to situations in which both donors and recipients have a genuine *concern* to solve the problems (Do key people really want the project to be successful, or do they just want the money?); when mutually satisfactory and enforceable *contracts* for the project can be agreed on (Will the work actually be done once the money is given out? Will money be siphoned off into private hands?); where there is the *capacity* to undertake the project, in terms of institutions, personnel, and infrastructure (Do people have the skills to do the work, and do they have the necessary resources, such as vehicles, research equipment, buildings, and access to information, to carry out the work?); and when the *causes* of the problem are addressed (Will the project treat the underlying causes of the problem, such as poverty, overpopulation, and mismanagement, or just provide temporary relief of the symptoms?). Despite these problems, international funding of conservation projects continues. Past experiences are informing new projects, which are often more effective because of additional rules but with the result that the application and accounting processes can be extremely cumbersome and time-consuming.

The need for increased funding for biodiversity remains great at the local, national, and international levels. At present, about $6 billion is spent each year on budgets for terrestrial protected areas, yet it would take $13 billion to expand and effectively manage systems protecting terrestrial biological diversity in the tropics alone (Brooks et al. 2009). While $13 billion is an enormous amount of money, it is less than the $20 billion spent annually just on agricultural subsidies in the United States, and it is dwarfed by the whopping $738 billion spent on U.S. military defense in 2010 (**Figure 9.10**). Similarly, while the conservation funds provided by the World Bank seem large, they are small compared with the other activities supported by the World Bank, national governments, and

While the recent increased funding for the protection of biodiversity is welcome, further funding is needed to accomplish the task.

(A)

(B)

FIGURE 9.10 (A) Countries spend huge amounts of money on military defense and agricultural subsidies. (B) Far less money is spent on biodiversity conservation and environmental protection. Around $6 billion per year is spent on the budgets of the world's national parks and protected areas, which is around the cost of a single modern aircraft carrier. (A, photograph courtesy of Micah P. Blechner/U.S. Navy; B, photograph © Steve Bloom Images/Alamy.)

related organizations. Certainly the world's priorities could be modestly adjusted to give more resources to the protection of biological diversity (Rand et al. 2010). Instead of countries rushing forward in a race to supply themselves with the next generation of fighter aircraft, missiles, ships, drones, and other weapons systems, what about spending what it takes to protect biological diversity? Instead of the world's affluent consumers buying the latest round of consumer luxuries and electronic gadgets to replace things that still work, what about contributing more money to conservation organizations and causes?

There is also a role to be played by conservation organizations and businesses working together to market "green products" to environmentally aware consumers. Already the Forest Stewardship Council and similar organizations are certifying wood products from sustainably managed forests, and coffee companies are marketing shade-grown coffee. If consumers are educated about the benefits of these products, and willing to buy them at a somewhat higher price, this could be a strong force in international conservation efforts.

Finally, a potentially huge new funding source to protect tropical forests is being adopted as part of the 2009 Copenhagen Accord. Because about 20% of global greenhouse gas emission results from tropical forest destruction, a funding mechanism called **REDD—Reducing Emissions from Deforestation and Forest Degradation**—could pay to protect tropical forests (Ghazoul et al. 2010). REDD would reward poorer nations for preserving forests by paying them for the carbon that is stored in their forests. There are huge concerns about whether this money will be well spent in protecting forests and reducing poverty in developing countries, or whether it will be diverted to other purposes or cause worse deforestation in other places. Organizations at all scales will be involved in designing and implementing projects, and monitoring what happens as REDD becomes a reality.

Conservation Education

Conservation biologists, ecologists, and other scientists acquire large amounts of information about the projects that they work on. Yet, often this information is only communicated at specialized scientific meetings and published in the form of technical papers. Conservation biologists need to communicate with a wider audience by writing about their work for local audiences, speaking about their work at public meetings, and especially, working with children who are learning about nature.

One of the best ways to educate people about conservation is for scientists and educators to involve groups of citizens in local conservation projects, especially those that include fieldwork. Such efforts involving direct outreach to ordinary citizens require creativity and attention to popular concerns, yet sometimes they can be very successful (Jacobson et al. 2006). These citizen scientists provide valuable contributions to scientific projects and often become strong advocates for conservation. Another rapidly expanding feature of such projects is the use of websites that allow citizens to enter their data online and to track the results of the project.

Many of these **citizen science conservation projects** involve birds. Project FeederWatch (www.birds.cornell.edu/pfw) is a Cornell Lab of Ornithology and Bird Studies Canada initiative that recruits people for annual winter surveys of bird populations at their feeders. Another venture developed by Cornell and the National Audubon Society is eBird (www.ebird.org), a giant database on bird presence and abundance as observed by bird-watchers throughout North America (Sullivan et al. 2009). Finally, Journey North (www.learner.org/jnorth) is a global study of migration and seasonal change in which students record when they observe various biological events, such as the first ruby-throated hummingbird or monarch butterfly in the

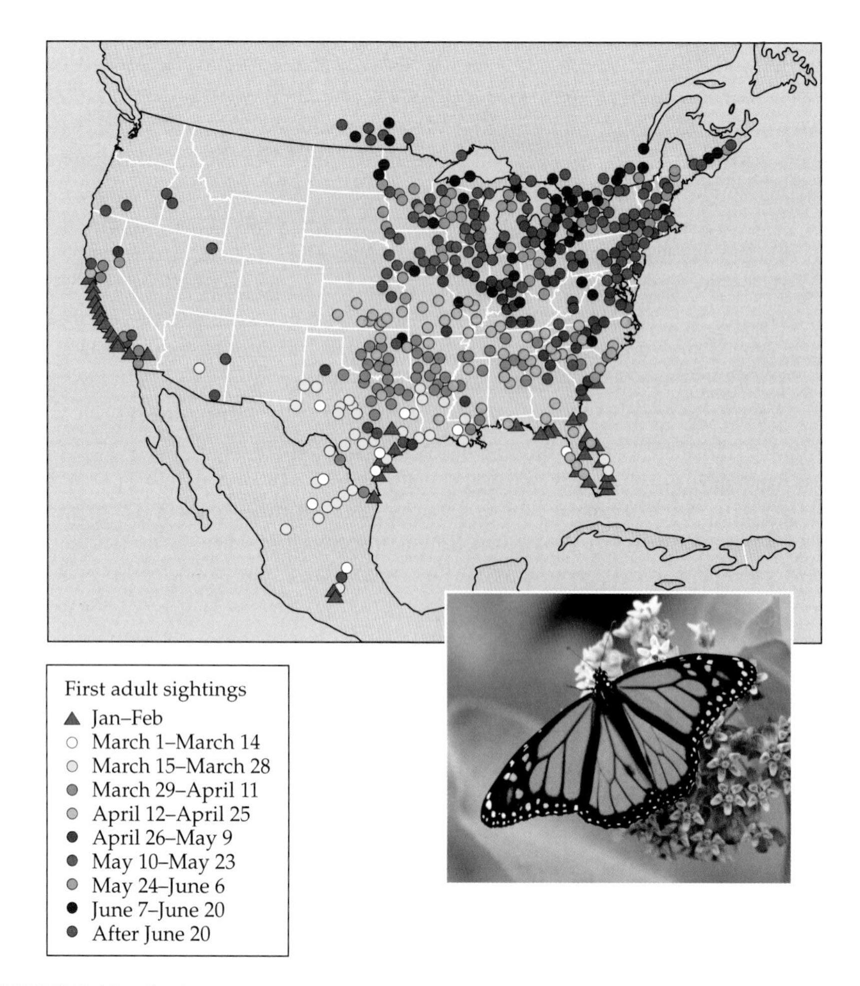

FIGURE 9.11 Each spring, thousands of citizen observers, especially students, contribute their observations to the Journey North website. This map presents the observations of the first appearance of monarch butterflies (*Danaus plexippus*) across North America in the spring of 2009. The butterflies primarily overwinter in central Mexico and start migrating into southern United States in March, arriving in northern United States and southern Canada in late May and June. There are secondary overwintering sites in California, Florida, and Texas. (After www.learner.org/jnorth; butterfly photograph by David McIntyre.)

spring; students then enter their data online, and they can see maps of data submitted by observers across North America (**Figure 9.11**).

Many conservation education projects involve working with school groups. The Global Rivers Environmental Education Network (GREEN) exemplifies an effective action project that uses a practical hands-on approach to teach environmental messages (www.earthforce.org). Students and teachers identify water and air quality problems in their local environments, acquire the skills and equipment needed to investigate the problems, carry out and analyze field research, recommend solutions, and then lobby for those changes. This program has allowed children in over 60 countries to learn about pollution, help protect their environment, and then share what they have learned on the GREEN website.

The West Indian Whistling-Duck and Wetlands Conservation Project illustrates how conservation biologists can reach hundreds of teachers and tens of thousands of students using teacher workshops and educational materials (www.scscb.org). The goal of the project is to raise awareness of endangered Caribbean birds and the wetland habitats in which they live. The leaders of the project decided to focus on teacher training and support in order to reach the widest possible audience. The project has produced and distributed a wide range of educational materials, including teachers' manuals, slide shows, puppet shows, posters, coloring books, and bird identification guides emphasizing conservation themes and the value of wetlands. Materials are printed in the various languages of the region and adapted to each island. These educational materials, workshops, and annual environmental festivals have contributed to species and wetlands protection; this has, in turn, positively affected bird populations, which on several islands are now stable or increasing.

The Role of Conservation Biologists

When a species is driven to extinction, it is usually by a combination of factors acting simultaneously or sequentially. Blaming a specific group of people or a certain industry for the destruction of biodiversity is a simplistic and usually ineffective strategy. The challenge for conservation biologists is to understand the national and international links that promote the destruction and to find viable alternatives (Sutherland et al. 2011). These alternatives must include finding ways to stabilize the size of the human population; supporting livelihoods for people (in both developing and developed countries) that do not damage the environment; providing incentives and penalties that will convince industries to value the environment; and restricting trade in products that are obtained by damaging the environment. Also crucial is a willingness on the part of people in developed countries to reduce their consumption of the world's resources and to pay fair prices for products that are produced in a sustainable, nondestructive manner. If these challenges are to be met, conservation biologists should strongly consider taking on several active roles (**Figure 9.12**).

First, they can become *effective educators* in the public forum as well as in the classroom. Studies show that students care more strongly about conservation issues after attending a conservation biology course (Caro et al. 2003). Conservation biologists need to educate as broad a range of people as possible about the problems that stem from the loss of biodiversity (Pace et al. 2010). Groups such as fishermen, hunters, bird-watchers, hikers, and church leaders may be motivated to help conservation efforts

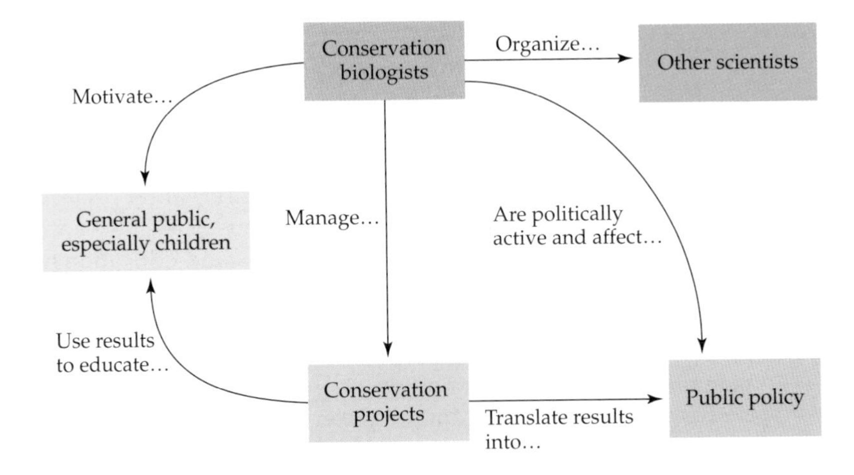

FIGURE 9.12 Conservation biologists need to be active in various ways to achieve the goals of conservation biology and the protection of biological diversity. Not every conservation biologist can be active in each role, but all of the roles are important.

once they become aware of the issues or recognize that their special area of concern or emotional well-being is dependent on conservation. The efforts of Merlin Tuttle and Bat Conservation International illustrate how public attitudes can be changed through education. When the government of Austin, Texas, began plans to exterminate hundreds of thousands of Mexican free-tailed bats (*Tadarida brasiliensis*) that lived under a downtown bridge, Tuttle and his colleagues conducted a successful publicity campaign to convince people that bats were both fun to watch and critical in controlling populations of noxious insects over a wide area. The situation has changed so radically that the government now protects the bats as a matter of civic pride, and citizens and tourists gather every night to watch the bats emerge (**Figure 9.13**).

The goal of conservation biology is not just to reveal new knowledge but to use that knowledge to protect biodiversity. Conservation biologists must learn to show the practical application of their work.

Conservation biologists can also become *politically active leaders* in order to influence public policy and the passage of new laws (Fleishman et al. 2011; Scott and Rachlow 2011). An important first step in this process is joining conservation organizations or political parties to learn more about the issues and to be more effective in advocating for change. Though much of the political process is time-consuming and tedious, it is often the only way to accomplish major conservation goals such as acquiring new land for reserves or preventing overexploitation of old-growth forests. Conservation biologists need to master the language and methods of the legal process and form effective alliances with environmental lawyers,

FIGURE 9.13 Citizens and tourists gather in the evening to watch Mexican free-tailed bats emerge from their roosts beneath the Congress Avenue bridge in Austin, Texas. (Photograph © Merlin D. Tuttle, Bat Conservation International, www.batcon.org.)

citizen groups, and politicians. To be effective, conservation biologists have to demonstrate the relevance of their research to public policy and concerns and show that their findings are unbiased and respectful of the values and concerns of all stakeholders (Wilhere 2012).

Finally, and most important, conservation biologists can become effective *managers* and *practitioners* of conservation projects (Neff 2011). They should be willing to walk on the land and go out on the water to find out what is really happening, to get dirty, to talk with local people, to knock on doors, and to take risks. Conservation biologists must learn everything they can about the species and ecosystems that they are trying to protect and then make that knowledge available to others in a form that can be readily understood and can affect decision making.

If conservation biologists are willing to put their ideas into practice, and to work with park managers, land-use planners, politicians, and local people, then progress will follow. Getting the right mix of models, new theories, innovative approaches, and practical examples is necessary for the success of the discipline. Once this balance is found, conservation biologists, working with energized citizens and government officials, will be in a position to protect the world's biological diversity during this unprecedented era of change.

Summary

- Sustainable development is economic development that satisfies present and future needs of human society while minimizing its impact on biodiversity. Achieving sustainable development is a challenge for conservation biology and society.
- Legal efforts to protect biodiversity occur at local, regional, and national levels and regulate activities affecting both private and public lands. Governments and private land trusts buy land for conservation purposes or acquire conservation easements and development rights for future protection. Associated laws limit pollution, regulate or ban certain types of development, and set rules for hunting and other recreational activities—all with the aim of preserving biodiversity and protecting human health.
- International agreements and conventions that protect biological diversity are needed for the following reasons: species migrate across borders, there is an international trade in biological products, the benefits of biological diversity are of international importance, and the threats to diversity are often international in scope and require international cooperation.
- Conservation groups, governments in developed countries, and the World Bank are increasing funding to protect biodiversity, especially in developing countries. National environmental funds and debt-for-nature swaps are also used to fund conservation activities. However, the amount of money is still inadequate to deal with the problems.
- Conservation biologists must demonstrate the practical value of the theories and approaches of their new discipline and must actively work with different elements of society to protect biodiversity. Conservation education is an important part of this activity.

For Discussion

1. What are the roles of government agencies, private conservation organizations, businesses, community groups, and individuals in the conservation of biodiversity? Can they work together, or are their interests necessarily opposed to each other?
2. How can conservation biologists provide links between basic science and a public environmental movement? What suggestions can you make for ways in which conservation biologists and environmental activists can energize and enrich each other in working toward an economically and environmentally stable world?
3. Sutherland and colleagues (2009) posed 100 questions for conservation biology (see Suggested Readings). Provide answers for the questions you consider to be the most urgent and important.

Suggested Readings

Bode, M., W. Probert, W. R. Turner, K. A. Wilson, and O. Venter. 2011. Conservation planning with multiple organizations and objectives. *Conservation Biology* 25: 295–304. There are both negative and positive aspects of having many conservation organizations working in the same place.

Fleishman, E. and 28 others. 2011. Top 40 priorities for science to inform US conservation and management policy. *BioScience* 61: 290–300. Outline of research priorities of greatest use to the government and society in reaching management decisions.

Jacobson, S. K., M. McDuff, and M. Monroe. 2006. *Conservation Education and Outreach Techniques*. Oxford University Press, Oxford, UK. Practical ways for enlarging public support for conservation.

Kapos, V. and 11 others. 2008. Calibrating conservation: New tools for measuring success. *Conservation Letters* 1: 155–164. A methodology for evaluating whether conservation projects have achieved their goals.

Neff, N. W. 2011. What research should be done and why? Four competing visions among ecologists. *Frontiers in Ecology and the Environment* 9: 462–469. Ecologists identify different reasons for carrying out research and different relationships to society.

Rands, M.R. and 12 others. 2010. Biodiversity conservation: Challenges beyond 2010. *Science* 329: 1298–1303. A radical change is required in how societies view and fund biodiversity conservation.

Sullivan, B., C. L. Wood, M. J. Iliff, R. E. Bonney, D. Fink, and S. Kelling. 2009. eBird: A citizen-based bird observation network in the biological sciences. *Biological Conservation* 142: 2282–2292. eBird is engaging tens of thousands of bird-watchers due to its user-friendly approach.

Sutherland, W. J. and 43 others. 2009. One hundred questions of importance to the conservation of global biological diversity. *Conservation Biology* 23: 557–567. Presents 100 scientific questions whose answers would have the greatest impact on conservation practice and policy.

Sutherland, W. J. and 14 others. 2010 A horizon scan of global conservation issues for 2011. *Trends in Ecology and Evolution* 26: 10–16. A review of key emerging topics in conservation biology, including new methods of pest control and social attitudes.

Wilhere, G. F. 2012. Inadvertent advocacy. *Conservation Biology* 26: 39–46. Conservation biologists need to be aware of their two separate roles as scientists providing information about the threats to biodiversity and as advocates for threatened biodiversity.

Appendix

Selected Environmental Organizations and Sources of Information

The best single reference on conservation activities is the *Conservation Directory* (www.nwf.org/conservationdirectory). This directory lists over 4200 local, national, and international conservation organizations. Online searches, especially using Google, provide a powerful way to search for information concerning people, organizations, places, and topics. Various conservation and environmental organizations have websites with employment opportunities and internships throughout the world. Two places to start are: www.webdirectory.com/Employment and www.ecojobs.com. Also look at the websites of the organizations listed below. Two publications of interest are: *The Eco Guide to Careers that Make a Difference; Environmental Work for a Sustainable World* (2004), published by Island Press, and *Careers in the Environment* (2007) by Mike Fasulo and Paul Walker, published by McGraw-Hill.

Searchable databases on species and countries are now rapidly expanding. These two are worth looking at:

Encyclopedia of Life
www.eol.org
Developing resource for species biology.

Global Biodiversity Information Facility
www.gbif.org
Free and open access to biodiversity data.

The following is a list of some major conservation organizations and resources:

Association of Zoos and Aquariums (AZA)
8403 Colesville Road, Suite 710
Silver Spring, MD 20910-3314 U.S.A.
www.aza.org
Preservation and propagation of captive wildlife.

BirdLife International
Wellbrook Court
Girton Road
Cambridge, CB3 0NA U.K.
www.birdlife.org
Determines status, priorities, and conservation plans for birds throughout the world.

CITES Secretariat of Wild Fauna and Flora
International Environment House
15 Chemin des Anémones
CH-1219 Châtelaine, Geneva, Switzerland
www.cites.org
Regulates trade in endangered species.

Conservation International (CI)
2011 Crystal Drive, Suite 500
Arlington, VA 22202 U.S.A.
www.conservation.org
Active in international conservation efforts and developing conservation strategies; home of Center for Applied Biodiversity Science.

Convention on Biological Diversity Secretariat
413 Rue Saint-Jacques, Suite 800
Montreal, Quebec H2Y 1N9 Canada
www.cbd.int
Promotes the goals of the CBD: sustainable development, biodiversity conservation, and equitable sharing of genetic resources.

Earthwatch Institute
14 Western Avenue
Boston, MA 02134 U.S.A.
www.earthwatch.org
Clearinghouse for international conservation projects in which volunteers can work with scientists.

Environmental Defense Fund (EDF)
257 Park Avenue South
New York, NY 10010 U.S.A.
www.edf.org
Involved in scientific, legal, and economic issues.

European Center for Nature Conservation (ECNC)
P.O. Box 90154
5000 LG Tilburg
The Netherlands
www.ecnc.org
Provides the scientific expertise that is required for making conservation policy.

Fauna & Flora International
Jupiter House, 4th Floor
Station Road
Cambridge, CB1 2JD U.K.
www.fauna-flora.org
Long-established international conservation body acting to protect species and ecosystems.

Food and Agriculture Organization of the United Nations (FAO)
Viale delle Terme di Caracalla
00513 Rome, Italy
www.fao.org
A UN agency supporting sustainable agriculture, rural development, and resource management.

Friends of the Earth
1100 15th St. N.W., 11th floor
Washington, DC 20005 U.S.A.
www.foe.org
Attention-grabbing organization working to improve and expand environmental policy.

Global Environment Facility Secretariat (GEF)
1818 H Street N.W.
Washington, D.C. 20433 U.S.A.
www.thegef.org
Funds international biodiversity and environmental projects.

Greenpeace International
Ottho Heldringstraat 5
1006 AZ Amsterdam
The Netherlands
www.greenpeace.org/international
Activist organization, known for grassroots efforts and dramatic protests against environmental damage.

Missouri Botanical Garden/Center for Plant Conservation
4344 Shaw Boulevard
St. Louis, MO 63110 U.S.A.
www.missouribotanicalgarden.org
Major center for worldwide plant conservation activities.

National Audubon Society
225 Varick Street, 7th floor
New York, NY 10014 U.S.A.
www.audubon.org
Wildlife conservation, public education, research, and political lobbying, with emphasis on birds.

National Council for Science and the Environment (NCSE)
1101 17th Street N.W., Suite 250
Washington, D.C. 20036 U.S.A.
www.ncseonline.org
Works to improve the scientific basis for environmental decision making; their website provides extensive environmental information.

National Wildlife Federation (NWF)
11100 Wildlife Center Drive
Reston, VA 20190-5362 U.S.A.
www.nwf.org
Advocates for wildlife conservation. Publishes the *Conservation Directory 2005–2006*, as well as children's publications *Ranger Rick* and *Your Big Backyard*.

Natural Resources Defense Council (NRDC)
40 West 20th Street
New York, NY 10011 U.S.A.
www.nrdc.org
Uses legal and scientific methods to monitor and influence government actions and legislation.

The Nature Conservancy (TNC)
4245 North Fairfax Drive, Suite 100
Arlington, VA 22203-1606 U.S.A.
www.nature.org
Emphasizes land preservation.

NatureServe
4600 N. Fairfax Drive, 7th floor
Arlington, VA 22209 U.S.A
www.natureserve.org
Maintains databases of endangered species for North America.

New York Botanical Garden/Institute for Economic Botany
200th Street and Kazimiroff Boulevard
Bronx, NY 10458 U.S.A.
www.nybg.org
Conducts research and conservation programs involving plants that are useful to people.

Ocean Conservancy
1300 19th Street N.W., 8th floor
Washigton, D.C. 20036 U.S.A.
www.oceanconservancy.org
Focuses on marine wildlife and ocean and coastal habitats.

Rainforest Action Network
221 Pine Street, 5th floor
San Francisco, CA 94104 U.S.A.
www.ran.org
Works for rain forest conservation and human rights.

Royal Botanic Gardens, Kew
Richmond Surrey TW9 3AB
United Kingdom
www.rbgkew.org.
The famous "Kew Gardens" are home to a leading botanical research institute and an enormous plant collection.

Sierra Club
85 Second Street, 2nd floor
San Francisco, CA 94105-3441 U.S.A.
www.sierraclub.org
Leading advocate for the preservation of wilderness and open space.

Smithsonian Institution/National Zoological Park
3001 Connecticut Avenue N.W.
Washington, D.C. 20008 U.S.A.
www.nationalzoo.si.edu
The National Zoo and the nearby U.S. National Museum of Natural History represent a vast resource of literature, biological materials, and skilled professionals.

Society for Conservation Biology (SCB)
4245 N. Fairfax Drive
Arlington, VA 22203 U.S.A.
www.conbio.org
Leading scientific society for the field. Develops and publicizes new ideas and scientific results through the journal *Conservation Biology* and annual meetings.

Student Conservation Association (SCA)
689 River Road
P.O. Box 550
Charlestown, NH 03603 U.S.A.
www.thesca.org
Places volunteers and interns with conservation organizations and public agencies.

United Nations Development Programme (UNDP)
1 United Nations Plaza
New York, NY 10017 U.S.A.
www.undp.org
Funds and coordinates international economic development activities.

United Nations Environment Programme (UNEP)
United Nations Avenue, Gigiri
P.O. Box 30552
Nairobi, Kenya
www.unep.org
International program of environmental research and management.

United States Fish and Wildlife Service
Department of the Interior
1849 C. Street N.W.
Washington, D.C. 20240 U.S.A.
www.fws.gov
The leading U.S. government agency concerned with conservation research and management; with connections to state governments and other government units, including the National Marine Fisheries Service, the U.S. Forest Service, and the Agency for International Development, which is active in developing nations.

Wetlands International
Horapark 9
6717 LZ Ede
The Netherlands
www.wetlands.org
Focus on the conservation and management of wetlands.

Wilderness Society
1615 M. Street N.W.
Washington, D.C. 20036 U.S.A.
www.wilderness.org
Devoted to preserving wilderness and wildlife.

Wildlife Conservation Society/New York Zoological Society (WCS)
2300 Southern Boulevard
Bronx, NY 10460-1099 U.S.A.
www.wcs.org
Leaders in wildlife conservation and research.

World Bank
1818 H. Street N.W.
Washington, D.C. 20433 U.S.A.
www.worldbank.org
Multinational bank involved in economic development; increasingly concerned with environmental issues.

World Conservation Monitoring Centre (WCMC)
219 Huntingdon Road
Cambridge CB3 0DL U.K.
www.unep-wcmc.org
Monitors global wildlife trade, the status of endangered species, natural resource use, and protected areas.

World Conservation Union (IUCN)
Rue Mauverney 28
CH-1196, Gland, Switzerland
www.iucn.org
Coordinating body for international conservation efforts. Produces directories of specialists and the Red Lists of endangered species.

World Resources Institute (WRI)
10 G. Street N.E., Suite 800
Washington, D.C. 20002 U.S.A.
www.wri.org
Produces environmental, conservation, and development reports.

World Wildlife Fund (WWF)
1250 24th Street N.W.
P.O. Box 97180
Washington, D.C. 20077-7180 U.S.A.
www.worldwildlife.org or www.wwf.org
Major conservation organization, with branches throughout the world. Active in national park management.

Xerces Society
628 NE Broadway, Suite 200
Portland, OR 97232 U.S.A.
www.xerces.org
Focuses on the conservation of insects and other invertebrates.

Zoological Society of London
Regent's Park, Outer Circle
London NW1 4RY U.K.
www.zsl.org
Center for worldwide activities to preserve nature.

Glossary

A

adaptive management Implementing a management plan and monitoring how well it works, then using the results to adjust the management plan.

adaptive restoration Using monitoring data to adjust management plans to achieve restoration goals.

Allee effect Inability of a species' social structure to function once a population of that species falls below a certain number or density of individuals.

alleles Different forms of the same gene (e.g., different alleles of the genes for certain blood proteins produce the different blood types found among humans).

alpha diversity The number of different species in a community or specific location; species richness.

amenity value Recreational value of biodiversity, including ecotourism.

arboretum Specialized botanical garden focusing on trees and other woody plants.

artificial incubation Conservation strategy that involves humans taking care of eggs or newborn animals.

artificial insemination Introduction of sperm into a receptive female animal by humans; used to increase the reproductive output particularly of endangered species.

assisted colonization People establish populations of rare and endangered species at new, suitable localities outside of their current range; may be done because climate change is rendering their current habitats unsuitable.

B

beneficiary value *See* bequest value.

bequest value The benefit people receive by preserving a resource or species for their children and descendants or future generations, and quantified as the amount people are willing to pay for this goal. Also known as beneficiary value.

beta diversity Rate of change of species composition along a gradient or transect.

binomial The unique two-part Latin name taxonomists bestow on a species, such as *Canis lupus* (gray wolf) or *Homo sapiens* (humans).

biodiversity The complete range of species, biological communities, and their ecosystem interactions and genetic variation within species. Also known as biological diversity.

biodiversity indicators Species or groups of species that provide an estimate of the biodiversity in an area when data on the whole community is unavailable. Also known as surrogate species.

biological community A group of species that occupies a particular locality.

biological definition of species Among biologists, the most generally used of several definitions of "species." A group of individuals that can potentially breed among themselves in the wild and that do not breed with individuals of other groups. Compare with morphological definition of species.

biological diversity *See* biodiversity.

biomagnification Process whereby toxins become more concentrated in animals at higher levels in the food chain.

biophilia The postulated predisposition in humans to feel an affinity for the diversity of the living world.

biopiracy Collecting and using biological materials for commercial, scientific, or personal use without obtaining the necessary permits.

bioprospecting Collecting biological materials as part of a search for new products.

bioregional management Management system that focuses on a single large ecosystem or a series of linked ecosystems, particularly where they cross political boundaries.

biosphere reserve One of the protected areas established as part of a United Nations program to demonstrate the compatibility of biodiversity conservation and sustainable development to benefit local people.

biota A region's plants and animals.

bushmeat Meat from any wild animal.

bushmeat crisis The sharp decline in wild animal populations caused by humans hunting for food.

bycatch Animals, including marine mammals, sea turtles, and fish of no commercial value, killed unintentionally during large-scale fishing operations.

C

carnivore An animal species that consumes other animals to survive. Also called a secondary consumer or predator. Compare with primary consumer.

carrying capacity The number of individuals or biomass of a species that an ecosystem can support.

census A count of the number of individuals in a population.

CITES *See* Convention on International Trade in Endangered Species.

citizen science conservation project Conservation project that involves public participation.

co-management Local people working as partners with government agencies and conservation organizations in protected areas.

commodity value *See* direct use value.

community conserved area Protected areas managed and sometimes established by local people.

compensatory mitigation When a new site is created or rehabilitated in compensation for a site damaged or destroyed elsewhere.

conservation banking A system involving developers paying landowners for the preservation of an endangered species or protected habitat type (or even restoration of a degraded habitat) to compensate for a species or habitat that is destroyed elsewhere.

conservation biology Scientific discipline that draws on diverse fields to carry out research on biodiversity, identify threats to biodiversity, and play an active role in the preservation of biodiversity

conservation concession Method of protecting land whereby a conservation organization pays a government or other landowner to preserve habitat rather than allow an extractive industry to damage the habitat.

conservation corridor *See* corridor.

conservation development *See* limited development.

conservation easement Method of protecting land in which landowners give up the right to develop or build on their property, often in exchange for financial or tax benefit.

conservation leasing Providing payments to private landowners who actively manage their land for biodiversity protection.

conservation units Species, ecosystems, and physical features of a region; data about them are gathered and stored by conservation organizations.

consumptive use value Value assigned to goods that are collected and consumed locally.

Convention Concerning the Protection of the World Cultural and Natural Heritage *See* World Heritage Convention.

Convention on Biological Diversity (CBD) A treaty that obligates countries to protect the biodiversity within their borders, and gives them the right to receive economic benefits from the use of that biodiversity.

Convention on International Trade in Endangered Species (CITES) The international treaty that establishes lists (known as Appendices) of species for which international trade is to be prohibited, regulated, or monitored.

corridor Connection between protected areas that allows for dispersal and migration. Also known as conservation corridor, habitat corridor, or movement corridor.

cost-benefit analysis Comprehensive analysis that compares values gained against the costs of a project or resource use.

cross-fostering Conservation strategy in which individuals from a common species raise the offspring of a rare, related species.

cultural eutrophication Algal blooms and associated impacts caused by excess mineral nutrients released into the water from human activity.

D

debt-for-nature swap Agreement in which a developing country agrees to fund additional conservation activities in exchange for a conservation organization canceling some of its discounted debt.

decomposer A species that feeds or grows on dead plant and animal material. Also called a detritivore.

deep ecology Philosophy emphasizing biodiversity protection, personal lifestyle changes, and working towards political change.

demographic stochasticity Random variation in birth, death, and reproductive rates in small populations, sometimes causing further decline in population size. Also called demographic variation.

demographic study Study in which individuals and populations are monitored over time to determine rates of growth, reproduction, and survival.

demographic variation *See* demographic stochasticity.

desertification Process by which ecosystems are degraded by human activities into man-made deserts.

detritivore *See* decomposer.

direct use value Value assigned to products, such as timber and animals, that are harvested and directly used by the people who harvest them. Also known as commodity value or private goods.

E

Earth Summit An international conference held in 1992 in Rio de Janeiro that resulted in new environmental agreements. Also known as the Rio Summit.

ecocolonialism Practice of governments and conservation organizations disregarding the land rights and traditions of local people in order to establish new conservation areas.

ecological economics Discipline that includes valuations of biodiversity in economic analyses.

ecological restoration Altering a site to reestablish an indigenous ecosystem.

ecologically extinct A species that has been so reduced in numbers that it no longer has a significant ecological impact on the biological community.

economic development Economic activity focused on improvements in efficiency and organization but not necessarily on increases in resource consumption.

economic growth Economic activity characterized by increases in the amount of resources used and in the amount of goods and services produced.

ecosystem A biological community together with its associated physical and chemical environment.

ecosystem diversity The variety of ecosystems present in a place or geographic area.

ecosystem management Large-scale management that often involves multiple stakeholders, the primary goal of which is the preservation of ecosystem components and processes.

ecosystem services Range of benefits provided to people from ecosystems, including flood control, clean water, and reduction of pollution.

ecotourism Tourism, especially in developing countries, focused on viewing unusual and/or especially charismatic biological communities and species that are unique to a country or region.

edge effect Altered environmental and biological conditions at the edges of a fragmented habitat.

effective population size (N_e) The number of breeding individuals in a population.

embryo transfer The surgical implantation of embryos into a surrogate mother; used to increase the number of individuals of a rare species, with a common species used as the surrogate mother.

endangered species A species that has a high risk of extinction in the wild in the near future; a category in the IUCN system and under the U.S. Endangered Species Act.

Endangered Species Act (ESA) An important U.S. law passed to protect endangered species and the ecosystems in which they live.

endemic Occurring in a place naturally, without the influence of people (e.g., gray wolves are endemic to Canada).

endemic species Species found in one place and nowhere else (e.g., the many lemur species found only on the island of Madagascar).

environmental economics Discipline that examines the economic impacts of environmental policies and decisions.

environmental ethics Discipline of philosophy that articulates the intrinsic value of the natural world and people's responsibility to protect the environment.

environmental impact assessment Evaluation of a project which considers its possible present and future impacts on the environment.

environmental justice Movement that seeks to empower and assist poor and politically weak people in protecting their own environments; their well-being and the protection of biological diversity are enhanced in the process.

environmental stochasticity Random variation in the biological and physical environment. Can increase the risk of extinction in small populations.

environmentalism A widespread movement, characterized by political activism, with the goal of protecting the natural environment.

eutrophication Process of degradation in aquatic environments caused by nitrogen and phosphorus pollution and characterized by algal blooms and oxygen depletion.

ex situ conservation Preservation of species under artificial conditions, such as in zoos, aquariums, and botanical gardens.

existence value The benefit people receive from knowing that a habitat or species exists and quantified as the amount that people are willing to pay to prevent species from being harmed or going extinct, habitats from being destroyed, and genetic variation from being lost

exotic species A species that occurs outside of its natural range due to human activity. Compare with endemic.

extant Presently alive; not extinct.

externalities Hidden costs or benefits that result from an economic activity to individuals or a society not directly involved in that activity.

extinct The condition in which no members of a species are currently living.

extinct in the wild A species no longer found in the wild, but individuals may remain alive in zoos, botanical gardens, or other artificial environments.

extinction cascade A series of linked extinctions whereby the extinction of one species leads to the extinction of one or more other species.

extinction debt The inevitable extinction of many species in coming years as the result of current human activities.

extinction vortex Tendency of small populations to decline toward extinction.

extirpated Local extinction of a population, even though the species may still exist elsewhere.

extractive reserve Protected area in which sustainable extraction of certain natural products is allowed.

F

fitness An individual's ability to grow, survive, and reproduce.

flagship species A species that captures public attention, aids in conservation efforts, such as establishing a protected area, and may be crucial to ecotourism.

focal species A species that provides a reason for establishing a protected area.

food chain Specific feeding relationships between species at different trophic levels.

food web A network of feeding relationships among species.

founder effect Reduced genetic variability that occurs when a new population is established ("founded") by a small number of individuals.

G

Gaia hypothesis An idea that the Earth's biological, physical, and chemical components interact to regulate the atmosphere and climate.

gamma diversity The number of species in a large geographic area.

gap analysis Comparing the distribution of endangered species and biological communities with existing and proposed protected areas to determine gaps in protection.

gap species A species that is not protected in any part of its range.

gene A unit (DNA sequence) on a chromosome that codes for a specific protein.

gene pool The total array of genes and alleles in a population.

genetic diversity The range of genetic variation found within a species.

genetic drift Loss of genetic variation and change in allele frequencies that occur by chance in small populations.

genetic swamping Differences between species becoming blurred when invasive species hybridize with native species.

genetic variation Genetic differences among individuals in a population.

genetically modified organism (GMO) An organism whose genetic code has been altered by scientists using recombinant DNA technology.

genome resource bank (GRB) Frozen collection of DNA, eggs, sperm, embryos, and other tissues of species that can be used in breeding programs and scientific research.

genotype Particular combination of alleles that an individual possesses.

geographic information systems (GIS) Computer analyses that integrate and display spatial data; relating in particular to the natural environment, ecosystems, species, protected areas, and human activities.

global climate change Climate characteristics that are changing now and will continue to change in the future, resulting in part from human activity.

Global Environment Facility (GEF) A large international program involved in funding conservation activities in developing countries.

global warming The current and future increases in temperatures caused by higher atmospheric concentrations of carbon dioxide and other greenhouse gases produced by human activities.

globalization The increasing interconnectedness of the world's economy.

globally extinct No individuals are presently alive anywhere.

greenhouse effect Warming of the Earth caused by carbon dioxide and other "greenhouse gases" in the atmosphere that allow the sun's radiation to penetrate and warm the Earth but prevent the heat generated by sunlight from re-radiating. Heat is thus trapped near the surface, raising the planet's temperature.

greenhouse gases Gases in the atmosphere, primarily carbon dioxide, that are transparent to sunlight but that trap heat near the Earth's surface.

guild A group of species at the same trophic level that use approximately the same environmental resources.

H

habitat conservation plans (HCPs) Regional plans that allow development in designated areas while protecting biodiversity in other areas.

habitat corridor *See* corridor.

habitat fragmentation Process whereby a continuous area of habitat is both reduced in area and divided into two or more fragments.

hard release In the establishment of a new population, when individuals from an outside source are released in a new location without assistance.

healthy ecosystem Ecosystem in which processes are functioning normally, whether or not there are human influences.

herbivore A species that eats plants or other photosynthetic organisms. Also called a primary consumer.

heterozygous Condition of an individual having two different allele forms of the same gene.

homozygous Condition of an individual having two identical allele forms of the same gene.

hotspot Region with numerous species, many of which are endemic, that is also under immediate threat from human activity.

hybrid Intermediate offspring resulting from mating between individuals of two different species.

hybrid vigor Increased fitness of offspring resulting from mating of unrelated individuals.

I

in situ conservation Preservation of natural communities and populations of endangered species in the wild.

inbreeding Self-fertilization or mating among close relatives.

inbreeding depression Lowered reproduction or production of weak offspring following mating among close relatives or self-fertilization.

indicator species Species used in a conservation plan to identify and often protect a biological community or set of ecosystem processes.

indirect use values Values provided by biodiversity that do not involve harvesting or destroying the resource (such as water quality, soil protection, recreation, and education). Also known as public goods.

integrated conservation and development project (ICDP) Conservation project that also provides for the economic needs and welfare of local people.

Intergovernmental Panel on Climate Change (IPCC) A group of leading scientists organized by the United Nations to study the impacts and implications of human activity on climate and ecosystems.

International Union for Conservation of Nature (IUCN) *See* IUCN.

intrinsic value Value of a species and other aspects of biodiversity for their own sake, unrelated to human needs.

introduction program Moving individuals to areas outside their historical range in order to create a new population of an endangered species.

invasive species An introduced species that increases in abundance at the expense of native species.

island biogeography model Formula for the relationship between island size and the number of species living on the island; the model can be used to predict the impact of habitat destruction on species extinctions, viewing remaining habitat as an "island" in the "sea" of a degraded ecosystem.

IUCN The World Conservation Union, a major international conservation organization; previously known as the International Union for the Conservation of Nature.

K

keystone resource Any resource in an ecosystem that is crucial to the survival of many species; for example, a watering hole.

keystone species A species that has a disproportionate impact (relative to its numbers or biomass) on the organization of a biological community. Loss of a keystone species may have far-reaching consequences for the community.

L

land trust Conservation organization that protects and manages land.

landscape ecology Discipline that investigates patterns of habitat types and their influence on species distribution and ecosystem processes.

legal title The right of ownership of land, recognized by a government and/or judicial system; traditional people often struggle to achieve this recognition.

limited development Compromise involving a landowner, a property developer, and a conservation organization that combines some development with protection of the remaining land.

limiting resource Any requirement for existence whose presence or absence limits a population's size. In the desert, for example, water is a limiting resource.

locally extinct A species that no longer exists in a place where it used to occur, but still exists elsewhere.

Long-Term Ecological Research (LTER) A U.S. government program to study gradual changes in ecosystems

M

management plan A statement of how to protect biodiversity in an area, along with methods for implementation.

marine protected area (MPA) Protected area of ocean and/or coastline established to rebuild and maintain marine biodiversity.

market failure Misallocation of resources in which certain individuals or businesses benefit from using a common resource, such as water, the atmosphere, or a forest, but other individuals, businesses or the society at large bears the cost.

maximum sustainable yield Greatest amount of a resource that can be harvested each year and replaced through population growth without detriment to the population.

metapopulation Shifting mosaic of populations of the same species linked by some degree of migration; a "population of populations."

minimum dynamic area (MDA) Area needed for a population to have a high probability of surviving into the future.

minimum viable population (MVP) Number of individuals necessary to ensure a high probability that a population will survive a certain number of years into the future.

mitigation Process by which a new population or habitat is created to compensate for a habitat damaged or destroyed elsewhere.

morphological definition of species A group of individuals, recognized as a species, that is morphologically, physiologically, or biochemically distinct from other groups. Compare with biological definition of species.

morphospecies Individuals that are probably a distinct species based on their appearance but that do not currently have a scientific name.

movement corridor *See* corridor.

multiple-use habitat An area managed to provide a variety of goods and services.

mutations Changes that occur in genes and chromosomes, sometimes resulting in new allele forms and genetic variation.

mutualistic relationship When two species benefit each other by their relationship.

N

national environmental fund (NEF) A trust fund or foundation that uses its annual income to support conservation activities.

natural history The ecology and distinctive characteristics of a species.

natural resources Commodities and qualities found in nature that are used and valued by people.

non-use value Value of something that is not presently used; for example, existence value.

nonconsumptive use value Value assigned to benefits provided by some aspect of biodiversity that does not involve harvesting or destroying the resource (such as water quality, soil protection, recreation, and education).

nongovernmental organization (NGO) A private organization that acts to benefit society in some way; many conservation organizations are NGOs.

normative discipline A discipline that embraces ethical commitment rather than ethical neutrality.

O

omnivore A species that eats both plants and animals.

open-access resources Natural resources that are not controlled by individuals but are collectively owned by society.

option value Value of biodiversity in providing possible future benefits for human society (such as new medicines).

outbreeding Mating and production of offspring by individuals that are not closely related, as in individuals from different populations of the same species. In general outbreeding leads to heterosis, a level of genetic variation that improves individual evolutionary fitness.

outbreeding depression Lowered fitness that occasionally occurs when individuals of different species or widely different populations mate and produce offspring.

overexploitation Intense harvest of a resource or species that results in its decline or loss.

P

payment for ecosystem services (PES) Direct payment to individual landowners and local communities that protect species or critical ecosystem characteristics.

perverse subsidies Government payments or other financial incentives to industries that result in environmentally destructive activities.

phenotype The morphological, physiological, anatomical, and biochemical characteristics of an individual that result from the expression of its genotype in a particular environment.

photochemical smog Visible air pollution resulting from chemicals released from human activities being transformed in sunlight.

polymorphic gene Within a population, a gene that has more than one form or allele.

population A geographically defined group of individuals of the same species that mate and otherwise interact with one another. Compare with metapopulation.

population biology Study of the ecology and genetics of populations, often with a focus on population numbers.

population bottleneck A radical reduction in population size (e.g., following an outbreak of infectious disease), sometimes leading to the loss of genetic variation.

population viability analysis (PVA) Demographic analysis that predicts the probability of a population persisting in an environment for a certain period of time; sometimes linked to various management scenarios.

precautionary principle Principle stating that it may be better to avoid taking a particular action due to the possibility of causing unexpected harm.

predator *See* carnivore.

prey An animal that is eaten as food by another species.

primary consumer *See* herbivore.

primary producer Organisms such as green plants, algae, and seaweeds that obtain their energy directly from the sun via photosynthesis. Also known as autotrophs.

private goods *See* direct use value.

productive use value Values assigned to products that are sold in markets.

protected area A habitat managed primarily or in large part for biodiversity.

R

Ramsar Convention on Wetlands A treaty that promotes the protection of wetlands of international importance.

rapid biodiversity assessments Species inventories and vegetation maps made by teams of biologists when urgent decisions must be made on where to establish new protected areas. Also known as *r*apid *a*ssessment *p*lans (RAPs).

recombination Mixing of the genes on the two copies of a chromosome that occurs during meiosis (i.e., in the formation of egg and sperm, which contain only one copy of each chromosome). Recombination is an important source of genetic variation.

reconciliation ecology The science of developing urban places in which people and biodiversity can coexist.

Red Data Books Compilations of lists ("Red Lists") of endangered species prepared by the IUCN and other conservation organizations.

Red List criteria Quantitative measures of threats to species based on the probability of extinction.

Red Lists Lists of endangered species prepared by the IUCN.

Reducing Emissions from Deforestation and Forest Degradation (REDD) Program using financial incentives to reduce the emissions of greenhouse gases from deforestation.

reference site Control site that provides goals for restoration in terms of species composition, community structure, and ecosystem processes.

reintroduction program The release of captive bred or wild-collected individuals at a site within their historical range where the species does not presently occur.

replacement cost approach How much people would have to pay for an equivalent product if what they normally use is unavailable.

representative site Protected area that includes species and ecosystem properties characteristic of a larger area.

resilience The ability of an ecosystem to return to its original state following disturbance.

resistance The ability of an ecosystem to remain in the same state even with ongoing disturbance.

restocking program The release of additional individuals into an existing population to increase population size and introduce genetic variation. Also referred to as augmentation.

restoration ecology The scientific study of restored populations, communities, and ecosystems.

Rio Summit *See* Earth Summit.

S

secondary consumer *See* carnivore.

seed bank Collection of stored seeds, collected from wild and cultivated plants; used in conservation and agricultural programs.

shifting cultivation Farming method in which farmers cut down trees, burn them, plant crops for a few years, and then abandon the site when soil fertility declines. Also called "slash-and-burn" agriculture.

sink population A population that receives an influx of new individuals from a source population.

sixth extinction episode The present mass extinction event which is just beginning.

SLOSS debate Controversy concerning the relative advantages of a *s*ingle *l*arge *o*r *s*everal *s*mall conservation areas.

soft release In the establishment of a new population, when individuals are given assistance during or after the release to increase the chance of success. Compare with hard release.

source population An established population from which individuals disperse to new locations.

species diversity The entire range of species found in a particular place.

species richness The number of species found in a community

stable ecosystem An ecosystem that is able to remain in roughly the same compositional state despite human intervention or stochastic events such as unseasonable weather.

stochasticity Random variation; variation happening by chance.

succession The gradual process of change in species composition, vegetation structure, and ecosystem characteristics following natural or human-caused disturbance.

survey Repeatable sampling method to estimate population size or density, or some other aspect of biodiversity.

sustainable development Economic development that meets present and future human needs without damaging the environment and biodiversity.

symbiotic relationship A mutualistic relationship in which neither of the two species involved can survive without the other.

T

taxonomist Scientist involved in the identification and classification of species.

threatened Species that fall into the endangered or vulnerable to extinction categories in the IUCN system. Under the U.S. Endangered Species Act, refers to species at risk of extinction, but at a lower risk than endangered species.

tragedy of the commons The unregulated use of a public resource that results in its degradation.

trophic cascade Major changes in vegetation and biodiversity resulting from the loss of a keystone species.

trophic levels Levels of biological communities representing ways in which energy is captured and moved through the ecosystem by the various types of species. *See* primary producer; herbivore; predator; detritivore.

U

umbrella species Protecting an umbrella species results in the protection of other species.

UNESCO Biosphere Reserves Program *See* biosphere reserve.

use value The direct and indirect value provided by some aspect of biodiversity.

W

World Bank International bank established to support economic development in developing countries.

World Heritage Convention A treaty that protects cultural and natural areas of international significance.

World Heritage site A cultural or natural area officially recognized as having international significance.

Z

zoning A method of managing protected areas that allows or prohibits certain activities in designated places.

Chapter Opening Photograph Credits

CHAPTER 1 A primatologist observes endangered gelada baboons in Ethiopia. Photograph © Suzanne Long/Alamy.

CHAPTER 2 A fire-tailed sunbird (*Aethopyga ignicauda*) visits flowers of the cinnabar rhododendron (*Rhododendron cinnabarinum*) in the eastern Himalayas of northeastern India. Photograph courtesy of Kamal Bawa.

CHAPTER 3 Ecotourists viewing a tiger in Ranthambhore National Park, Rajasthan, India. Photograph © Aditya "Dicky" Singh/Alamy.

CHAPTER 4 An experiment to test how a changing climate affects an ecosystem. Photograph by J. S. Dukes.

CHAPTER 5 The South Island giant moa (*Dinornis robustus*), a giant (up to 3.6 m [over 11 feet] in height) flightless bird, went extinct around 700 years ago after people arrived in New Zealand. Photograph © National Geographic Society/Corbis.

CHAPTER 6 A wildlife veterinarian and a conservation biologist examine a Magellanic penguin chick on the Argentinian coast. Photograph courtesy of Sue Moore.

CHAPTER 7 Banff National Park, the oldest national park in Canada and a World Heritage Site, is known for its great natural beauty, such this view of Bow Lake and Crowfoot Mountain. Photograph © John E Marriott/All Canada Photos/Corbis.

CHAPTER 8 Farmers and nursery workers plant tree seedlings on degraded land in Costa Rica. Photograph © Gary Braasch/Corbis.

CHAPTER 9 An extensive solar power station at the foot of an undeveloped mountain region. Photograph © Yu Lan/shutterstock.

Bibliography

Abesamis, R. A. and G. R. Russ. 2005. Density-dependent spillover from a marine reserve: Long-term evidence. *Ecological Applications* 15: 1798–1812. (7)

Abson, D. J. and M. Termansen. 2011. Valuing ecosystem services in terms of ecological risks and returns. *Conservation Biology* 25: 250–258. (3)

Aburto-Oropeza, O., B. Erisman, G. R. Galland, I. Mascareñas-Osorio, et al. 2011. Large recovery of fish biomass in a no-take marine reserve. *PLoS ONE* 6(8): e23601. (7)

Adams, J. S. 2006. *The Future of the Wild: Radical Conservation for a Crowded World*. Beacon Press, Boston. (8)

Alexander, S. (ed.). 2009. *Voluntary Simplicity: The Poetic Alternative to Consumer Culture*. Stead and Daughters, Whanganui, New Zealand. (3)

Allan, B. F., H. P. Dutra, L. S. Goessling, K. Barnett, et al. 2010. Invasive honeysuckle eradication reduces tick-borne disease risk by altering host dynamics. *Proceedings of the National Academy of Sciences USA* 107: 18523–18527. (4)

Allen, C., R. S. Lutz, and S. Demarais. 1995. Red imported fire ant impacts on northern bobwhite populations. *Ecological Applications* 5: 632–638. (4)

Allendorf, F. W. and G. Luikart. 2007. *Conservation and the Genetics of Populations*. Blackwell Publishing, Oxford, UK. (5)

Almeida, A. P. and S. L. Mendes. 2007. An analysis of the role of local fishermen in the conservation of the loggerhead turtle (*Caretta caretta*) in Pontal do Ipiranga, Linhares, ES, Brazil. *Biological Conservation* 134: 106–112. (1)

Altieri, M. A. 2004. Linking ecologists and traditional farmers in the search for sustainable agriculture. *Frontiers in Ecology and the Environment* 2: 35–42. (6)

Altman, I., A. M. H. Blakeslee, G. C. Osio, C. B. Rillahan, et al. 2011. A practical approach to implementation of ecosystem-based management: A case study using the Gulf of Maine marine ecosystem. *Frontiers in Ecology and the Environment* 9: 183–189. (8)

Anadón, J. D., A. Gimenez, R. Ballestar, and I. Pérez. 2009. Evaluation of local ecological knowledge as a method for collecting extensive data on animal abundance. *Conservation Biology* 23: 617–625. (7)

Andam, K. S., P. J. Ferraro, K. R. E. Sims, A. Healy, and M. B. Holland. 2010. Protected areas reduced poverty in Costa Rica and Thailand. *Proceedings of the National Academy of Sciences USA* 107: 9996–10001. (7)

Anderegg, W. R. L., J. W. Prall, J. Harold, and S. H. Schneider. 2010. Expert credibility in climate change. *Proceedings of the National Academy of Sciences USA* 107: 12107–12109. (4)

Anderson, E. P. and J. A. Maldonado-Ocampo. 2011. A regional perspective on the diversity and conservation of tropical Andean fishes. *Conservation Biology* 25: 30–39. (2)

Andersson, E., S. Barthel, and K. Ahrne. 2007. Measuring social-ecological dynamics behind the generation of ecosystem services. *Ecological Applications*. 17: 1267–1287.

Andrew-Essien, E. and F. Bisong. 2009. Conflicts, conservation and natural resource use in protected area systems: An analysis of recurrent issues. *European Journal of Scientific Research* 25: 118–129. (8)

Armstrong, D. P. and P. J. Seddon. 2008. Directions in reintroduction biology. *Trends in Ecology and Evolution* 23: 20–25. (6)

Armsworth, P. R., L. Cantu-Salazar, M. Parnell, Z. G. Davies, and R. Stoneman. 2011. Management costs for small protected areas and economies of scale in habitat conservation. *Biological Conservation* 144: 423–429. (7)

Arnold, A. E. and F. Lutzoni. 2007. Diversity and host range of foliar fungal endophytes: Are tropical leaves biodiversity hotspots? *Ecology* 88: 541–549. (2)

Association of Zoos & Aquariums. 2009. *http://www.aza.org* (6)

Aukema, J. E., B. Leung, K. Kovacs, C. Chivers, et al. 2011. Economic impacts of non-native forest insects in the continental United States. *PLoS ONE* 6(9): e24587. (4)

Ausband, D. E. and K. R. Foresman. 2007. Swift fox reintroductions on the Blackfeet Indian Reservation, Montana, USA. *Biological Conservation* 136: 423–430. (6)

Azam, F. and A. Z. Worden. 2004. Oceanography: Microbes, molecules, and marine ecosystems. *Science* 303: 1622–1624. (2)

Azzurro, E., P. Moschella, and F. Maynou. 2011. Tracking signals of change in Mediterranean fish diversity based on local ecological knowledge. *PLoS ONE* 6(9): e24885. (6)

Bagla, P. 2010. Hardy cotton-munching pests are latest blow to GM crops. *Science* 327: 1439. (4)

Baillie, J. E. M., C. Hilton-Taylor, and S. N. Stuart (eds.). 2004. *2004 IUCN Red List of threatened species: A global species assessment*. IUCN/SSC Red List Programme, Cambridge, UK. (5)

Baker, J. D. and P. M. Thompson. 2007. Temporal and spatial variation in age-specific survival rates of a long-lived mammal, the Hawaiian monk seal. *Proceedings of the Royal Society B* 274: 407–415. (6)

Balmford, A., J. Beresford, J. Green, R. Naidoo, et al. 2009. A global perspective on trends in nature-based tourism. *PLoS Biology* 7: e1000144. (3)

Balmford, A., A. Bruner, P. Cooper, R. Costanza, et al. 2002. Economic reasons for conserving wild nature. *Science* 297: 950–953. (3)

Balmford, A., R. E. Green, and J. P. W. Scharlemann. 2005. Sparing land for nature: Exploring the potential of changes in agricultural yield on the area needed for crop production. *Global Change Biology* 11: 1594–1605. (3)

Barber-Meyer, S. M. 2010. Dealing with the clandestine nature of wildlife-trade market surveys. *Conservation Biology* 24: 918–923. (3)

Barnosky, A. D., N. Matzke, S. Tomiya, G. O. U. Wogan, et al. 2011. Has Earth's sixth mass extinction already arrived? *Nature* 471: 51–57. (1, 5)

Barrett, M. A., T. A. Bouley, A. H. Stoertz, and R. W. Stoertz. 2011. Integrating a One Health approach in education to address global health and sustainability challenges. *Frontiers in Ecology and the Environment* 9: 239–235. (4)

Barriopedro, D., E. M. Fischer, J. Luterbacher, R. M. Trigo, and R. Garcia-Herrera. 2011. The hot summer of 2010: Redrawing the temperature record map of Europe. *Science* 332: 220–224. (4)

Beattie, A. and P. Ehrlich. 2010. The missing link in biodiversity conservation. *Science* 328: 307–308. (3)

Beck, M. W., R. D. Brumbaugh, L. Airoldi, A. Carranza, et al. 2011. Oyster reefs at risk and recommendations for conservation, restoration, and management. *BioScience* 61: 107–116. (5)

Becker, C. G., C. R. Fonseca, C. F. B. Haddad, and P. I. Prado. 2010. Habitat split as a cause of local population declines of amphibians with aquatic larvae. *Conservation Biology* 24: 287–294. (4)

Beier, P., W. Spencer, R. F. Baldwin, and B. H. McRae. 2011. Towards best practices for developing regional connectivity maps. *Conservation Biology* 25: 879–892. (7)

Beissinger, S. R., E. Nicholson, and H. P. Possingham. 2009. Application of population viability analysis to landscape conservation planning. *In* J. J. Millspaugh and F. R. Thompson, III (eds.), *Models for Planning Wildlife Conservation in Large Landscapes*, pp. 33–50. Academic Press, San Diego, CA. (6)

Bell, C. D., J. M. Blumenthal, A. C. Broderick, and B. J. Godley. 2010. Investigating potential for depensation in marine turtles: How low can you go? *Conservation Biology* 24: 226–235. (5)

Bell, T. J., M. L. Bowles, and K. A. McEachern. 2003. Projecting the success of plant population restoration with viability analysis. *In* C. A. Brigham and M. M. Schwartz (eds.), *Population Viability in Plants*, pp. 313–348. Springer-Verlag, Heidelberg. (6)

Bender, M. A., T. R. Knudson, R. E. Tuleya, J. J. Sirutis, et al. 2010. Modeled impact of anthropogenic warming on the frequency of intense Atlantic hurricanes. *Science* 327: 454–8. (4)

Bennett, E. L., E. Blencowe, K. Brandon, D. Brown, et al. 2007. Hunting for consensus: Reconciling bushmeat harvest, conservation, and development policy in West and Central Africa. *Conservation Biology* 21: 884–887. (4)

Berger, J. 1990. Persistence of different-sized populations: An empirical assessment of rapid extinctions in bighorn sheep. *Conservation Biology* 4: 91–98. (5)

Berger, J. 1999. Intervention and persistence in small populations of bighorn sheep. *Conservation Biology* 13: 432–435. (5)

Berkes, F., T. P. Hughes, R. S. Steneck, J. A. Wilson, et al. 2006. Globalization, roving bandits, and marine resources. *Science* 311: 1557–1558. (4)

Beyer, H. L., E. H. Merrill, N. Varley, and M. S. Boyce. 2007. Willow on Yellowstone's northern range: Evidence for a trophic cascade? *Ecological Applications* 17: 1563–1571. (2)

Bhagwat, S. A., N. Dudley, and S. R. Harrop. 2011. Religious following in biodiversity hotspots: Challenges and opportunities for conservation and development. *Conservation Letters* 4: 234–240. (3)

Bhagwat, S. A. and C. Rutte. 2006. Sacred groves: Potential for biodiversity management. *Frontiers in Ecology and the Environment* 4: 519–524. (8)

Bhatti, S., S. Carrizosa, P. McGuire, and T. Young (eds.). 2009. *Contracting for ABS: The Legal and Scientific Implications of Bioprospecting Contracts.* IUCN, Gland, Switzerland. (3)

Bickford, D., D. J. Lohman, N. S. Sodhi, P. K. L. Ng, et al. 2007. Cryptic species as a window on diversity and conservation. *Trends in Ecology and Evolution* 22: 148–155. (2)

Bisht, I. S., P. S. Mehta, and D. C. Bhandari. 2007. Traditional crop diversity and its conservation on-farm for sustainable agricultural production in Kumaon Himalaya of Uttaranchal state: A case study. *Genetic Resources and Crop Evolution* 54: 345–357. (8)

Black, S. A. 2011. Leadership and conservation effectiveness: Finding a better way to lead. *Conservation Letters* 4: 329–339. (1)

Blanco, G., J. A. Lemus, and M. Garía-Montijano. 2011. When conservation management becomes contraindicated: Impact of food supplementation on health of endangered wildlife. *Ecological Applications* 21: 2469–2477. (6)

Blue Lists. 2009. *http://www.bluelists.ethz.ch* (6)

Bobbink, R., K. Hicks, J. Galloway, T. Spranger, et al. 2010. Global assessment of nitrogen deposition effects on terrestrial plant diversity. *Ecological Applications* 20: 30–59. (4)

Bode, M., W. Probert, W. R. Turner, K. A. Wilson, and O. Venter. 2011. Conservation planning with multiple organizations and objectives. *Conservation Biology* 25: 295–304. (7)

Boersma, P. D. 2006. Landscape-level conservation for the sea. *In* M. J. Groom, G. K. Meffe, and C. R. Carroll (eds.), *Principles of Conservation Biology*, 3rd ed, pp. 447–448. Sinauer Associates, Sunderland, MA. (6)

Bohlen, P. J., S. Lynch, L. Shabman, M. Clark, et al. 2009. Paying for environmental services from agricultural lands: An example from the northern Everglades. *Frontiers in Ecology and the Environment* 7: 46–55. (9)

Borrini-Feyerabend, G., M. Pimbert, T. Farvar, A. Kothari, and Y. Renard. 2004. *Sharing Power: Learning by Doing in Co-Management of Natural Resources throughout the World.* IIED and IUCN/CEESP/CMWG, Cenesta, Tehran. (8)

Botanic Gardens Conservation International (BGCI). 2005. *http://www.bgci.org* (6)

Bouché, P., I. Douglas-Hamilton, G. Wittemyer, A. J. Nianogo, et al. 2011. Will elephants soon disappear from West African savannahs? *PLoS ONE* 6(6): e20619. (5)

Bouzat, J. L., J. A. Johnson, J. E. Toepfer, S. A. Simpson, et al. 2008. Beyond the beneficial effects of translocations as an effective tool for the genetic restoration of isolated populations. *Conservation Genetics* 10: 191–201. (5)

Bowkett, A. E. 2009. Recent captive-breeding proposals and the return of the ark concept to global species conservation. *Conservation Biology* 23: 773–776. (6)

Bowles, M. L., J. L. McBride, and R. F. Betz. 1998. Management and restoration ecology of Mead's milkweed. *Annals of the Missouri Botanical Garden* 85: 110–125. (6)

Boyles, J. G., P. M. Cryan, G. F. McKracken, and T. H. Kunz. 2011. Economic importance of bats in agriculture. *Science* 332: 41–42. (3)

Bradley, B. A., D. M. Blumenthal, R. Early, E. D. Grosholz, et al. 2012. Global change, global trade, and the next wave of plant invasions. *Frontiers in Ecology and the Environment* 10: 20–28. (4)

Bradshaw, A. D. 1990. The reclamation of derelict land and the ecology of ecosystems. *In* W. R. Jordan III, M. E. Gilpin, and J. D. Aber (eds.), *Restoration Ecology: A Synthetic Approach to Ecological Research*, pp. 53–74. Cambridge University Press, Cambridge. (8)

Bradshaw, C. J. A., N. S. Sodhi, and B. W. Brook. 2009. Tropical turmoil: A biodiversity tragedy in progress. *Frontiers in Ecology and the Environment* 7: 79–87. (4)

Braithwaite, R. W. 2001. Tourism, role of. *In* S. A. Levin (ed.), *Encyclopedia of Biodiversity*, vol. 5, pp. 667–679. Academic Press, San Diego, CA. (3)

Brandt, A. A., A. J. Gooday, S. N. Brandão, S. Brix, et al. 2007. First insights in the biodiversity and biogeography of the Southern Ocean deep sea. *Nature* 447: 307–311. (2)

Branton, M. and J. S. Richardson. 2011. Assessing the value of the umbrella-species concept for conservation planning with meta-analysis. *Conservation Biology* 25: 9–20. (7)

Briant, G., V. Gond, and S. G. W. Laurance. 2010. Habitat fragmentation and the desiccation of forest canopies: A case study from eastern Amazonia. *Biological Conservation* 143: 2763–2769. (4)

Briggs, S. V. 2009. Priorities and paradigms: Directions in threatened species recovery. *Conservation Letters* 2: 101–108. (6)

Brinson, M. M. and S. D. Eckles. 2011. U.S. Department of Agriculture conservation program and practice effects on wetland ecosystem services: A synthesis. *Ecological Applications* 21: S116–S127. (8)

Brodie, J. F., O. E. Helmy, W. Y. Brockelman, and J. L. Maron. 2009. Bushmeat poaching reduces the seed dispersal and population growth rate of a mammal-dispersed tree. *Ecological Applications* 19: 854–863. (7)

Brook, A., M. Zint, and R. DeYoung. 2003. Landowner's response to an Endangered Species Act listing and implications for encouraging conservation. *Conservation Biology* 17: 1638–1649. (6)

Brooks, T. M., S. J. Wright, and D. Sheil. 2009. Evaluating the success of conservation actions in safeguarding tropical forest biodiversity. *Conservation Biology* 23: 1448–1457. (9)

Brush, S. B. 2007. Farmers' rights and protection of traditional agricultural knowledge. *World Development* 35: 1499–1514. (6)

Bryant, D., L. Burke, J. McManus, and M. Spalding. 1998. *Reefs at Risk: A Map-Based Indicator of Threats to the World's Coral Reefs*. World Resources Institute, Washington, D.C. (4)

Buchholz, R. 2007. Behavioral biology: An effective and relevant conservation tool. *Trends in Ecology and Evolution* 22: 401–407. (6)

Buckley, R. 2009. Parks and tourism. *PLoS Biology* 7: e1000143. (3)

Bull, A. T. 2004. *Microbial Diversity and Bioprospecting*. ASM Press, Washington, D.C. (3)

Bulman, C. R., R. J. Wilson, A. R. Holt, A. L. Galvez-Bravo, et al. 2007. Minimum viable metapopulation size, extinction debt, and the conservation of declining species. *Ecological Applications* 17: 1460–1473. (6)

Burke, L., K. Reytar, M. Spalding, and A. Perry. 2011. *Reefs at Risk Revisited*. World Resources Institute, Washington DC, USA. (4)

Butchart, S. H. and J. P. Bird. 2010. Data deficient birds on the IUCN Red List: What don't we know and why does it matter? *Biological Conservation* 143: 239–247. (6)

Butler, R. A. and W. F. Laurance. 2008. New strategies for conserving tropical forests. *Trends in Ecology and Evolution* 23: 469–472. (4, 8)

Cain, M. L., W. D. Bowman, and S. D. Hacker. 2008. *Ecology*. Sinauer Associates, Sunderland, MA. (2)

Cameron, S. A., J. D. Lozier, J. P. Strange, J. B. Koch, et al. 2011. Patterns of widespread decline in North American bumble bees. *Proceedings of the National Academy of Sciences USA* 108: 662–667. (3)

Carnicer, J., M. Coll, M. Ninyerola, X. Pons, et al. 2011. Widespread crown condition decline, food web disruption, and amplified tree mortality with increased climate change-type drought. *Proceedings of the National Academy of Sciences USA* 108: 1474–1478. (4)

Caro, T., M. Borgerhoff Mulder, and M. Moore. 2003. Effects of conservation education on reasons to conserve biological diversity. *Biological Conservation* 114: 143–152. (9)

Carraro, C., J. Eyckmans, and M. Finus. 2006. Optimal transfers and participation decisions in international environmental agreements. *The Review of International Organizations* 1: 379–396. (6)

Carruthers, E. H., D. C. Schneider, and J. D. Neilson. 2009. Estimating the odds of survival and identifying mitigation opportunities for common bycatch in pelagic longline fisheries. *Biological Conservation* 142: 2620–2630. (4)

Castelletta, M., N. S. Sodhi, and R. Subaraj. 2000. Heavy extinctions of forest avifauna in Singapore: Lessons for biodiversity conservation in Southeast Asia. *Conservation Biology* 14: 1870–1880. (5)

Cawardine, J., J. K. Carissa, K. A. Wilson, R. L. Pressey, and H. P. Possingham. 2009. Hitting the target and missing the point: Target-based conservation planning in context. *Conservation Letters* 2: 4–11. (7)

Ceballos, G. 2007. Conservation priorities for mammals in megadiverse Mexico: The efficiency of reserve networks. *Ecological Applications* 17: 569–578. (7)

Chan, K. M. A. 2008. Value and advocacy in conservation biology: Crisis discipline or discipline in crisis? *Conservation Biology* 22: 1–3. (1)

Chape, S., M. D. Spalding, and M. D. Jenkins (eds.). 2008. *The World's Protected Areas: Status, Values, and Prospects in the Twenty-First Century*. University of California Press, CA. (7)

Chapin, F. S., S. R. Carpenter, G. P. Kofinas, C. Folke, et al. 2010. Ecosystem stewardship: Sustainability strategies for a rapidly changing planet. *Trends in Ecology and Evolution* 25: 241–249. (8)

Chapman, J. W., T. W. Miller, and E. V. Coan. 2003. Live seafood species as recipes for invasion. *Conservation Biology* 17: 1386–1395. (4)

Chen, I., J. K. Hill, R. Ohlemüller, D. B. Roy, and C. D. Thomas. 2011. Rapid range shifts of species associated with high levels of climate warming. *Science* 333: 1024–1026. (4)

Chiba, S. 2010. Invasive non-native species provision of refugia for endangered native species. *Conservation Biology* 24: 1141–1147. (4)

Chittaro, P. M., I. C. Kaplan, A. Keller, and P. S. Levin. 2010. Trade-offs between species conservation and the size of marine protected areas. *Conservation Biology* 24: 197–206. (5)

Chivian, E. and A. Bernstein (eds.). 2008. *Sustaining Life: How Human Health Depends on Biodiversity*. Oxford University Press, New York. (3)

Chornesky, E. A., A. M. Bartuska, G. H. Aplet, K. O. Britton, et al. 2005. Science priorities for reducing the threat of invasive species to sustainable forestry. *BioScience* 55: 335–348. (4)

Christian-Smith, J. and A. Merenlender. 2010. The disconnect between restoration goals and practices: A case study of watershed restoration in the Russian River basin. *Restoration Ecology* 18: 95–102. (8)

Christie, M. R., B. N. Tissot, M. A. Albins, J. P. Beets, et al. 2010. Larval connectivity in an effective network of marine protected areas. *PLoS ONE* 5(12): e15715. (7)

Christy, B. 2010. Asia's wildlife trade. *National Geographic Magazine* 217(January): 78–107. (4)

Cinner, J. E. and S. Aswani. 2007. Integrating customary management into marine conservation. *Biological Conservation* 140: 201–216. (4)

Clark, P. and D. E. Johnson. 2009. *Wolf-cattle interactions in the Northern Rocky Mountains*. In: Range Field Data 2009 Progress Report. Special Report 1092. Corvallis, OR: Oregon State University, Agricultural Experiment Station, pp. 1–7. (6)

Clausen, R. and R. York. 2008. Economic growth and marine diversity: Influence of human social structure on decline of marine trophic levels. *Conservation Biology* 22: 458–466. (4)

Clausnitzer, V., V. J. Kalkman, M. Ram, B. Collen, et al. 2009. Odonata enter the biodiversity crisis debate: The first global assessment of an insect group. *Biological Conservation* 142: 1864–1869. (6)

Clavero, M., L. Brotons, P. Pons, and D. Sol. 2009. Prominent role of invasive species in avian biodiversity loss. *Biological Conservation* 142: 2043–2049. (5)

Clavero, M. and E. García-Berthou. 2005. Invasive species are a leading cause of animal extinctions. *Trends in Ecology and Evolution* 20: 110–110. (4)

Clewell, A. F. and J. Aronson. 2006. Motivations for the restoration of ecosystems. *Conservation Biology* 20: 420–428. (8)

Cockle, K. L., K. Martin, and T. Wesolowski. 2011. Woodpeckers, decay, and the future of cavity-nesting vertebrate communities worldwide. *Frontiers in Ecology and Conservation* 9: 377–382. (2)

Cole, T. 1965. Essay on American Scenery. *In* J. W. McCoubrey(ed.), *American Art, 1700–1960*, pp. 98–109. Prentice-Hall, Englewood Cliffs, NJ. (1)

Coleman, J. M., O. K. Huh, and D. Braud. 2008. Wetland loss in world deltas. *Journal of Coastal Research 24* (1A): 1–14. (4)

Commission for the Conservation of Antarctic Marine Living Resources (CCAMLR). *http://www.ccamlr.org* (6)

Common, M. and S. Stagl. 2005. *Ecological Economics: An Introduction*. Cambridge University Press, New York. (3)

Conservation International and K. J. Caley. 2008. Biological diversity in the Mediterranean Basin. *In* C. J. Cleveland (ed.), *Encyclopedia of Earth*. Environmental Information Coalition, National Council for Science and the Environment, Washington, D.C. *http://www.eoearth.org/article/Biological_diversity_in_the_Mediterranean_Basin* (2)

Convention on International Trade in Endangered Species of Wild Flora and Fauna (CITES). *http://www.cites.org* (6)

Cork, S. J., T. W. Clark, and N. Mazur. 2000. Introduction: An interdisciplinary effort for koala conservation. *Conservation Biology* 14: 606–609. (7)

Corlatti, L., K. Hacklander, and F. Frey-Roos. 2009. Ability of wildlife overpasses to provide connectivity and prevent genetic isolation. *Conservation Biology* 23: 548–556. (7)

Corlett, R. T. 2011. Impacts of warming on tropical lowland rainforests. *Trends in Ecology and Evolution* 26: 606–613. (4)

Corlett, R. T. and R. B. Primack. 2010. *Tropical Rain Forests: An Ecological and Biogeographical Comparison*, 2nd ed. Blackwell Publishing, Malden, MA. (2, 3, 4)

Corlett, R. T. and I. M. Turner. 1996. The conservation value of small, isolated fragments of lowland tropical rain forest. *Trends in Ecology and Evolution* 11: 330–333. (7)

Corral-Verdugo, V., M. Bonnes, C. Tapia-Fonllem, B. Fraijo-Sing, et al. 2009. Correlates of pro-sustainability orientation: The affinity towards diversity. *Journal of Environmental Psychology* 29: 34–43. (1)

Costanza, R., R. d'Arge, R. de Groot, S. Farber, et al. 1997. The value of the world's ecosystem services and natural capital. *Nature* 387: 253–260. (3)

Costello, C., J. M. Drake, and D. M. Lodge. 2007. Evaluating an invasive species policy: Ballast water exchange in the Great Lakes. *Ecological Applications* 17: 655–662. (4)

Cousins, S. A. O. and D. Vanhoenacker. 2011. Detection of extinction debt depends on scale and specialization. *Biological Conservation* 144: 782–787. (5)

Cox, P. A. and T. Elmqvist. 1997. Ecocolonialism and indigenous-controlled rainforest preserves in Samoa. *Ambio* 26: 84–89. (8)

Cox, R. L. and E. C. Underwood. 2011. The Importance of Conserving Biodiversity Outside of Protected Areas in Mediterranean Ecosystems. *PLoS ONE* 6(1): e14508. (8)

Cox, T. M., R. L. Lewison, R. Zydelis, L. B. Crowder, et al. 2007. Comparing effectiveness of experimental and implemented bycatch reduction measures: The ideal and the real. *Conservation Biology* 21: 1155–1164. (4)

Crowl, T. A., T. O. Crist, R. R. Parmenter, G. Belovsky, and A. E. Lugo. 2008. The spread of invasive species and infectious disease as drivers of ecosystem change. *Frontiers in Ecology and the Environment* 6: 238–246. (4)

Czech, B. 2008. Prospects for reconciling the conflict between economic growth and biodiversity conservation with technological progress. *Conservation Biology* 22: 1389–1398. (1)

Dahles, H. 2005. A trip too far: Ecotourism, politics, and exploitation. *Development Change* 36: 969–971. (3)

Daly, G. L., Y. D. Lei, C. Teixeira, D. C. G. Muir, et al. 2007. Accumulation of current-use pesticides in neotropical montane forests. *Environmental Science and Technology* 41: 1118–1123. (4)

Daszak, P., A. A. Cunningham, and A. D. Hyatt. 2000. Emerging infectious diseases of wildlife—threats to biodiversity and human health. *Science* 287: 443–449. (4)

Davis, M. A. 2009. *Invasion Biology*. Oxford University Press, Oxford, UK. (4, 7)

Dawson, M. R., L. Marivaux, C. K. Li, K. C. Beard, and G. Métais. 2006. *Laonastes* and the "Lazarus effect" in recent mammals. *Science* 311: 1456–1458. (2)

de Bello, F., S. Lavorel, P. Gerhold, Ü. Reier, and M. Pärtel. 2010. A biodiversity monitoring framework for practical conservation of grasslands and shrublands. *Biological Conservation* 143: 9–17. (7)

De Grammont, P. C. and A. D. Cuarón. 2006. An evaluation of threatened species categorization systems used on the American continent. *Conservation Biology* 20: 14–27. (6)

De Merode, E. and G. Cowlishaw. 2006. Species protection, the changing informal economy, and the politics of access to the bushmeat trade in the Democratic Republic of Congo. *Conservation Biology* 20: 1262–1271. (4)

De Roos, A. M. 2008. Demographic analysis of continuous-time life-history models. *Ecology Letters* 11: 1–15. (6)

Diamond, J. 2005. *Collapse: How Societies Choose to Fail or Succeed*. Penguin Books, New York. (1, 3)

Dias, M. S., W. E. Magnusson, and J. Zuanon. 2010. Effects of reduced-impact logging on fish assemblages in Central Amazonia. *Conservation Biology* 24: 278–286. (8)

Dickman, A. J., E. A. Macdonald, and D. W. Macdonald. 2011. A review of financial instruments to pay for predator conservation and encourage human-carnivore coexistance. *Proceedings Of The National Academy Of Sciences USA* 108: 13937–13944. (8)

Dietz, R. W. and B. Czech. 2005. Conservation deficits for the continental United States: An ecosystem gap analysis. *Conservation Biology* 19: 1478–1487. (7)

Dobson, A. 2005. Monitoring global rates of biodiversity change: Challenges that arise in meeting the Convention on Biological Diversity (CBD) 2010 goals. *Philosophical Transactions of the Royal Society B-Biological Sciences* 360: 229–241. (3)

Donald, P. F., F. J. Sanderson, I. J. Burfield, S. M. Bierman, et al. 2007. International conservation policy delivers benefits for birds in Europe. *Science* 317: 810–813. (9)

Donath, T. W., S. Bissels, N. Hölzel, and A. Otte. 2007. Large scale application of diaspore transfer with plant material in restoration practice: Impact of seed and microsite limitation. *Biological Conservation* 138: 224–234. (6)

Doney, S. C. 2010. The growing human footprint on coastal and open-ocean biogeochemistry. *Science* 328: 1512–1516. (4)

Donlan C. J., D. K. Wingfield, L. B. Crowder, and C. Wilcox. 2010. Using expert opinion surveys to rank threats to endangered species: A case study with sea turtles. *Conservation Biology* 24: 1586–1595. (6)

Donlan, J., H. W. Greene, J. Berger, C. E. Bock, et al. 2006. Re-wilding North America. *Nature* 436: 913–914. (6)

Donovan, T. M. and C. W. Welden. 2002. *Spreadsheet Exercises in Conservation Biology and Landscape Ecology*. Sinauer Associates, Sunderland, MA. (6)

Drayton, B. and R. B. Primack. 1999. Experimental extinction of garlic mustard (*Alliaria petiolata*) populations: Implications for weed science and conservation biology. *Biological Invasions* 1: 159–167. (5)

Dreschler, M. and F. Wätzold. 2009. Applying tradable permits to biodiversity conservation: Effects of space-dependent conservation benefits and cost heterogeneity on habitat allocation. *Ecological Economics* 68: 1083–1092. (9)

Driscoll, C. T., Y. J. Han, C. Y. Chen, D. C. Evers, et al. 2007. Mercury contamination in forest and freshwater ecosystems in the northeastern United States. *BioScience* 57: 17–28. (4)

Driscoll, D. A. 1999. Genetic neighbourhood and effective population size for two endangered frogs. *Biological Conservation* 88: 221–229. (5)

Duchelle, A. E., M. R. Guariguata, G. Less, M. A. Albornoz, et al. 2012. Evaluating the opportunities and limitations to multiple use of Brazil nuts and timber in Western Amazonia. *Forest Ecology and Management* 268: 39–48. (8)

Dudley, N., L. Higgins-Zogib, and S. Mansourian. 2009. The links between protected areas, faiths, and sacred natural sites. *Conservation Biology* 23: 568–577. (1)

Dudley, R. K. and S. P. Platania. 2007. Flow regulation and fragmentation imperil pelagic-spawning riverine fishes. *Ecological Applications* 17: 2074–2086. (4)

Dukes, J. S., N. R. Chiariello, S. R. Loarie, and C. B. Field. 2011. Strong response of an invasive plant species (*Centaurea solstitialis* L.) to global environmental changes. *Ecological Applications* 21: 1887–1894. (4)

Ehrlich, P. R. and L. H. Goulder. 2007. Is current consumption excessive? A general framework and some indications for the United States. *Conservation Biology* 21: 1145–1154. (4)

Eigenbrod, F., B. J. Anderson, P. R. Armsworth, A. Heinemeyer, et al. 2009. Ecosystem service benefits of contrasting conservation strategies in a human-dominated region. *Proceedings of the Royal Society B* 276: 2903–2911. (7)

Elkinton, J. S., D. Parry, and G. H. Boettner. 2006. Implicating an introduced generalist parasitoid in the invasive browntail moth's enigmatic demise. *Ecology* 87: 2664–2672. (4)

Emerton, L. 1999. Balancing the opportunity costs of wildlife conservation for communities around Lake Mburo National Park, Uganda. Evaluating Eden Series Discussion Paper No. 5, International Institute for Environment and Development, London. (3)

Encyclopedia of the Nations. 2009. India. *http://www.nationsencyclopedia.com/economies/Asia-and-the-Pacific/India.html* (1)

Epps, C. W., P. J. Palsboll, J. D. Wehausen, G. K. Goderick, et al. 2005. Highways block gene flow and cause a rapid decline in genetic diversity of desert bighorn sheep. *Ecology Letters* 8: 1029–1038. (6)

Epps, C. W., J. D. Wehausen, V. C. Bleich, S. G. Torres, and J. S. Brashares. 2007. Optimizing dispersal and corridor models using landscape genetics. *Journal of Applied Ecology* 44: 714–724. (6)

Estes, J. A., J. Terbough, J. S. Brashares, M. E. Power, et al. 2011. Trophic downgrading of planet earth. *Science* 333: 301–306. (2)

Esty, D. C., M. Levy, T. Srebotnjak, and A. de Sherbinin. 2005. *Environmental Sustainability Index: Benchmarking National Environmental Stewardship*. Yale Center for Environmental Law & Policy, New Haven, CT. (3)

Evans, S. R. and B. C. Sheldon. 2008. Interspecific patterns of genetic diversity in birds: Correlations with extinction risk. *Conservation Biology* 22: 1016–1025. (5)

Faeth, S. H., P. S. Warren, E. Shochat, and W. A. Marussich. 2005. Trophic dynamics in urban communities. *BioScience* 55: 399–407. (2)

Faith, D. P. 2008. Threatened species and the potential loss of phylogenetic diversity: Conservation scenarios based on estimated extinction probabilities and phylogenetic risk analysis. *Conservation Biology* 22: 1461–1470. (7)

Falk, D. A., M. A. Palmer, and J. B. Zedler (eds.). 2006. *Foundations of Restoration Ecology: The Science and Practice of Ecological Restoration*. Island Press, Washington, D.C. (8)

Farmer, J. R., D. Knapp, V. J. Meretsky, C. Chancellor, and B. C. Fischer. 2011. Motivations influencing the adoption of conservation easements. *Conservation Biology* 25: 827–834. (9)

Feist, B. E., E. R. Buhle, P. Arnold, J. W. Davis, and N. L. Scholz. 2011. Landscape Ecotoxicology of Coho Salmon Spawner Mortality in Urban Streams. *PLoS ONE* 6(8): e23424. (4)

Fennell, D. A. 2007. *Ecotourism*, 3rd ed. Routledge, New York. (3)

Ferraz, G., G. J. Russell, P. C. Stouffer, R. O. Bierregaard, et al. 2003. Rates of species loss from Amazonian forest fragments. *Proceedings of the National Academy of Sciences USA* 100: 14069–14073. (5)

Ferrer. M., I. Newton, and M. Pandolfi. 2009. Small populations and offspring sex-ratio deviations in eagles. *Conservation Biology* 23: 1017–1025. (5)

Ferro, P. J., M. M. Hanauer, and K. R. E. Sims. 2011. Conditions associated with protected area success in conservation and poverty reduction. *Proceedings of the National Academy of Sciences USA* 108: 13913–13918. (7)

Fischer, J. and D. B. Lindenmayer. 2000. An assessment of published results of animal relocations. *Biological Conservation* 96: 1–11. (6)

Fisher, B., D. P. Edwards, X. Giam, and D. S. Wilcove. 2011. The high costs of conserving Southeast Asia's lowland rainforests. *Frontiers in Ecology and the Environment* 9: 329–334. (7)

Fisher, D. O. and S. P. Blomberg. 2012. Inferring extinction of mammals from sighting records, threats, and biological traits. *Conservation Biology* 26: 57–67. (5)

Fitzpatrick, B. M., J. R. Johnson, D. K. Kump, J. J. Smith, et al. 2010. Rapid spread of invasive genes into threatened native species. *Proceedings of the National Academy of Sciences USA*: 107: 3388–3393. (2)

Flather, C. H., G. D. Hayward, S. R. Beissinger, and P. A. Stephens. 2011. Minimum viable populations: Is there a "magic number" for conservation practitioners? *Trends in Ecology and Evolution* 26: 307–316. (5)

Fleishman, E., D. E. Blockstein, J. A. Hall, M. B. Mascia, et al. 2011. Top 40 priorities for science to inform US conservation and management policy. *BioScience* 61: 290–300. (9)

Flohre, A., C. Fischer, T. Aavik, J. Bengtsson, et al. 2011. Agricultural intensification and biodiversity partitioning in European landscapes comparing plants, carabids, and birds. *Ecological Applications* 21: 1772–1780.

Foley, J., D. Clifford, K. Castle, P. Cryan, and R. S. Ostfeld. 2011. Investigating and managing the rapid emergence of white-nose syndrome, a novel, fatal, infectious disease of hibernating bats. *Conservation Biology* 25: 223–231. (4)

Fontaine, B., P. Bouchet, K. Van Achterberg, M. A. Alonso-Zarazaga, et al. 2007. The European Union's 2010 target: Putting rare species in focus. *Biological Conservation* 139: 167–185. (6)

Forister, M. L., A. C. McCall, N. J. Sanders, J. A. Fordyce, et al. 2010. Compounded effects of climate change and habitat alteration shift patterns of butterfly diversity. *Proceedings of the National Academy of Sciences USA* 107: 2088–2092. (4)

Forzza, R. C., J. F. A. Baumgratz, C. E. M. Bicudo, D. A. L. Canhos, et al. 2012. New Brazilian floristic list highlights conservation challenges. *Bioscience* 62: 39–45. (2)

Foster, B. L., K. Kindscher, G. R. Houseman, and C. A. Murphy. 2009. Effects of hay management and native species sowing on grassland community structure, biomass, and restoration. *Ecological Applications* 19: 1884–1896. (8)

Foster, S. J. and A. C. J. Vincent. 2005. Enhancing sustainability of the international trade in seahorses with a single minimum size limit. *Conservation Biology* 19: 1044–1050. (4)

Fox, H. E., M. B. Mascia, X. Basurto, A. Costa, et al. 2012. Reexamining the science of marine protected areas: Linking knowledge to action. *Conservation Letters* 5: 1–10. (7)

Foxcroft, L. C., V. Jarošík, P. Pyšek, D. M. Richardson, and M. Rouget. 2011. Protected-area boundaries as filters of plant invasions. *Conservation Biology* 25: 400–405. (4)

Francis, C. M., A. V. Borisenko, N. V. Ivanova, J. L. Eger, et al. 2010. The role of DNA barcodes in understanding and conservation of mammal diversity in southeast Asia. *PLoS ONE* 5(9): e12575. (2, 7)

Frankham, R. 2005. Genetics and extinction. *Biological Conservation* 126: 131–140. (5)

Frankham, R., J. D. Ballou, and D. A. Briscoe. 2009. *Introduction to Conservation Genetics*, 2nd ed. Cambridge University Press, Cambridge, UK. (2, 5)

Galbraith, C. A., P. V. Grice, G. P. Mudge, S. Parr, and M. W. Pienkowski. 1998. The role of statutory bodies in ornithological conservation. *Ibis* 137: S224–S231. (1)

Gallant, A. L., R. W. Klaver, G. S. Casper, and M. J. Lannoo. 2007. Global rates of habitat loss and implications for amphibian conservation. *Copeia* 2007: 967–979. (4)

Gardiner, M. M., D. A. Landis, C. Gratton, C. D. DiFonzo, et al. 2009. Landscape diversity enhances biological control of an introduced crop pest in the north-central USA. *Ecological Applications* 19: 143–154. (3)

Gardiner, S., S. Caney, D. Jamieson, and H. Shue. 2010. *Climate Ethics: Essential Readings*. Oxford University Press, New York. (3)

Garzón-Machado, V., J. M. González-Mancebo, A. Palomares-Martínez, A. Acevedo-Rodríguez, et al. 2010. Strong negative effect of alien herbivores on endemic legumes of the Canary pine forest. *Biological Conservation* 143: 2685–2694. (4)

Gascoigne, J., L. Berec, S. Gregory, and F. Courchamp. 2009. Dangerously few liaisons: A review of mate-finding Allee effects. *Population Ecology* 51: 355–372. (5)

Gaston, K. J. and J. I. Spicer. 2004. *Biodiversity: An Introduction*, 2nd ed. Blackwell Publishing, Oxford, UK. (2)

Germano, J. M. and P. J. Bishop. 2009. Suitability of amphibians and reptiles for translocation. *Conservation Biology* 23: 7–15. (6)

Gerrodette, T. and W. G. Gilmartin. 1990. Demographic consequences of changing pupping and hauling sites of the Hawaiian monk seal. *Conservation Biology* 4: 423–430. (6)

Geyer, J., I. Kiefer, S. Kreft, V. Chavez, et al. 2011. Classification of climate-change-induced stresses on biological diversity. *Conservation Biology* 35: 708–715. (4)

Ghazoul, J., R. A. Butler, J. Mateo-Vega, and L. P. Koh. 2010. REDD: A reckoning of environment and development implications. *Trends in Ecology and Evolution* 25: 396–402. (8)

Gibson, L., T. M. Lee, L. P. Koh, B. W. Brook, et al. 2011. Primary forests are irreplaceable for sustaining tropical biodiversity. *Nature* 478: 378–381. (4, 8)

Gigon, A., R. Langenauer, C. Meier, and B. Nievergelt. 2000. Blue Lists of threatened species with stabilized or increasing abundance: A new instrument for conservation. *Conservation Biology* 14: 402–413. (6)

Gilpin, M. E. and M. E. Soulé. 1986. Minimum viable populations: Processes of species extinction. *In* M. E. Soulé (ed.), *Conservation Biology: The Science of Scarcity and Diversity*, pp. 19–34. Sinauer Associates, Sunderland, MA. (5)

Global Footprint Network: Advancing the Science of Sustainability. 2009. *http://www.footprint.org* (4)

Godefroid, S., C. Piazza, G. Rossi, S. Buord, et al. 2011a. How successful are plant species reintroductions? *Biological Conservation* 144: 672–682. (6)

Godefroid, S., S. Rivière, S. Waldren, N. Boretos, R. Eastwood, et al. 2011b. To what extent are threatened European plant species conserved in seed banks? *Biological Conservation* 144: 1494–1498. (6)

Godfray, H. C. J., J. R. Beddington, I. R. Crute, L. Haddad, et al. 2010. Food security: The challenge of feeding 9 billion people. *Science* 327: 812–818. (4)

Gooden, B., K. French, P. J. Turner, and P. O. Downey. 2009. Impact threshold for an alien plant invader, *Lantana camara* L., on native plant communities. *Biological Conservation* 142: 2631–2641. (4)

Gordon, D. R. and C. A. Gantz. 2008. Screening new plant introductions for potential invasiveness. *Conservation Letters* 1: 227–235. (4)

Gore, A. 2006. *An Inconvenient Truth: The Planetary Emergency of Global Warming and What We Can Do About It*. Rodale Books, New York. (4)

Granek, E. F., S. Polasky, C. V. Kappel, D. J. Reeds, et al. 2010. Ecosystem services as a common language for coastal ecosystem-based management. *Conservation Biology* 24: 207–216. (3)

Grassle, J. F. 2001. Marine ecosystems. *In* S. A. Levin (ed.), *Encyclopedia of Biodiversity*, vol. 4, pp. 13–26. Academic Press, San Diego, CA. (2)

Gray, L. K., T. Gylander, M. S. Mbogga, P. Chen, A. Hamann. 2011. Assisted migration to address climate change: Recommendations for aspen reforestation in western Canada. *Ecological Applications* 21: 1591–1603. (6)

Greene, R. M., J. C. Lehrter, and J. D. Hagy. 2009. Multiple regression models for hindcasting and forecasting midsummer hypoxia in the Gulf of Mexico. *Ecological Applications* 19: 1161–1175. (4)

Greiner, R. and A. Lankester. 2007. Supporting on-farm biodiversity conservation through debt-for-conservation swaps: Concept and critique. *Land Use Policy* 24: 458–471. (9)

Grenier, M. B., D. B. McDonald, and S. W. Buskirk. 2007. Rapid population growth of a critically endangered carnivore. *Science* 317: 779. (6)

Griffiths, R. A. and L. Pavajeau. 2008. Captive breeding, reintroduction, and the conservation of amphibians. *Conservation Biology* 22: 852–861. (6)

Grimm, N. B., D. Foster, P. Groffman, J. M. Grove, et al. 2008. The changing landscape: Ecosystem responses to urbanization and pollution across climatic and societal gradients. *Frontiers in Ecology and the Environment* 6: 264–272. (4, 9)

Groom, M. J., G. K. Meffe, and C. R. Carroll (eds.). 2006. *Principles of Conservation Biology*, 3rd ed. Sinauer Associates, Sunderland, MA. (2, 3, 4, 5)

Groombridge, B. and M. D. Jenkins. 2010. *World Atlas of Biodiversity; Earth's living resources in the 21st century*. University of California Press, Berkeley. (1)

Gross, L. 2008. Can farmed and wild salmon coexist? *PLoS Biology* 6: e46. (3)

Grouios, C. P. and L. L. Manne. 2009. Utility of measuring abundance versus consistent occupancy in predicting biodiversity persistence. *Conservation Biology* 23: 1260–1269. (5)

Grumbine, R. E. 2007. China's emergences and the prospects for global sustainability. *BioScience* 57: 249–255. (4)

Grumbine, R. E. and J. Xu. 2011. Creating a "Conservation with Chinese Characteristics." *Biological Conservation* 144: 1347–1355. (7)

Guerrant, E. O. 1992. Genetic and demographic considerations in the sampling and reintroduction of rare plants. *In* P. L. Fiedler and S. K. Jain (eds.), *Conservation Biology: The Theory and Practice of Nature Conservation, Preservation and Management*, pp. 321–344. Chapman and Hall, New York. (5)

Guerrant, E. O. Jr., K. Havens, and M. Maunder. 2004. *Ex Situ Conservation: Supporting Species Survival in the Wild*. Island Press, Washington, D.C. (6)

Gullison, R. E., P. C. Frumhoff, J. G. Canadell, C. B. Field, et al. 2007. Tropical forests and climate policy. *Science* 316: 985–986. (9)

Gusset, M., S. J. Ryan, M. Hofmeyr, G. V. Dyk, et al. 2007. Efforts going to the dogs? Evaluating attempts to re-introduce endangered wild dogs in South Africa. *Journal of Applied Ecology* 45: 100–108. (6)

Gutiérrez, D. 2005. Effectiveness of existing reserves in the long-term protection of a regionally rare butterfly. *Conservation Biology* 19: 1586–1597. (6)

Gutierrez, N. L., R. Hilborn, and O. Defeo. 2011. Leadership, social capital and incentives promote successful fisheries. *Nature* 470: 386–389. (4)

Haig, S. M., E. A. Beever, S. M. Chambers, H. M. Draheim, et al. 2006. Taxonomic considerations in listing subspecies under the U.S. Endangered Species Act. *Conservation Biology* 20: 1584–1594. (2)

Halfar, J. and R. M. Fujita. 2007. Danger of deep-sea mining. *Science* 316: 987. (4)

Hall, J. A. and E. Fleishman. 2010. Demonstration as a means to translate conservation science into practice. *Conservation Biology* 24: 120–127. (1)

Hambler, C., P. A. Henderson, and M. R. Speight. 2011. Extinction rates, extinction-prone habitats, and indicator groups in Britain and at larger scales. *Biological Conservation* 144: 713–721. (4, 5)

Hamer, A. J. and K. M. Parris. 2011. Local and landscape determinants of amphibian communities in urban ponds. *Ecological Applications* 21: 378–390. (7)

Hamlin, K. L., R. A. Garrott, P. J. White, and J. A. Cunningham. 2008. Contrasting wolf-ungulate interactions in the greater Yellowstone ecosystem. *Terrestrial Ecology* 3: 541–577. (6)

Haney, J. C., T. Kroeger, F. Casey, A. Quarforth, et al. 2007. Wilderness discounts on livestock compensation costs for imperiled gray wolf *Canis lupus*. *USDA Forest Service Proceedings* RMRS-P-49. (6)

Hannah, L. 2010. A global conservation system for climate-change adaptation. *Conservation Biology* 24: 70–77. (7)

Hansen, A. J., C. R. Davis, N. Piekielek, J. Gross, et al. 2011. Delineating the ecosystems containing protected areas for monitoring and management. *BioScience* 61: 363–373. (7, 8)

Hansen, M. C., S. V. Stehman, P. V. Potapov, B. Arunarwati, et al. 2009. Quantifying changes in the rates of forest clearing in Indonesia from 1990 to 2005 using remotely sensed data sets. *Environmental Research Letters* 4: 034001. (7)

Hanson, T., T. M. Brooks, G. A. B. da Fonseca, M. Hoffmann, et al. 2009. Warfare in biodiversity hotspots. *Conservation Biology* 23: 578–587. (7)

Hardwick, K. A., P. Fiedler, L. C. Lee, B. Pavlik, et al. 2011. The role of botanic gardens in the science and practice of ecological restoration. *Conservation Biology* 25: 265–275. (6)

Hart, M. M. and J. T. Trevors. 2005. Microbe management: Application of mycorrhizal fungi in sustainable agriculture. *Frontiers in Ecology and the Environment* 10: 533–539. (3)

Harvell, D., R. Aronson, N. Baron, J. Connell, et al. 2004. The rising tide of ocean diseases: Unsolved problems and research priorities. *Frontiers in Ecology and the Environment* 2: 375–382. (4)

Hayward, M. W. 2009. Conservation management for the past, present, and future. *Biodiversity and Conservation* 18: 765–775. (8)

Heber, S. and J. V. Briskie. 2010. Population bottlenecks and increased hatching failure in endangered birds. *Conservation Biology* 24: 1674–1678. (5)

Hedges, S., M. J. Tyson, A. F. Sitompul, M. F. Kinnaird, et al. 2005. Distribution, status, and conservation needs of Asian elephants (*Elephas maximus*) in Lampung Province, Sumatra, Indonesia. *Biological Conservation* 124: 35–48. (5)

Hedrick, P. 2005. Large variance in reproductive success and the N_e/N ratio. *Evolution* 59: 1596–1599. (5)

Helfield, J. M., S. J. Capon, C. Nilsson, R. Jansson, and D. Palm. 2007. Restoration of rivers used for timber floating: Effects on riparian plant diversity. *Ecological Applications* 17: 840–851. (8)

Henwood, W. D. 2010. Toward a strategy for the conservation and protection of the world's temperate grasslands. *Great Plains Research* 20: 121–134. (4)

Hinz, H., V. Prieto, and M. J. Kaiser. 2009. Trawl disturbance on benthic communities: Chronic effects and experimental predictions. *Ecological Applications* 19: 761–773. (4)

Hobbs, R. J., D. N. Cole, L. Yung, E. S. Zavaleta, et al. 2010. Guiding concepts for park and wilderness stewardship in an era of global environmental change. *Frontiers in Ecology and the Environment* 8: 483–490. (7, 8)

Hoeinghaus, D. J., A. A. Agostinho, L. C. Gomes, F. M. Pelicice, et al. 2009. Effects of river impoundment on ecosystem services of large tropical rivers: Embodied energy and market value of artisanal fisheries. *Conservation Biology* 23: 1222–1231. (3)

Hoffmann, M., T. M. Brooks, G. A. B. da Fonseca, C. Gascon, et al. 2008. Conservation planning and the IUCN Red List. *Endangered Species Research* 6: 113–125. (7)

Hoffmann, M., C. Hilton-Taylor, A. Angulo, M. Böhm, et al. 2010. The impact of conservation on the status of the world's vertebrates. *Science* 330: 1503–1509. (5)

Hogan, C. M., World Wildlife Fund, S. Sarkar, and M. McGinley. 2008. Madagascar dry deciduous forests. *In* C. J. Cleveland (ed.), *Encyclopedia of Earth*. Environmental Information Coalition, National Council for Science and the Environment, Washington, D.C. *http://www.eoearth.org/article/Madagascar_dry_deciduous_forests* (4)

Holden, E. and K. Linnerud. 2007. The sustainable development area: Satisfying basic needs and safeguarding ecological sustainability. *Sustainable Development* 15: 174–187. (9)

Hole, D. G., B. Huntley, J. Arinaitwe, S. H. M. Butchart, et al. 2011. Toward a management framework for networks of protected areas in the face of climate change. *Conservation Biology* 25: 305–315. (4)

Holland, G. J., J. S. A. Alexander, P. Johnson, A. H. Arnold, et al. 2012. Conservation cornerstones: Capitalising on the endeavours of long-term monitoring projects. *Biological Conservation* 145: 95–101. (6)

Holt, R. D. and M. Barfield. 2010. Metapopulation perspectives on the evolution of species' niches. *In* S. Cantrell, C. Cosner, and S. Ruan (eds.),

Spatial Ecology, pp. 189–212. Chapman and Hall, Boca Raton, FL. (6)

Houston, D., K. Mcinnes, G. Elliott, D. Eason, et al. 2007. The use of a nutritional supplement to improve egg production in the endangered kakapo. *Biological Conservation* 138: 248–255. (6)

Huang, D., R. A. Haack, and R. Zhang. 2011. Does global warming increase establishment rates of invasive alien species? A centurial time series analysis. *PLoS ONE* 6(9): e24733. (4)

Hughes, A. R., S. L. Williams, C. M. Duarte, K. L. Heck, Jr., and M. Waycott. 2009. Associations of concern: Declining seagrasses and threatened dependent species. *Frontiers in Ecology and the Environment* 7: 242–246. (2)

Hughes, J. B. and J. Roughgarden. 2000. Species diversity and biomass stability. *American Naturalist* 155: 618–627. (5, 7)

Ibarra-Macias, A., W. D. Robinson, and M. S. Gaines. 2011. Experimental evaluation of bird movements in a fragmented Neotropical landscape. *Biological Conservation* 144: 703–712. (4)

International Union for Conservation of Nature (IUCN). 2010. *http://iucn.org* (6)

International Whaling Commission (IWC). 2009. *http://www.iwcoffice.org* (6)

IPCC. 2007. *Climate Change 2007: The physical science basis. Contribution of Working Group I to the Fourth Assessment Report of the Intergovernmental Panel on Climate Change*. S. Solomon, D. Qin, M. Manning, Z. Chen, et al. (eds.). Cambridge University Press, Cambridge, UK. (4)

Isbell, F., V. Calcagno, A. Hector, J. Connolly, et al. 2011. High plant diversity is needed to maintain ecosystem services. *Nature* 477: 199–202. (3)

IUCN. 2001. IUCN Red List Categories and Criteria: Version 3.1. IUCN Species Survival Commission. IUCN, Gland, Switzerland. *http://www.iucnredlist.org/technical-documents/categories-and-criteria/2001–categories-criteria* (6)

IUCN. 2004. 2004 IUCN Red List of Threatened Species. *http://www.iucnredlist.org* (4)

IUCN. 2009. IUCN Red List of Threatened Species. Version 2009.2. *http://www.iucnredlist.org* (6)

IUCN/SSC Re-introduction Specialist Group. 2007. *http://iucnsscrsg.org* (6)

Jackson, J. B. C. 2008. Ecological extinction and evolution in the brave new ocean. *Proceedings of the National Academy of Sciences USA* 105: 11458–11465. (5)

Jackson, S. T., J. L. Betancourt, R. K. Booth, and S. T. Gray. 2009. Ecology and the ratchet of events: Climate variability, niche dimensions, and species distributions. *Proceedings of the National Academy of Sciences USA* 106: 19685–19692. (4)

Jackson, S. F., K. Walker, and K. J. Gaston. 2009. Relationship between distributions of threatened plants and protected areas in Britain. *Biological Conservation* 142: 1515–1522. (7)

Jacob, J., E. Jovic, and M. B. Brinkerhoff. 2009. Personal and planetary well-being: Mindfulness meditation, pro-environmental behavior and personal quality of life in a survey from the social justice and ecological sustainability movement. *Social Indicators Research* 93: 275–294. (3)

Jacobson, S. K. 2006. The importance of public education for biological conservation. *In* M. J. Groom, G. K. Meffe, and C. R. Carroll (eds.), *Principles of Conservation Biology*, 3rd ed, pp. 681–683. Sinauer Associates, Sunderland, MA. (9)

Jacobson, S. K., M. D. McDuff, and M. C. Monroe. 2006. *Conservation Education and Outreach Techniques*. Oxford University Press, Oxford. (9)

Jacquemyn, H., C. Van Mechelen, R. Brys, and O. Honnay. 2011. Management effects on the vegetation and soil seed bank of calcareous grasslands: An 11-year experiment. *Biological Conservation* 144: 416–422. (7)

Jaeger, I., H. Hop, and G. W. Gabrielsen. 2009. Biomagnification of mercury in selected species from an Arctic marine food web in Svalbard. *Science of the Total Environment* 407: 4744–4751. (4)

Jähnig, S. C., A. W. Lorenz, D. Hering, C. Antons, et al. 2011. River restoration success: A question of perception. *Ecological Applications* 21: 2007–2015. (7)

Jamieson, I. G. 2011. Founder effects, inbreeding, and loss of genetic diversity in four avian reintroduction programs. *Conservation Biology* 25: 115–123. (5)

Janzen, D. H. 2001. Latent extinctions—the living dead. *In* S. A. Levin (ed.), *Encyclopedia of Biodiversity*, pp. 689–700. Academic Press, San Diego, CA. (5)

Jarošík, V., M. Konvička, P. Pyšek, T. Kadlec, and J. Beneš. 2011. Conservation in a city: Do the same principles apply to different taxa? *Biological Conservation* 144: 490–499. (7)

Johnson, C. 2009. Megafaunal decline and fall. *Science* 326: 1072–1073. (7)

Jones, H. P. 2010. Seabird islands take mere decades to recover following rat eradication. *Ecological Applications* 20: 2075–2080. (4)

Jones, H. L. and J. M. Diamond. 1976. Short-time-base studies of turnover in breeding birds of the California Channel Islands. *Condor* 76: 526–549. (5)

Jones, K. E., N. G. Patel, M. A. Levy, A. Storeygard, et al. 2008. Global trends in emerging infectious diseases. *Nature* 451: 990–994. (4)

Joppa, L. N., S. R. Loarie, and S. L. Pimm. 2008. On the protection of "protected areas." *Proceedings of the National Academy of Sciences USA* 105: 6673–6678. (7)

Joppa, L. N., D. L. Roberts, N. Myers, and S. L. Pimm. 2011. Biodiversity hotspots house most undiscovered plant species. *Proceedings of the National Academy of Sciences USA* 108: 13171–13176. (7)

Joppa, L. N., D. L. Roberts, and S. L. Pimm. 2011. The population ecology and social behavior of taxonomists. *Trends in Ecology and Evolution* 26: 551–553. (2)

Jovan, S. and B. McCune. 2005. Air-quality bioindication in the greater central valley of California, with epiphytic macrolichen communities. *Ecological Applications* 15: 1712–1726. (3)

Kadoya, T., S. Suda, and I. Washitani. 2009. Dragonfly crisis in Japan: A likely consequence of recent agricultural habitat degradation. *Biological Conservation* 142: 1899–1905. (7)

Kannan, R. and D. A. James. 2009. Effects of climate change on global biodiversity: A review of key literature. *Tropical Ecology* 50: 31–39. (4)

Kapos, V., A. Balmford, R. Aveling, P. Bubb, et al. 2008. Calibrating conservation: New tools for measuring success. *Conservation Letters* 1: 155–164. (8, 9)

Karanth, K. and R. DeFries. 2011. Nature-based tourism in Indian protected areas: New challenges for park management. *Conservation Letters* 4: 137–149. (3)

Karesh, W. B., R. A. Cook, E. L. Bennett, and J. Newcomb. 2005. Wildlife trade and global disease emergence. *CDC Emerging Infectious Diseases* 11: 1000–1002. (4)

Karl, T. R. 2006. Written statement for an oversight hearing: Introduction to Climate Change before the Committee on Government Reform, U.S. House of Representatives, Washington, D.C. (4)

Keatley, B. E., E. M. Bennett, G. K. MacDonald, Z. E. Taranu, and I. Gregory-Eaves. 2011. Land-use legacies are important determinants of lake eutrophication in the Anthropocene. *PLoS ONE* 6(1): e15913. (4)

Keeton, W. S., C. E. Kraft, and D. R. Warren. 2007. Mature and old-growth riparian forests: Structure, dynamics, and effects on Adirondack stream habitats. *Ecological Applications* 17: 852–868. (7)

Keller, R. P., K. Frang, and D. M. Lodge. 2008. Preventing the spread of invasive species: Economic benefits of intervention guided by ecological predictions. *Conservation Biology* 22: 80–88. (4)

Kelm, D. H., K. R. Wiesner, and O. von Helversen. 2008. Effects of artificial roosts for frugivorous bats on seed dispersal in a Neotropical forest pasture mosaic. *Conservation Biology* 22: 733–741. (2)

Kirkby, C. A., R. Giudice-Granados, B. Day, K. Turner, et al. 2010. The market triumph of ecotourism: An economic investigation of the private and social benefits of competing land uses in the Peruvian Amazon. *PLoS ONE* 5(9): e13015. (3)

Kissui, B. M. and C. Packer. 2004. Top-down population regulation of a top predator: Lions in the Ngorongoro Crater. *Proceedings of the Royal Society B* 271: 1867–1874. (4)

Knapp, R. A., C. P. Hawkins, J. Ladau, and J. G. McClory. 2005. Fauna of Yosemite National Park lakes has low resistance but high resilience to fish introductions. *Ecological Applications* 15: 835–847. (2)

Knowlton, N. and J. B. C. Jackson. 2008. Shifting baselines, local impacts, and global change on coral reefs. *PLoS Biology* 6: e54. (2)

Kobori, H. and R. Primack. 2003. Participatory conservation approaches for Satoyama: The traditional forest and agricultural landscape of Japan. *Ambio* 32: 307–311. (7)

Kociolek, A. V., A. P. Clevenger, C. C. St. Clair, and D. S. Proppe. 2011. Effects of road networks on bird populations. *Conservation Biology* 25: 241–249. (4)

Koh, L. P. and J. Ghazoul. 2010. A matrix-calibrated species-area model for predicting biodiversity losses due to land-use change. *Conservation Biology* 24: 994–1001. (5)

Koontz, T. M. and J. Bodine. 2008. Implementing ecosystem management in public agencies: Lessons from the U.S. Bureau of Land Management and the Forest Service. *Conservation Biology* 22: 60–69. (8)

Kremen, C. and R. S. Ostfeld. 2005. A call to ecologists: Measuring, analyzing and managing ecosystem services. *Frontiers in Ecology and the Environment* 10: 539–548. (3)

Kross, S. M., J. M. Tylianakis, and X. J. Nelson. 2012. Effects of introducing threatened falcons into vineyards on abundance of passeriformes and bird damage to grapes. *Conservation Biology* 26: 142–149. (3)

Kulkarni, M. V., P. M. Groffman, and J. B. Yavitt. 2008. Solving the global nitrogen problem: It's a gas! *Frontiers in Ecology and the Environment* 6: 199–206. (4)

Laikre, L., F. W. Allendorf, L. C. Aroner, C. S. Baker, et al. 2010. Neglect of genetic diversity in implementation of the Convention on Biological Diversity. *Conservation Biology* 24: 86–88. (2)

Laikre, L., M. K. Schwartz, R. S. Waples, and N. Ryman. 2010. Compromising genetic diversity in the wild: Unmonitored large-scale release of plants and animals. *Trends in Ecology and Evolution* 25: 520–529. (4)

Lalas, C., H. Ratz, K. McEwan, and S. D. McConkey. 2007. Predation by New Zealand sea lions (*Phocarctos hookeri*) as a threat to the viability of yellow-eyed penguins (*Megadyptes antipodes*) at Otago Peninsula, New Zealand. *Biological Conservation* 135: 235–246. (7)

Lamb, J. B. and B. L. Willis. 2011. Using coral disease prevalence to assess the effects of concentrating tourism activities on offshore reefs in a tropical marine park. *Conservation Biology* 25: 1044–1052. (3)

Lant, C. L., J. B. Ruhl, and S. E. Kraft. 2008. The tragedy of ecosystem services. *BioScience* 58: 969–974. (3)

Laurance, W. F. 2007. Have we overstated the tropical biodiversity crisis? *Trends in Ecology and Evolution* 22: 65–70. (4, 5)

Laurance, W. F., J. L. C. Camargo, R. C. C. Luizão, S. G. Laurance, et al. 2011. The fate of Amazonian forest fragments: A 32-year investigation. *Biological Conservation* 144: 56–67. (4)

Laurance, W. F., M. Goosem, and S. G. W. Laurance. 2009. Impacts of roads and linear clearings on tropical forests. *Trends in Ecology and Evolution* 24: 659–679. (4)

Laurance, S. G. and W. F. Laurance. 1999. Tropical wildlife corridors: Use of linear rainforest remnants by arboreal mammals. *Biological Conservation* 91: 231–239. (7)

Laurance, W. F., T. E. Lovejoy, H. L. Vasconcelos, E. M. Bruna, et al. 2002. Ecosystem decay of Amazonian forest fragments: A 22–year investigation. *Conservation Biology* 16: 605–618. (4)

Laurance, W. F. and R. C. Luizão. 2007. Driving a wedge into the Amazon. *Nature* 448: 409–10. (4)

Lawler, J. J., S. P. Campbell, A. D. Guerry, M. B. Kolozsvary, et al. 2002. The scope and treatment of threats in endangered species recovery plans. *Ecological Applications* 12: 663–667. (4)

Lawler, J. J., S. L. Shafer, B. A. Bancroft, and A. R. Blaustein. 2010. Projected climate impacts for the amphibians of the Western Hemisphere. *Conservation Biology* 24: 38–50. (5)

Lawrence, A. J., R. Afif, M. Ahmed, S. Khalifa, and T. Paget. 2010. Bioactivity as an options value of sea cucumbers in the Egyptian Red Sea. *Conservation Biology* 24: 217–225. (3)

Lawrence, D. J., E. R. Larson, C. A. R. Liermann, M. C. Mims, et al. 2011. National parks as protected areas for U.S. freshwater fish diversity. *Conservation Letters* 4: 364–371. (7)

Leidner, A. K. and N. M. Haddad. 2011. Combining measures of dispersal to identify conservation strategies in fragmented landscapes. *Conservation Biology* 25: 1022–1031. (6)

Leidner, A. K. and M. C. Neel. 2011. Taxonomic and geographic patterns of decline for threatened and endangered species in the United States. *Conservation Biology* 25: 716–725. (5)

Leopold, A. 1949. *A Sand County Almanac and Sketches Here and There*. Oxford University Press, New York. (1, 3)

Lepczyk, C. A., A. G. Mertig, and J. Liu. 2003. Landowners and cat predation across rural-to-urban landscapes. *Biological Conservation* 115: 191–201. (4)

Letnic, M., F. Koch, C. Gordon, M. S. Crowther, and C. R. Dickman. 2009. Keystone effects of an alien top-predator stem extinctions of native mammals. *Proceedings of the Royal Society B* 276: 3249–3256. (2)

Leu, M., S. E. Hanser, and S. T. Knick. 2008. The human footprint in the west: A large-scale analysis of anthropogenic impacts. *Ecological Applications* 18: 1119–1139. (4)

Levin, S. A. (ed.). 2001. *Encyclopedia of Biodiversity*. Academic Press, San Diego, CA. (2)

Licht, D. S., J. J. Millspaugh, K. E. Kunkel, C. O. Kockanny, and R. O. Peterson. 2010. Using small populations of wolves for ecosystem restoration and stewardship. *BioScience* 60: 147–153. (8)

Lindenmayer, D. and M. Hunter. 2010. Some guiding concepts for conservation biology. *Conservation Biology* 24: 1459–1468. (1)

Lindenmayer, D. B., G. E. Likens, A. Haywood, and L. Miezis. 2011. Adaptive monitoring in the real world: Proof of concept. *Trends in Ecology and Evolution* 26: 641–646. (7)

Linder, J. M. and J. F. Oates. 2011. Differential impact of bushmeat hunting on monkey species and implications for primate conservation in Korup National Park, Cameroon. *Biological Conservation* 144: 738–745. (4)

Lindsey, P. A., R. Alexander, J. T. duToit, and M. G. L. Mills. 2005. The cost efficiency of wild dog conservation in South Africa. *Conservation Biology* 19: 1205–1214. (6)

Lindsey, P. A., P. A. Roulet, and S. S. Romañach. 2007. Economic and conservation significance of the trophy hunting industry in sub-Saharan Africa. *Biological Conservation* 134: 455–469. (3, 8)

Linkie, M., R. Smith, Y. Zhu, D. J. Martyr, et al. 2008. Evaluating biodiversity conservation around a large Sumatran protected area. *Conservation Biology* 22: 683–690. (8)

Lloyd, P., T. E. Martin, R. L. Redmond, U. Langer, and M. M. Hart. 2005. Linking demographic effects of habitat fragmentation across landscapes to continental source-sink dynamics. *Ecological Applications* 15: 1504–1514. (4)

Lobell, D. B., M. B. Burke, C. Tebaldi, M. D. Mastrandrea, et al. 2008. Prioritizing climate change adaptation needs for food security in 2030. *Science* 319: 607–610. (4)

Loss, S. R. and R. B. Blair. 2011. Reduced density and nest survival of ground-nesting songbires relative to earthworm invasions in northern hardwood forests. *Conservation Biology* 25: 983–993. (4)

Loss, S. R., L. A. Terwilliger, and A. C. Peterson. 2011. Assisted colonization: Integrating conservation strategies in the face of climate change. *Biological Conservation* 144: 92–100. (6)

Lotze, H. K., M. Coll, A. M. Magera, C. Ward-Paige, and L. Airoldi. 2011. Recovery of marine animal populations and ecosystems. *Trends in Ecology and Evolution* 26: 595–605. (1)

Lotze, H. K. and B. Worm. 2009. Historical baselines for large marine mammals. *Trends in Ecology and Evolution* 24: 254–262. (4)

Loucks, C., M. B. Mascia, A. Maxwell, K. Huy, et al. 2009. Wildlife decline in Cambodia, 1953–2005: Exploring the legacy of armed conflict. *Conservation Letters* 2: 82–92. (4)

Lovelock, J. 1988. *The Ages of Gaia: A bibliography of our living earth*. Oxford University Press, New York and Oxford. (1)

Lowman, M. D., E. Burgess, and J. Burgess. 2006. *It's a Jungle Up There: More Tales from the Treetops*. Yale University Press, New Haven, CT. (2)

MacArthur, R. H. and E. O. Wilson. 1967. *The Theory of Island Biogeography*. Princeton University Press, Princeton, NJ. (5)

Mace, G. M. and J. E. M. Baillie. 2007. The 2010 Biodiversity Indicators: Challenges for science and policy. *Conservation Biology* 21: 1406–1413. (6)

Maezono, Y., R. Kobayashi, M. Kusahara, and T. Miyashita. 2005. Direct and indirect effects of exotic bass and bluegill on exotic and native organisms in farm ponds. *Ecological Applications* 15: 638–650. (4)

Maiorano, L., A. Falcucci, and L. Boitani. 2008. Size-dependent resistance of protected areas to land-use change. *Proceedings of the Royal Society B* 275: 1297–1304. (7)

Mangel, M. and C. Tier. 1994. Four facts every conservation biologist should know about persistence. *Ecology* 75: 607–614. (5)

Marcovaldi, M. A. and M. Chaloupka. 2007. Conservation status of the loggerhead sea turtle in Brazil: An encouraging outlook. *Endangered Species Research* 3: 133–143. (1)

Margules, C. and S. Sarkar. 2007. *Systematic Conservation Planning*. Cambridge University Press, Cambridge, U.K. (7)

Marquard, E. A. Weigelt, V. M. Temperton, C. Roscher, et al. 2009. Plant species richness and functional composition drive overyielding in a six-year grassland experiment. *Ecology* 90: 3290–3302. (2)

Martinuzzi, S., W. A. Gould, A. E. Lugo, and E. Medina. 2009. Conversion and recovery of Puerto Rican mangroves: 200 years of change. *Forest Ecology and Management* 257: 75–84. (4)

Mascia, M. B. and C. A. Claus. 2009. A property rights approach to understanding human displacement from protected areas: The case of Marine Protected Areas. *Conservation Biology* 23: 16–23. (7)

Mascia, M. B. and S. Pallier. 2011. Protected areas downgrading ,downsizing, and degazettement (PADD) and its conservation implications. *Conservation Letters* 4: 9–20. (7, 9)

Mawdsley, J. R., R. O'Malley, and D. S. Ojima. 2009. A review of climate change adaptation strategies for wildlife management and biodiversity conservation. *Conservation Biology* 23: 1080–1089. (4)

Maxted, N. 2001. Ex situ, in situ conservation. *In* S. A. Levin (ed.), *Encyclopedia of Biodiversity* 2: 683–696. Academic Press, San Diego, CA. (6)

McCarthy, M. A., C. J. Thompson, and N. S. G. Williams. 2006. Logic for designing nature reserves for multiple species. *American Naturalist* 167: 717–727. (6)

McClelland, E. K. and K. A. Naish. 2007. What is the fitness outcome of crossing unrelated fish populations? A meta-analysis and an evaluation of future research directions. *Conservation Genetics* 8: 397–416. (5)

McClenachan, L., A. B. Cooper, K. E. Carpenter, and N. K. Dulvy. 2012. Extinction risk and bottlenecks in the conservation of charismatic marine species. *Conservation Letters* 5: 73–80. (5)

McCook, L. J., T. Ayling, M. Cappo, J. H. Choat, et al. 2010. Adaptive management of Great Barrier Reef: A globally significant demonstration of the benefits of networks of marine reserves. *Proceedings of the National Academy of Sciences USA* 107: 18278–18285. (3, 7)

McDonald, R. I. and T. M. Boucher. 2011. Global development and the future of the protected area strategy. *Biological Conservation* 144: 383–392. (7)

McDonald-Madden, E., M. C. Runge, H. P. Possingham, T. G. Martin. 2011. Optimal timing for managed relocation of species faced with climate change. *Nature Climate Change* 1: 261–265. (6)

McKinley, D. C., M. G. Ryan, R. A. Birdsey, C. P. Giardina, et al. 2011. A synthesis of current knowledge on forests and carbon storage in the United States. *Ecological Applications* 21: 1902–1924. (3)

McLachlan, J. S., J. J. Hellmann, and M. W. Schwartz. 2007. A framework for debate of assisted migration in an era of climate change. *Conservation Biology* 21: 297–302. (6)

McShane, T. O., P. D. Hirsch, T. C. Trung, A. N. Songorwa, et al. 2011. Hard choices: Making trade-offs between biodiversity conservation and human well-being. *Biological Conservation* 144: 966–972. (8)

Meffe, G. C., C. R. Carroll, and contributers. 1997. *Principles of Conservation Biology*, 2nd ed. Sinauer Associates, Sunderland, MA. (5)

Meijaard, E., G., G. Albar, Nardiyono, Y. Rayadin, et al. 2010. Unexpected ecological resilience in Bornean orangutans and implications for pulp and paper plantation management. *PLoS ONE* 5(9): e12813. (8)

Merchant, C. 2002. *The Columbia guide to American environmental history*. Columbia University Press. (1)

Messina, J. P. and M. A. Cochrane. 2007. The forests are bleeding: How land use change is creating a new fire regime in the Ecuadorian Amazon. *Journal of Latin American Geography* 6.1: 85–100. (4)

Meyer, C. K., M. R. Whiles, and S. G. Baer. 2010. Plant community recovery following restoration in temporarily variable riparian wetlands. *Restoration Ecology* 18: 52–64. (8)

Milder, J. C., J. P. Lassoie, and B. L. Bedford. 2008. Conserving biodiversity and ecosystem function through limited development: An empirical evaluation. *Conservation Biology* 22: 70–79. (9)

Millennium Ecosystem Assessment (MEA). 2005. *Ecosystems and Human Well-being*. 4 volumes. Island Press, Covelo, CA. (1, 3, 4, 5)

Miller, B., W. Conway, R. P. Reading, C. Wemmer, et al. 2004. Evaluating the conservation mission of zoos, aquariums, botanical gardens, and natural history museums. *Conservation Biology* 18: 86–93. (6)

Miller, J. K., J. M. Scott, C. R. Miller, and L. P Waits. 2002. The Endangered Species Act: Dollars and sense? *BioScience* 52: 163–168. (6)

Miller, K. R. 1996. *Balancing the Scales: Guidelines for Increasing Biodiversity's Chances through Bioregional Management*. World Resources Institute, Washington, D.C. (8)

Miller-Rushing, A. J. and R. B. Primack. 2008. Global warming and flowering times in Thoreau's Concord: A community perspective. *Ecology* 89: 332–341. (5)

Min, S., X. Zhang, F. W. Zwiers, and G. C. Hegerl. 2011. Human contribution to more-intense precipitation extremes. *Nature* 470: 378–381. (4)

Minteer, B. A. and T. R. Miller. 2011. The new conservation debate: Ethical foundations, strategic trade-offs, and policy opportunities. *Biological Conservation* 144: 945–947. (8)

Mitchell, A. M., T. I. Wellicome, D. Brodie, and K. M. Cheng. 2011. Captive-reared burrowing owls show higher site-affinity, survival, and reproductive performance when reintroduced using a soft-release. *Biological Conservation* 144: 1382–1391. (6)

Mittermeier, R. A., P. R. Gil, M. Hoffman, J. Pilgrim, et al. 2005. *Hotspots Revisited: Earth's Biologically Richest and Most Endangered Terrestrial Ecoregions*. Conservation International, Washington, D.C. (7)

Molano-Flores, B. and T. J. Bell. 2012. Projected population dynamics for a federally endangered plant under different climate change emission scenarios. *Biological Conservation* 145: 130–138. (6)

Molnar, J. L., R. L Gamboa, C. Revenga, and M. D. Spalding. 2008. Assessing the global threat of invasive species to marine biodiversity. *Frontiers in Ecology and the Environment* 9: 485–492. (4)

Morgan, J. L., S. E. Gergel, and N. C. Coops. 2010. Aerial photography: A rapidly evolving tool for ecological management. *BioScience* 60: 47–59. (7)

Moseley, L. (ed.). 2009. *Holy Ground: A Gathering of Voices on Caring for Creation*. Sierra Club Books, San Francisco, CA. (3)

Moyle, P. B., J. V. E. Katz, and R. M. Quinones. 2011. Rapid decline of California's native inland fishes: A status assessment. *Biological Conservation* 144: 2414–2423. (5)

Mueller, J. G., I. H. B. Assanou, I. D. Guimbo, and A. M. Almedom. 2010. Evalutating rapid participatory rural appraisal as an assessment of ethnoecological knowledge and local biodiversity patterns. *Conservation Biology* 24: 140–150. (6)

Munson, L., K. A. Terio, R. Kock, T. Mlengeya, et al. 2008. Climate extremes promote fatal co-infections during canine distemper epidemics in African lions. *PLoS ONE* 3: e2545. (5)

Murray-Smith, C., N. A. Brummitt, A. T. Oliveira-Filho, S. Bachman, et al. 2009. Plant diversity hotspots in the Atlantic coastal forests of Brazil. *Conservation Biology* 23: 151–163. (7)

Musiani, M., C. Mamo, L. Boitani, C. Callaghan, et al. 2003. Wolf depredation trends and the use of fladry barriers to protect livestock in western North America. *Conservation Biology* 17: 1538–1547. (6)

Myers, N., N. Golubiewski, and C. J. Cleveland. 2007. Perverse subsidies. *In* C. J. Cleveland (ed.), *Encyclopedia of Earth*. Environmental Information Coalition, National Council for Science and the Environment, Washington, D.C. *http://www.eoearth.org/article/Perverse_subsidies* (3)

Naess, A. 2008. *The Ecology of Wisdom: Writings by Arne Naess*. A. Drengson and B. Devall (eds.). Counterpoint, Berkeley, CA. (3)

NatureServe: A Network Connecting Science with Conservation. 2009. *http://www.natureserve.org* (6)

NatureServe Explorer. 2009. *http://www.natureserve.org/explorer* (6)

Naughton-Treves, L., M. B. Holland, and K. Brandon. 2005. The role of protected areas in conserving biodiversity and sustaining local livelihoods. *Annual Review of Environmental Resources* 30: 219–252. (9)

Nee, S. 2003. Unveiling prokaryotic diversity. *Trends in Ecology and Evolution* 18: 62–63. (2)

Neff, J. C., R. L. Reynolds, J. Belnap, and P. Lamothe. 2005. Multi-decadal impacts of grazing on soil physical and biogeochemical properties in southeast Utah. *Ecological Applications* 15: 87–95. (4)

Neff, N. W. 2011. What research should be done and why? Four competing visions among ecologists. *Frontiers in Ecology and the Environment* 9: 462–469. (9)

Nelson, A. and K. M. Chomitz. 2011. Effectiveness of strict vs. multiple use protected areas in reducing tropical forest fires: A global analysis using matching methods. *PLoS ONE* 6(8): e22722. (7)

Nelson, M. P. and J. A. Vucetich. 2009. On advocacy by environmental scientists: What, whether, why, and how. *Conservation Biology* 23: 1090–1101. (1)

Nepstad, D., S. Schwartzman, B. Bamberger, M. Santilli, et al. 2006. Inhibition of Amazon deforestation and fire by parks and indigenous lands. *Conservation Biology* 20: 65–73. (8)

Newmark, W. D. 1995. Extinction of mammal populations in western North American national parks. *Conservation Biology* 9: 512–527. (7)

Newmark, W. D. 2008. Isolation of African protected areas. *Frontiers in Ecology and the Environment* 6: 321–328. (8)

Nicholson, T. E., K. A. Mayer, M. M. Staedler, and A. B. Johnson. 2007. Effects of rearing methods on survival of released free-ranging juvenile southern sea otters. *Biological Conservation* 138: 313–320. (6)

Nijman, V., K. A. I. Nekaris, G. Donati, M. Bruford, and J. Fa. 2011. Primate conservation: Measuring and mitigating trade in primates. *Endangered Species Research* 13: 159–161. (4)

Niles, L. J. 2009. Effects of horseshoe crab harvest in Delaware Bay on red knots: Are harvest restrictions working? *BioScience* 59: 153–164. (3)

Norden, N., J. Chave, P. Belbenoit, A. Caubère, et al. 2009. Interspecific variation in seedling responses to seed limitation and habitat conditions for 14 neotropical woody species. *Journal of Ecology* 97: 186–197. (2)

Norris, K., A. Asase, B. Collen, J. Gockowski, et al. 2010. Biodiversity in a forest-agriculture mosaic—The changing face of West African rainforests. *Biological Conservation* 143: 2341–2350. (8)

Norris, S. 2007. Ghosts in our midst: Coming to terms with amphibian extinctions. *BioScience* 57: 311–316. (4)

Noss, R. F. 1992. Essay: Issues of scale in conservation biology. *In* P. L. Fiedler and S. K. Jain (eds.), *Conservation Biology: The Theory and Practice of Nature Conservation, Preservation and Management*, pp. 239–250. Chapman and Hall, New York. (3)

Noss, R. F., E. T. La Roe III, and J. M. Scott. 1995. *Endangered Ecosystems of the United States: A Preliminary Assessment of Loss and Degradation. Biological Report 28.* U.S. Department of the Interior, National Biological Service, Washington, D.C. (4)

Nunez-Iturri, G., O. Olsson, and H. F. Howe. 2008. Hunting reduces recruitment of primate-dispersed trees in Amazonian Peru. *Biological Conservation* 141: 1536–1546. (2)

O'Meilla, C. 2004. Current and reported historical range of the American burying beetle. U.S. Fish and Wildlife Services, Oklahoma Ecological Services Field Office, OK. (5)

O'Neill, B. C., M. Dalton, R. Fuchs, L. Jiang, et al. 2010. Global demographic trends and future carbon emissions. *Proceedings of the National Academy of Sciences USA* 107: 17521–17526. (4)

Odell, J., M. E. Mather, and R. M. Muth. 2005. A biosocial approach for analyzing environmental conflicts: A case study of horseshoe crab allocation. *BioScience* 55: 735–748. (3)

Oehlmann J., U. Schulte-Oehlmann, W. Kloas, O. Jagnytsch, et al. 2009. A critical analysis of the biological impacts of plasticizers on wildlife. *Philosophical Transactions of the Royal Society B* 364: 2047–2062. (4)

Okin, G. S., A. Parsons, J. Wainwright, J. E. Herrick, et al. 2009. Do changes in connectivity explain desertification? *BioScience* 59: 237–244. (4)

Olden, J. D., M. J. Kennard, J. J. Lawler, and N. L. Poff. 2011. Challenges and opportunities in implementing managed relocation for conservation of freshwater species. *Conservation Biology* 25: 40–47. (6)

Olds, A. D., R. M. Connolly, K. A. Pitt, and P. S. Maxwell. 2012. Habitat connectivity improves reserve performance. *Conservation Letters* 5: 56–63. (7)

Oppel, S., B. M. Beaven, M. Bolton, J. Vickery, and T. W. Body. 2011. Eradication of invasive mammals on islands inhabited by humans and domestic animals. *Conservation Biology* 25: 232–240. (4)

Orr, D. W. 2007. Optimism and hope in a hotter time. *Conservation Biology* 21: 1392–1395. (1)

Osterlind, K. 2005. Concept formation in environmental education: 14–year olds' work on the intensified greenhouse. *International Journal of Science Education* 27: 891–908. (3)

Ostfeld, R. S. 2009. Climate change and the distribution and intensity of infectious diseases. *Ecology* 4: 903–905. (4)

Pace, M. L., S. E. Hampton, K. E. Limburg, E. M. Bennett, et al. 2010. Communicating with the public: Opportunities and rewards for individual ecologists. *Frontiers in Ecology and the Environment* 8: 292–298. (9)

Pan, Y., R. A. Birdsey, J. Fang, R. Houghton, et al. 2011. A large and persistent carbon sink in the world's forests. *Science* 333: 988–993. (3)

Pandolfi, J. M., S. R. Connoly, D. J. Marshall, and A. L. Cohen. 2011. Projecting coral reef futures under global warming and ocean acidification. *Science* 333: 418–442. (4)

Pardini, R., S. M. de Souza, R. Braga-Neto, and J. P. Metzger. 2005. The role of forest structure, fragment size and corridors in maintaining small mammal abundance and diversity in an Atlantic forest landscape. *Biological Conservation* 12: 253–266. (7)

Parry, L., J. Barlow, and C. A. Peres. 2009. Allocation of hunting effort by Amazonian smallholders:

Implications for conserving wildlife in mixed-use landscapes. *Biological Conservation* 142: 1777–1786. (4)

Peakall, R., D. Ebert, L. J. Scott, P. F. Meagher, and C. A. Offord. 2003. Comparative genetic study confirms exceptionally low genetic variation in the ancient and endangered relictual conifer, *Wollemia nobilis* (Araucariaceae). *Molecular Ecology* 12: 2331–2343. (5)

Peery, M. Z., S. R. Beissinger, S. H. Newman, E. B. Burkett, and T. D. Williams. 2004. Applying the declining population paradigm: Diagnosing causes of poor reproduction in the marbled murrelet. *Conservation Biology* 18: 1088–1098. (5)

Perry, G. and D. Vice. 2009. Forecasting the risk of brown tree snake dispersal from Guam: A mixed transport-establishment model. *Conservation Biology* 23: 992–1000. (4)

Persha, L, A. Agrowal, and A. Chhatre. 2011. Social and ecological synergy: Local rulemaking, forest livelihoods, and biodiversity conservation. *Science* 331: 1606–1608. (8)

Peterson, M. J., D. M. Hall, A. M. Feldpausch-Parker, and T. R. Peterson. 2010. Obscuring ecosystem function with the application of the ecosystem services concept. *Conservation Biology* 24: 113–119. (3)

Petrovan, S. O., A. I. Ward, P. Wheeler. 2011. Detectability counts when assessing populations for biodiversity targets. *PLoS ONE* 6(9): e24206. (6)

Philpott, S. M., P. Bichier, R. Rice, and R. Greenberg. 2007. Field-testing ecological and economic benefits of coffee certification programs. *Conservation Biology* 21: 975–985. (8)

Phua, M. H., S. Tsuyuki, N. Furuya, and J. S. Lee. 2008. Detecting deforestation with a spectral change detection approach using multitemporal Landsat data: A case study of Kinabalu Park, Sabah, Malaysia. *Journal of Environmental Management* 88: 784–795. (4)

Pimentel, D., L. Lach, R. Zuniga, and D. Morrison. 2000. Environmental and economic costs of nonindigenous species in the United States. *BioScience* 50: 53–65 (4).

Pimentel, D., R. Zuniga, and D. Morrison. 2005. Update on the environmental and economic costs associated with alien-invasive species in the United States. *Ecological Economics* 52: 273–288. (4)

Pimm, S. L. and C. Jenkins. 2005. Sustaining the variety of life. *Scientific American* 293(33): 66–73. (4, 5)

Polidoro, B. A., K. E. Carpenter, L. Collins, N. C. Duke, et al. 2010. The loss of species: Mangrove extinction risk and geographic areas of global concern. *PLoS ONE* 5(4): e10095. (4)

Posey, D. A. and M. J. Balick (eds.). 2006. *Human Impacts on Amazonia: The Role of Traditional Ecological Knowledge in Conservation and Development*. Columbia University Press, New York. (7)

Post, E., J. Brodie, M. Hebblewhite, A. D. Anders, J. A. K. Maier, and C. C. Wilmers. 2009. Global population dynamics and hot spots of response to climate change. *BioScience* 59: 489–497. (4)

Power, M. E., D. Tilman, J. A. Estes, B. A. Menge, et al. 1996. Challenges in the quest for keystones. *BioScience* 46: 609–620. (2)

Power, T. M. and R. N. Barret. 2001. *Post-Cowboy Economics: Pay and Prosperity in the New American West*. Island Press, Washington, D.C. (3)

Prato, T. 2005. Accounting for uncertainty in making species protection decisions. *Conservation Biology* 19: 806–814. (3)

Prescott-Allen, C. and R. Prescott-Allen. 1986. *The First Resource: Wild Species in the North American Economy*. Yale University Press, New Haven, CT. (3)

Primack, R. B. and A. J. Miller-Rushing. 2012. Uncovering, collecting, and analyzing records to investigate the ecological impacts of climate change: A template from Thoreau's Concord. *BioScience* 62: 170–181. (4)

Primack, R. B., A. J. Miller-Rushing, and K. Dharaneeswaran. 2009. Changes in the flora of Thoreau's Concord. *Biological Conservation* 142: 500–508. (5)

Ramakrishnan, U., J. A. Santosh, U. Ramakrishnan, and R. Sukumar. 1998. The population and conservation status of Asian elephants in the Periyar Tiger Reserve, southern India. *Current Science India* 74: 110–113. (5)

Rands, M. R., W. M. Adams, L. Bennun, S. H. M. Butchart, et al. 2010. Biodiversity conservation: Challenges beyond 2010. *Science* 329: 1298–1303. (9)

Randolph, J. and G. M. Masters. 2008. *Energy for Sustainability: Technology, Planning, Policy*. Island Press, Washington, D.C. (4)

Ranganathan, J., K. M. A. Chan and G. C. Daily. 2007. Satellite detection of bird communities in tropical countryside. *Ecological Applications* 17: 1499–1510. (7)

Rao, M. and P. McGowan. 2002. Wild-meat use, food security, livelihoods, and conservation. *Conservation Biology* 16: 580–583. (3)

Redford, K. H. and W. M. Adams. 2009. Payment for ecosystem services and the challenge of saving nature. *Conservation Biology* 23: 785–787. (3)

Redford, K. H., G. Amato, J. Baillie, P. Beldomenico, et al. 2011. What does it mean to successfully conserve a (vertebrate) species? *BioScience* 61: 39–48. (6)

Reed, J. M., C. S. Elphick, A. F. Zuur, E. N. Ieno, and G. M. Smith. 2007. Time series analysis of Hawaiian waterbirds. *In* A. F. Zuur, E. N. Ieno, and G. M. Smith (eds.), *Analysis of Ecological Data*. Springer-Verlag, the Netherlands. (6)

Régnier, C., B. Fontaine, and P. Bouchet. 2009. Not knowing, not recording, not listing: Numerous unnoticed mollusk extinctions. *Conservation Biology* 23: 1214–1221. (5, 6)

Relyea, R. A. 2005. The impact of insecticides and herbicides on the biodiversity and productivity of aquatic communities. *Ecological Applications* 15: 618–627. (4)

Reyers, B., D. J. Roux, R. M. Cowling, A. E. Ginsgurg, et al. 2010. Conservation planning as a transdisciplinary process. *Conservation Biology* 24: 957–965. (1)

Ricciardi, A. 2003. Predicting the impacts of an introduced species from its invasion history: An empirical approach applied to zebra mussel invasions. *Freshwater Biology* 48: 972–981. (4)

Rinella, M. F., B. D. Maxwell, P. K. Fay, T. Weaver, and R. L. Sheley. 2009. Control effort exacerbates invasive-species problem. *Ecological Applications* 19: 155–162. (4)

Ripple, W. J. and R. L. Beschta. 2012. Trophic cascades in Yellowstone: The first 15 years after wolf reintroduction. *Biological Conservation* 145: 205–213. (2)

Rissman, A. R. and V. Butsic. 2011. Land trust defense and enforcement of conserved areas. *Conservation Letters* 4: 31–37. (9)

Roark, E. B., T. P. Guilderson, R. B. Dunbar, and B. L. Ingram. 2006. Radiocarbon-based ages and growth rates of Hawaiian deep-sea corals. *Marine Ecology Progress Series* 327: 1–14. (2)

Roberts, C. M., C. J. McClean, J. E. N. Veron, J. P. Hawkins, et al. 2002. Marine biodiversity hotspots and conservation priorities for tropical reefs. *Science* 295: 1280–1284. (7)

Robertson, M. M. 2006. Emerging ecosystem service markets: Trends in a decade of entrepreneurial wetland banking. *Frontiers in Ecology and the Environment* 4: 297–302. (8)

Robinson, J. G. 2011. Ethical pluralism, pragmatism, and sustainability in conservation practice. *Biological Conservation* 144: 958–965. (3)

Robinson, R. A., H. P. Q. Crick, J. A. Learmonth, I. M. D. Maclean, et al. 2008. Travelling through a warming world: Climate change and migratory species. *Endangered Species Research* 7: 87–99. (4)

Robles, M. D., C. H. Flather, S. M Stein, M. D. Nelson, and A. Cutko. 2008. The geography of private forests that support at-risk species in the conterminous United States. *Frontiers in Ecology and the Environment* 6: 301–307. (8)

Rocha, M. F., M. Passamani, and J. Louzada. 2011. A small mammal community in a forest fragment, vegetation corridor and coffee matrix system in the Brazilian Atlantic forest. *PLoS ONE* 6(8): e23312. (7)

Rodrigues, A. S. L., R. M. Ewers, L. Parry, C. Souza, Jr., et al. 2009. Boom-and-bust development patterns across the Amazon deforestation frontier. *Science* 324: 1435–1437. (4)

Rodríguez, J. P., K. M. Rodríguez-Clark, J. E. M. Baillie, N. Ash, et al. 2011. Establishing IUCN Red List criteria for threatened ecosystems. *Conservation Biology* 25: 21–29. (7)

Rolston, H., III. 1989. *Philosophy Gone Wild: Essays on Environmental Ethics*. Prometheus Books, Buffalo, NY. (3)

Rosenzweig, C., D. Karoly, M. Vicarelli, P. Neofotis, et al. 2008. Attributing physical and biological impacts to anthropogenic climate change. *Nature* 453: 353–358. (9)

Roux, D. J., J. L. Nel, P. J. Ashton, A. R. Deacon, et al. 2008. Designing protected areas to conserve riverine biodiversity: Lessons from the hypothetical redesign of Kruger National Park. *Biological Conservation* 141: 100–117. (7)

Ruane, J. 2000. A framework for prioritizing domestic animal breeds for conservation purposes at the national level: A Norwegian case study. *Conservation Biology* 14: 1385–1393. (6)

Sachs, J. D. 2008. *Common Wealth: Economics for a Crowded Planet*. Penguin Press, New York. (1)

Sagoff, M. 2008. On the compatibility of a conservation ethic with biological science. *Conservation Biology* 21: 337–345. (3)

Sairam, R., S. Chennareddy, and M. Parani. 2005. OBPC Symposium: Maize 2004 & Beyond—plant regeneration, gene discovery, and genetic engineering of plants for crop improvement. *In Vitro Cellular and Developmental Biology—Plant* 41: 411. (3)

Sanderson, E., M. Jaiteh, M. A. Levy, K. H. Redford, et al. 2002. The human footprint and the last of the wild. *BioScience* 52: 891–904. (4)

Saraux, C., C. Le Bohec, J. M. Durant, V. A. Viblanc, et al. 2011. Reliability of flipper-banded penguins as indicators of climate change. *Nature* 469: 203–206. (6)

Scheckenbach, F., K. Hausmann, C. Wylezich, M. Weitere, and H. Arndt. 2010. Large-scale patterns in biodiversity of microbial eukaryotes from the abyssal sea floor. *Proceedings of the National Academy of Sciences USA* 107: 115–120. (2)

Schlenker, W. and M. J. Roberts. 2009. Nonlinear temperature effects indicate severe damages to US crop yields under climate change. *Proceedings of the National Academy of Sciences USA* 106: 15594–15598. (4)

Schleuning, M. and D. Matthies. 2009. Habitat change and plant demography: Assessing the extinction risk of a formerly common grassland perennial. *Conservation Biology* 23: 174–183. (5)

Schuldt, A. and T. Assmann. 2010. Invertebrate diversity and national responsibility for species conservation across Europe—A multi-taxon approach. *Biological Conservation* 143: 2747–2756. (2)

Schwartz, M. W. 2008. The performance of the Endangered Species Act. *Annual Review of Ecology, Evolution, and Systematics* 39: 279–299. (6)

Schwenk, W. S. and T. M. Donovan. 2011. A multi-species framework for landscape conservation planning. *Conservation Biology* 25: 1010–1021. (7)

Scott, J. M., B. Csuti, and F. Davis. 1991. Gap analysis: An application of Geographic Information Systems for wildlife species. *In* D. J. Decker, M. E. Krasny, G. R. Goff, C. R. Smith, and D. W. Gross (eds.), *Challenges in the Conservation of Biological Resources: A Practitioner's Guide*, pp. 167–179. Westview Press, Boulder, CO. (7)

Scott, J. M., D. D. Goble, J. A. Wiens, D. S. Wilcove, et al. 2005. Recovery of imperiled species under the Endangered Species Act: The need for a new approach. *Frontiers in Ecology and the Environment* 3: 383–389. (6)

Scott, J. M. and J. L. Rachlow. 2011. Refocusing the debate about advocacy. *Conservation Biology* 25: 1–3. (9)

Sebastián-González, E., J. A. Sánchez-Zapata, F. Botella, J. Figuerola, et al. 2011. Linking cost efficiency evaluation with population viability analysis to prioritize wetland bird conservation actions. *Biological Conservation* 144: 2354–2361. (6)

Seddon, P. J., D. P. Armstrong, and R. F. Maloney. 2007. Developing the science of reintroduction biology. *Conservation Biology* 21: 303–312. (6)

Sethi, P. and H. F. Howe. 2009. Recruitment of hornbill-dispersed trees in hunted and logged forests of the eastern Indian Himalaya. *Conservation Biology* 23: 710–718. (3)

Sewall, B. J., A. L. Freestone, M. F. E. Moutui, N. Toilibou, et al. 2011. Reorienting systematic conservation assessment for effective conservation planning. *Conservation Biology* 25: 688–696. (7)

Shafer, C. L. 1997. Terrestrial nature reserve design at the urban/rural interface. *In* M. W. Schwartz (ed.), *Conservation in Highly Fragmented Landscapes*, pp. 345–378. Chapman and Hall, New York. (7)

Shaffer, M. L. 1981. Minimum population sizes for species conservation. *BioScience* 31: 131–134. (5)

Shanley, P. and L. Luz. 2003. The impacts of forest degradation on medicinal plant use and implications for health care in eastern Amazonia. *BioScience* 53: 573–584. (3)

Siikamaki, J. 2011. Contributions of the US state park system to nature recreation. *Proceedings of the National Academy of Sciences USA* 108: 14031–14036. (3)

Singh, H. S. and L. Gibson. 2011. A conservation success story in the otherwise dire megafauna extinction crisis: The Asiatic lion (*Panthera leo persica*) of Gir forest. *Biological Conservation* 144: 1753–1757. (7)

Smith, M. J., H. B. Diaz, M. A. Clemente-Munoz, J. Donaldson, et al. 2011. Assessing the impacts of international trade on CITES-listed species: Current practices and opportunities for scientific research. *Biological Conservation* 144: 82–91. (6)

Snelgrove, P. V. R. 2001. Marine sediments. *In* S. A. Levin (ed.). *Encyclopedia of Biodiversity*. Academic Press, San Diego, CA. (4)

Snow, A. A., D. A. Andow, P. Gepts, E. M. Hallerman, et al. 2005. Genetically engineered organisms and the environment: Current status and recommendations. *Ecological Applications* 15: 377–404. (4)

Soares-Filho, B., P. Moutinho, D. Nepstad, A. Anderson, et al. 2010. Role of Brazilian Amazon protected areas in climate change mitigation. *Proceedings of the National Academy of Sciences USA* 107: 10821–10826. (7, 8)

Sodhi, S. N., R. Butler, W. F. Laurance, and L. Gibson. 2011. Conservation successes at micro-, meso- and macroscales. *Trends in Ecology and Evolution* 26: 585–594. (1)

Soulé, M. E. 1985. What is conservation biology? *BioScience* 35: 727–734. (1)

Soulé, M. E. and D. Simberloff. 1986. What do genetics and ecology tell us about the design of nature reserves? *Biological Conservation* 35: 19–40. (7)

Spalding, M. D., L. Fish, and L. J. Wood. 2008. Towards representative protection of the world's coasts and oceans—Progress, gaps, and opportunities. *Conservation Letters* 1: 217–226. (7)

Sponberg, A. F. 2009. Great Lakes: Sailing to the forefront of national water policy? *BioScience* 59: 372. (8)

Srinivasan, U. T., S. P. Carey, E. Hallstein, P. A. T. Higgins, et al. 2008. The debt of nations and the distribution of ecological impacts from human activities. *Proceedings of the National Academy of Sciences USA* 105: 1768–1773. (9)

Stearns, B. P. and S. C. Stearns. 2010. Still watching, from the edge of extinction. *BioScience* 60: 141–146. (1)

Stein, B. A., L. S. Kutner, and J. S. Adams (eds.). 2000. *Precious Heritage: The Status of Biodiversity in the United States*. Oxford University Press, New York. (4)

Stokes, E. J., S. Strindberg, P. C. Bakabana, P. W. Elkan, et al. 2010. Monitoring great ape and elephant abundance at large spatial scales: Measuring effectiveness of a conservation landscape. *PLoS ONE* 5(4): e10294. (8)

Stokstad, E. 2007. Gambling on a ghost bird. *Science* 317: 888–892. (5)

Stouffer, P. C., E. I. Johnson, R. O. Bierregaard Jr, and T. E. Lovejoy. 2011. Understory bird communities in Amazonian rainforest fragments: Species turnover through 25 Years post-isolation in recovering landscapes. *PLoS ONE* 6(6): e20543. (4)

Strain, D. 2011. 8.7 million: A new estimate for all the complex species on earth. *Science* 333: 1083. (2)

Suárez, E., M. Morales, R. Cueva, V. U. Bucheli, et al. 2009. Oil industry, wild meat trade and roads: Indirect effects of oil extraction activities in a protected area in north-eastern Ecuador. *Animal Conservation* 12: 364–373. (4)

Sullivan, B., C. L. Wood, M. J. Iliff, R. E. Bonney, et al. 2009. eBird: A citizen-based bird observation network in the biological sciences. *Biological Conservation* 142: 2282–2292. (9)

Sutherland, W. J., W. M. Adams, R. B. Aronson, R. Aveling, et al. 2009. One hundred questions of importance to the conservation of global biological diversity. *Conservation Biology* 23: 557–567. (9)

Sutherland, W. J., S. Bardsley, L. Bennun, M. Clout, et al. 2011. Horizon scan of global conservation issues for 2011. *Trends in Ecology and Evolution* 26: 10–16. (9)

Sutherland, W. J., M. Clout, I. M. Côté, P. Daszak, et al. 2010. A horizon scan of global conservation issues for 2010. *Trends in Ecology and Evolution* 25: 1–6. (9)

Swanson, M. E., J. F. Franklin, R. L. Beschta, C. M. Crisafulli, et al. 2011. The forgotten stage of forest succession: Early-successional ecosystems on forest sites. *Frontiers in Ecology and the Environment* 9: 117–125. (2)

Tallis, H. and S. Polasky. 2009. Mapping and valuing ecosystem services as an approach for conservation and natural-resource management. *Annals of the New York Academy of Sciences* 1162: 265–283. (7)

Taylor, M. F. J., K. F. Suckling, and J. J. Rachlinski. 2005. The effectiveness of the Endangered Species Act: A quantitative analysis. *BioScience* 55: 360–366. (6)

Temple, S. A. 1991. Conservation biology: New goals and new partners for managers of biological resources. *In* D. J. Decker, M. Krasny, G. R. Goff, C. R. Smith, and D. W. Gross (eds.), *Challenges in the Conservation of Biological Resources: A Practitioner's Guide*, pp. 45–54. Westview Press, Boulder, CO. (1)

Theobald, D. M. 2004. Placing exurban land-use change in a human modification framework.

Frontiers in Ecology and the Environment 2: 139–144. (8)

Thiollay, J. M. 1989. Area requirements for the conservation of rainforest raptors and game birds in French Guiana. *Conservation Biology* 3: 128–137. (5)

Thomas, J. A., M. G. Telfer, D. B. Roy, C. D. Preston, et al. 2004. Comparative losses of British butterflies, birds, and plants and the global extinction crisis. *Science* 303: 1879–1881. (4)

Thoreau, H. D. 1854. *Walden; or, Life in the Woods*. Ticknor and Fields, Boston. (3)

Thorp, J. H., J. E. Flotemersch, M. D. Delong, A. F. Casper, et al. 2010. Linking ecosystem services, rehabilitation, and river hydrogeomorphology. *BioScience* 60: 67–74. (3)

Tilman, D. 1999. The ecological consequences of change in biodiversity: A search for general principles. *Ecology* 80: 1455–1474. (3)

Timmer, V. and C. Juma. 2005. Biodiversity conservation and poverty reduction come together in the tropics: Lessons learned from the Equator Initiative. *Environment* 47: 25–44. (8)

Tittensor, D. P., C. Mora, W. Jetz, H. K. Lotze, et al. 2010. Global patterns and predictors of marine biodiversity across taxa. *Nature* 466: 1098–1101. (2)

Tognelli, M. F., M. Fernández, and P. A. Marquet. 2009. Assessing the performance of the existing and proposed network of marine protected areas to conserve marine biodiversity in Chile. *Biological Conservation* 142: 3147–3153. (7)

Towne, E. G., D. C. Hartnett, and R. C. Cochran. 2005. Vegetation trends in tallgrass prairie from bison and cattle grazing. *Ecological Applications* 15: 1550–1559. (7)

Traill, L. W., C. J. A. Bradshaw, and B. W. Brook. 2007. Minimum viable population size: A meta-analysis of 30 years of published estimates. *Biological Conservation* 139: 159–166. (5)

Traill, L. W., B. W. Brook, R. R. Frankham, and C. J. A. Bradshaw. 2010. Pragmatic population viability targets in a rapidly changing world. *Biological Conservation* 143: 28–34. (5, 6)

Tranquilli, S., M. Abedi-Lartey, F. Amsini, L. Arranz, et al. 2012. Lack of conservation effort rapidly increases African great ape extinction risk. *Conservation Letters* 5: 48–55. (7)

Troëng, S. and E. Rankin. 2005. Long-term conservation efforts contribute to positive green turtle *Chelonia mydas* nesting trend at Tortuguero, Costa Rica. *Biological Conservation* 121: 111–116. (6)

U.S. Census Bureau. *http://www.census.gov* (1)

United Nations Development Programme. 2006. *http://www.undp.org* (4)

Uthicke, S., B. Schaffelke, and M. Byrne. 2009. A boom-bust phylum? Ecological and evolutionary consequences of density variations in echinoderms. *Ecological Monographs* 79: 3–24. (4)

Vadeboncoeur, Y., P. B. McIntyre, and J. V. Zanden. 2011. Borders of biodiversity: Life at the edge of the world's large lakes. *BioScience* 61: 526–537. (2)

Valeila, I. and P. Martinetto. 2007. Changes in bird abundance in eastern North America: Urban sprawl and global footprint? *BioScience* 57: 360–370. (4)

van de Kerk, G., A. R. Manuel, and G. Douglas. 2009. Sustainable Society Index. *In* C. J. Cleveland (ed.), *Encyclopedia of Earth*. Environmental Information Coalition, National Council for Science and the Environment, Washington, D.C. *http://www.eoearth.org/article/Sustainable_Society_Index* (3)

Van Swaay, C., D. Maes, S. Collins, M. L. Munguira, et al. 2011. Applying IUCN criteria to invertebrates: How red is the Red List of European butterflies? *Biological Conservation* 144: 470–478. (6)

Vandermeer, J., I. Perfecto, and S. Philpott. 2010. Ecological complexity and pest control in organic coffee production: Uncovering an autonomous ecosystem service. *BioScience* 60: 527–537. (8)

Venevsky, S. and I. Venevskaia. 2005. Hierarchical systematic conservation planning at the national level: Identifying national biodiversity hotspots using abiotic factors in Russia. *Conservation Biology* 124: 235–251. (7)

Venter, O. and L. P. Koh. 2012. Reducing emissions from deforestation and forest degradation (REDD+): Game changer or just another quick fix? *Annals of the New York Academy of Sciences* 1249: 137–150. (8)

Venter, O., J. E. M. Watson, E. Meijaard, W. F. Laurance, and H. P. Possingham. 2010. Avoiding unintended outcomes from REDD. *Conservation Biology* 24: 5–6. (3)

Verstraete, M. M., R. J. Scholes, and M. S. Smith. 2009. Climate and desertification: Looking at an old problem through new lenses. *Frontiers in Ecology and the Environment* 7: 421–428. (4)

Vianna, G. M. S., M. G. Meekan, D. J. Pannell, S. P. Marsh, and J. J. Meeuwig. 2012. Socio-economic value and community benefits from shark-diving tourism in Palau: A sustainable use of reef shark populations. *Biological Conservation* 145: 267–277. (3)

Vilà, M., C. Basnou, P. Pyšek, M. Josefsson, et al. 2010. How well do we understand the impacts of alien species on ecosystem services? A pan-European cross-taxa assessment. *Frontiers in Ecology and the Environment* 8: 135–144. (4)

Vince, G. 2011. Embracing invasives. *Science* 331: 1383–1384. (4)

Vredenburg, V. T. 2004. Reversing introduced species effects: Experimental removal of introduced fish leads to rapid recovery of a declining frog. *Proceedings of the National Academy of Sciences USA* 101: 7646–7650. (4)

Vynne, C., J. R. Skalski, R. B. Machado, M. J. Groom, et al. 2011. Effectiveness of scat-detection dogs in determining species presence in a tropical savanna landscape. *Conservation Biology* 25: 154–162. (6)

Wake, D. B. and V. T. Vredenburg. 2008. Are we in the midst of the sixth mass extinction? A view from the world of amphibians. *Proceedings of the National Academy of Sciences USA* 105: 11466–11473. (5)

Wallach, A. D., B. Murray, and A. J. O'Neill. 2009. Can threatened species survive where the top predator is absent? *Biological Conservation* 142: 43–52. (2)

Warkentin, I. G., D. Bickford, N. S. Sodhi, and C. J. A. Bradshaw. 2009. Eating frogs to extinction. *Conservation Biology* 23: 1056–1059. (4)

Wasser, S., J. Poole, P. Lee, K. Lindsay, et al. 2010. Elephants, ivory and trade. *Science* 327: 1331–1332. (6)

Waters, S. S. and O. Ulloa. 2007. Preliminary survey on the current distribution of primates in Belize. *Neotropical Primates* 14: 80–82. (8)

Watson, J. E., M. C. Evans, J. Carwardine, R. A. Fuller, et al. 2011. The capacity of Australia's protected-area system to represent threatened species. *Conservation Biology* 25: 324–332. (7)

Weis, J. S. and C. J. Cleveland. 2008. DDT. *In* C. J. Cleveland (ed.), *Encyclopedia of Earth*. Environmental Information Coalition, National Council for Science and the Environment, Washington, D.C. *http://www.eoearth.org/article/DDT* (4)

West, P. and D. Brockington. 2006. An anthropological perspective on some unexpected consequences of protected areas. *Conservation Biology* 20: 609–616. (8)

West, P. C., G. R. Narisma, C. C. Barford, C. J. Kucharik, and J. A. Foley. 2011. An alternative approach for quantifying climate regulation by ecosystems. *Frontiers in Ecology and the Environment* 9: 126–133. (3)

Western and Central Pacific Fisheries Commission (WCPFC). 2009. *http://www.wcpfc.int* (6)

White, P. S. 1996. Spatial and biological scales in reintroduction. *In* D. A. Falk, C. I. Millar, and M. Olwell (eds.), *Restoring Diversity: Strategies for Reintroduction of Endangered Plants*, pp. 49–86. Island Press, Washington, D.C. (6)

Whittier, T. R., P. L. Ringold, A. T. Herlihy, and S. M. Pierson. 2008. A calcium-based invasion risk assessment for zebra and quagga mussels (*Driessena* spp). *Frontiers in Ecology and the Environment* 6: 180–184. (4)

Wiersma, Y. F., T. D. Nudds, and D. H. Rivard. 2004. Models to distinguish effects of landscape patterns and human population pressures associated with species loss in Canadian national parks. *Landscape Ecology* 19: 773–786. (7)

Wikramanayake, E., E. Dinerstein, J. Seidensticket, S. Lumpkin, et al. 2011. A landscape-based conservation strategy to double the wild tiger population. *Conservation Letters* 4: 219–227. (7)

Wikström, L., P. Milberg, and K. Bergman. 2008. Monitoring of butterflies in semi-natural grasslands: Diurnal variation and weather effects. *Journal of Insect Conservation* 13: 203–211. (6)

Wilcove, D. S. and J. Lee. 2004. Using economic and regulatory incentives to restore endangered species: Lessons learned from three new programs. *Conservation Biology* 18: 638–645. (6)

Wilcove, D. S. and L. L. Master. 2005. How many endangered species are there in the United States? *Frontiers in Ecology and the Environment* 3: 414–420. (6)

Wilcove, D. S. and M. Wikelski. 2008. Going, going, gone: Is animal migration disappearing? *PLoS Biology* 6: 1361–1364. (7)

Wildt, D. E., P. Comizzoli, B. Pukazhenthi, and N. Songsasen. 2009. Lessons from biodiversity—The value of nontraditional species to advance reproductive science, conservation, and human health. *Molecular Reproduction and Development* 77: 397–409. (6)

Wilhere, G. F. 2008. The how-much-is-enough myth. *Conservation Biology* 22: 514–517. (5)

Wilhere, G. F. 2012. Inadvertent advocacy. *Conservation Biology* 26: 39–46. (9)

Willi, Y., M. van Kleunen, S. Dietrich, and M. Fischer. 2007. Genetic rescue persists beyond first-generation outbreeding in small populations of a rare plant. *Proceedings of the Royal Society B* 274: 2357–2364. (5)

Willis, C. G., B. R. Ruhfel, R. B. Primack, A. J. Miller-Rushing, et al. 2010. Favorable climate change response explains non-native species' success in Thoreau's woods. *PLoS ONE* 5(1): e8878. (4)

Willis, C. G., B. R. Ruhfel, R. B. Primack, A. J. Miller-Rushing, and C. C. Davis. 2008. Phylogenetic patterns of species loss in Thoreau's woods are driven by climate change. *Proceedings of the National Academy of Sciences USA* 105: 17029–17033. (4, 5)

Wilson, E. O. 1989. Threats to biodiversity. *Scientific American* 261(3): 108–116. (5)

Wilson, E. O. 2003. The encyclopedia of life. *Trends in Ecology and Evolution* 18: 77–80. (2)

Wilson, E. O. 2010. *The Diversity of Life*. Harvard University Press, Cambridge. (2)

Wilson, J. R. U., E. E. Dormontt, P. J. Prentis, A. J. Lowe, and D. M. Richardson. 2009. Something in the way you move: Dispersal pathways affect invasion success. *Trends in Ecology and Evolution* 24: 136–144. (4)

Woodhams, D. C. 2009. Converting the religious: Putting amphibian conservation in context. *BioScience* 59: 463–464. (3)

Wooldridge, S. A. and T. J. Done. 2009. Improved water quality can ameliorate effects of climate change on corals. *Ecological Applications* 19: 1492–1499. (4)

World Resources Institute (WRI). 2003. *World Resources 2002–2004: Decisions for the Earth: Balance, voice, and power*. World Resources Institute, Washington, D.C. (9)

World Resources Institute (WRI). 2005. *World Resources 2005: The Wealth of the Poor—Managing Ecosystems to Fight Poverty*. World Resources Institute, Washington, D.C. (4).

World Wildlife Fund (WWF) and M. McGinley. 2009. Central American dry forests. *In* C. J. Cleveland (ed.), *Encyclopedia of Earth*. Environmental Information Coalition, National Council for Science and the Environment, Washington, D.C. *http://www.eoearth.org/article/Central_American_dry_forests* (4)

World Wildlife Fund (WWF) International. *http://www.panda.org* (6)

Worldwatch Institute. 2008. Making better energy choices. *http://www.worldwatch.org* (1)

Wright, S. 1931. Evolution in Mendelian populations. *Genetics* 16: 97–159. (5)

Wu, R., S. Zhang, D. W. Yu, P. Zhao, et al. 2011. Effectiveness of China's nature reserves in representing ecological diversity. *Frontiers in Ecology and the Environment* 9: 383–389. (7)

Yamaoko, K., H. Moriyama, and T. Shigematsu. 1977. Ecological role of secondary forests in the traditional farming area in Japan. *Bulletin of Tokyo University* 20: 373–384. (7)

Yamaura, Y., T. Kawahara, S. Iida, and K. Ozaki. 2008. Relative importance of the area and shape of patches to the diversity of multiple taxa. *Conservation Biology* 22: 1513–1522. (7)

Zander, K. K. and S. T. Garnett. 2011. The economic value of environmental services on indigenous-held lands in Australia. *PLoS ONE* 6(8): e23154. (3)

Zaradic, P. A., O. R. W. Pergams, and P. Kareiva. 2009. The impact of nature experience on willingness to support conservation. *PLoS ONE* 4: e7367. (9)

Zarin, D. J., M. D. Schulze, E. Vidal, and M. Lentini. 2007. Beyond reaping the first harvest: Management objectives for timber production in the Brazilian Amazon. *Conservation Biology* 21: 916–925. (8)

Zedler, J. B. 2005. Restoring wetland plant diversity: A comparison of existing and adaptive approaches. *Wetlands Ecology and Management* 13: 5–14. (8)

Zhu, Y. Y., Y. Y. Wang, H. R. Che, and B. R. Lu. 2003. Conserving traditional rice varieties through management for crop diversity. *BioScience* 53: 158–162. (8)

Zimmerer, K. S. 2006. Cultural ecology: At the interface with political ecology—The new geographies of environmental conservation and globalization. *Progress in Human Geography* 30: 63–78. (9)

Zimmermann, A., M. Hatchwell, L. Dickie, and C. D. West (eds.) 2008. *Zoos in the 21st Century: Catalysts for Conservation*. Cambridge University Press, Cambridge. (6)

Zydelis, R., B. P. Wallace, E. L. Gilman, and T. B. Werner. 2009. Conservation of marine megafauna through minimization of fisheries bycatch. *Conservation Biology* 23: 608–616. (4)

Index

The letter *f* after a page number indicates that the entry is included in a figure; *t* after a page number indicated that the entry is included in a table.

About the Author

Richard B. Primack is a Professor in the Biology Department at Boston University. He received his B.A. at Harvard University in 1972 and his Ph.D. at Duke University in 1976, and then was a postdoctoral fellow at the University of Canterbury. He has served as a visiting professor at the University of Hong Kong and Tokyo University, and has been awarded Bullard and Putnam Fellowships from Harvard University and a Guggenheim Fellowship. Dr. Primack was President of the Association for Tropical Biology and Conservation, and is currently Editor-in-Chief of the journal *Biological Conservation*. Twenty-seven foreign-language editions of his textbooks have been produced, with local coauthors adding in local examples. He is an author of rain forest books, most recently *Tropical Rain Forests: An Ecological and Biogeographical Comparison*, Second Edition (with Richard Corlett). Dr. Primack's research interests include: the biological impacts of climate change; the loss of species in protected areas; tropical forest ecology; and conservation education. He is currently writing a popular book about the effects of climate change in Concord since the time of Henry David Thoreau and Walden.

About the Book

Editor: Andrew D. Sinauer
Project Editor: Danna Niedzwiecki
Production Manager: Christopher Small
Book Design: Joanne Delphia
Book Layout: Ann Chiara and Janice Holabird
Cover Design: Joanne Delphia
Photo Research: David McIntyre
Manufacturer: Nordica International Ltd.